Advanced Concrete and Construction Materials

Advanced Concrete and Construction Materials

Editors

Mahdi Kioumarsi
Vagelis Plevris

MDPI • Basel • Beijing • Wuhan • Barcelona • Belgrade • Novi Sad • Cluj • Manchester

Editors

Mahdi Kioumarsi
Department of Built Environment
OsloMet—Oslo Metropolitan University
Oslo
Norway

Vagelis Plevris
Department of Civil and Environmental Engineering
Qatar University
Doha
Qatar

Editorial Office
MDPI
St. Alban-Anlage 66
4052 Basel, Switzerland

This is a reprint of articles from the Special Issue published online in the open access journal *Sustainability* (ISSN 2071-1050) (available at: www.mdpi.com/journal/sustainability/special_issues/advanced_concrete_construction_materials).

For citation purposes, cite each article independently as indicated on the article page online and as indicated below:

Lastname, A.A.; Lastname, B.B. Article Title. *Journal Name* **Year**, *Volume Number*, Page Range.

ISBN 978-3-7258-0446-7 (Hbk)
ISBN 978-3-7258-0445-0 (PDF)
doi.org/10.3390/books978-3-7258-0445-0

Contents

About the Editors

Mahdi Kioumarsi

Dr. Mahdi Kioumarsi holds the position of Associate Professor at the Department of Built Environment at OsloMet—Oslo Metropolitan University, as well as at the Department of Engineering at Ostfold University in Norway. He earned his Ph.D. in Structural Engineering from the Norwegian University of Science and Technology (NTNU), Norway.

With a rich background in structural engineering, Dr. Kioumarsi emerged as a preeminent figure in sustainable construction. His academic journey reflects a broad spectrum of scholarly pursuits, covering the analysis of large-scale structures and conducting thorough investigations at the material level, particularly focusing on concrete. The foundational principle of his research methodology is the seamless integration of modeling and experimental approaches.

Throughout his academic career, Dr. Kioumarsi has played a crucial role in coordinating and actively participating in numerous regional, national, and international research projects. These collaborative endeavors span diverse domains, involving both academic and industrial partners.

Vagelis Plevris

Dr. Vagelis (or Evangelos) Plevris is an Associate Professor at Qatar University's Department of Civil and Environmental Engineering in Doha, Qatar. Additionally, he serves as the Chief Editor for "Computational Methods in Structural Engineering" within the journal *Frontiers in Built Environment*, published by Frontiers media in Switzerland.

He earned his Diploma in Civil Engineering, specializing in Structural Engineering, from the National Technical University of Athens (NTUA). He furthered his education with an MSc in "Structural Design and Analysis of Structures" from NTUA, complemented by a Master of Business Administration (MBA) from the Athens University of Economics and Business (AUEB). He completed his academic journey with a PhD from NTUA's School of Civil Engineering. The title of his Doctoral Thesis was "Innovative Computational Techniques for the Optimum Structural Design Considering Uncertainties".

His research interests encompass the static and dynamic analysis of structures, earthquake engineering, concrete structures, the optimal design of structures, the reliability and probabilistic analysis of structures, as well as soft computing techniques and their engineering applications.

sustainability MDPI

Editorial

Advanced Concrete and Construction Materials for Sustainable Structures

Mahdi Kioumarsi [1] and Vagelis Plevris [2,*]

1 Department of Built Environment, OsloMet—Oslo Metropolitan University, 0166 Oslo, Norway; mahdi.kioumarsi@oslomet.no
2 Department of Civil and Environmental Engineering, Qatar University, Doha P.O. Box 2713, Qatar
* Correspondence: vplevris@qu.edu.qa

1. Introduction

Innovation in structural engineering has sparked remarkable advancements in the building materials sector and the construction industry in general [1,2]. Within this dynamic landscape, the traditional perception of concrete as a static, unchanging material has given way to a dynamic field marked by continuous innovation. The definition of durable and sustainable concrete has evolved over time, mirroring the shifting paradigms in modern construction. Today, sustainability is at the forefront of concrete science [3], driving the search for materials that not only meet the demands of construction but also tread lightly on our environment [4].

The sustainability of cementitious construction materials has emerged as a paramount concern, commanding the attention of researchers, engineers, and industry experts alike [5]. The production of cement accounts for approximately 8–10% of CO_2 emissions, thereby contributing to the global rise in environmental temperatures [6,7]. The imperative to reconcile the immense benefits of concrete structures with the pressing need to reduce their environmental footprint has catalyzed a wave of exploration into advanced concrete materials. These materials, collectively referred to as "advanced concretes", are characterized by their transformative microstructures, alternative binders to traditional Portland cement, and innovative additives designed to enhance durability and safety and minimize environmental impact throughout the life cycle of structures [8].

Among the noteworthy innovations are sustainable cementitious materials like alkali-activated binders and calcium sulphoaluminate cement, which hold the potential to revolutionize construction practices by reducing the carbon footprint of concrete [9]. However, as these novel materials emerge, the imperative for further research persists. The journey towards their full potential necessitates the optimization and standardization of mix designs, grounded in precise specifications that can guide the construction industry toward sustainable solutions.

This Special Issue, titled "Advanced Concrete and Construction Materials", aspired to provide a platform for groundbreaking research and critical reviews that transcend existing boundaries of knowledge in the field of advanced concrete construction materials. Embarking on this journey into the realm of advanced concrete and construction materials, we invited researchers, scholars, and industry professionals to contribute their expertise and insights. Our objective was to collect original research contributions that not only extend the frontiers of understanding but also pave the way for practical applications in the construction industry. Together, we aimed to shape the future of construction by harnessing the transformative potential of advanced materials and fostering a sustainable, resilient, and innovative built environment.

The scope of this collection of papers spans a wide spectrum of topics, offering a comprehensive exploration of cutting-edge developments in the field of construction materials. The studies delve into diverse facets of advanced concrete technology, addressing

Citation: Kioumarsi, M.; Plevris, V. Advanced Concrete and Construction Materials for Sustainable Structures. *Sustainability* **2024**, *16*, 1427. https://doi.org/10.3390/su16041427

Received: 4 February 2024
Accepted: 6 February 2024
Published: 8 February 2024

critical issues of sustainability, durability, and innovation. From enhancing the strength and durability of fly ash aggregate concrete through nanosilica incorporation to investigating the use of waste tires for reinforcing concrete columns under axial loads, these studies showcase innovative approaches that are reshaping the construction industry. Additionally, the collection delves into the world of numerical modeling and design methods for specialized cementitious composites, such as Polyvinyl-Alcohol-engineered cementitious composite beams. It evaluates the potential of water treatment sludge as a sustainable replacement for clay in fired clay bricks. Furthermore, topics like corrosion resistance, plant-based cementitious composites, the durability of alkali-activated slag concrete, and the use of biological agents like *Sporosarcina pasteurii* for sustainable construction materials expand the horizons of what is possible in construction technology.

The papers collectively highlight the urgency of advancing construction materials to meet the evolving demands of sustainability, resilience, and environmental responsibility in the construction industry. The research work presented in this collection paves the way for a more sustainable and efficient future in construction, addressing contemporary challenges while envisioning a more resilient and eco-friendly built environment. In the following section, we present a brief description of each of the ten articles in the collection.

2. Contributions

Azimi and Toufigh (Contribution 1) investigate the impact of incorporating slag into alkali-activated fly ash slag (AAFS) concrete. The study explores the mechanical properties, pore structure, and transport characteristics of AAFS concrete with varying slag content. Through unconfined compression and ultrasonic pulse velocity tests, it is observed that an increasing slag content improves the compressive strength, with the most significant enhancement occurring at a 20% slag replacement. The inclusion of slag leads to a noteworthy reduction in pore volume, particularly in gel pores, and an increase in surface area, indicating the presence of more geopolymeric products in AAFS gel. The microstructure of the AAFS gel is found to be more tortuous than the AAS gel, and the SEM/EDX analysis reveals the role of Ca ions in forming a highly compacted gel structure. Additionally, a higher slag content (up to 30%) significantly decreases transport properties due to microstructural changes. The study establishes an exponential relationship between ultrasonic pulse velocity (UPV) and compressive strength, highlighting the sensitivity of UPV to changes in compressive strength. Moreover, the research demonstrates that AAFS with equivalent strength levels exhibits a significantly lower water permeability compared to conventional concrete, emphasizing the influence of slag on pore structure modification. This paper offers valuable insights into the optimization of AAFS concrete by tailoring its composition to enhance both mechanical and transport properties.

The study of Peng et al. (Contribution 2) explores the enhancement of the strength and durability of cold-bonded fly ash aggregate (CFAA) concrete through the incorporation of nanosilica (NS). The study conducts various tests and analyses, revealing that the optimal NS dosage is 2 wt%, resulting in improved compressive and splitting tensile strength as well as early-age strength. NS's positive impact on the cementitious matrix reduces porosity but should be carefully dosed to avoid microscale deficiencies. Furthermore, the research demonstrates that CFAA concrete exhibits better resistance to carbonates than sulfates during dry–wet cycles, with NS mitigating deterioration in later cycles. Under freeze–thaw conditions, NS-modified CFAA concrete displays desirable spallation resistance and internal frost resistance but limited effectiveness against sulfate attacks. This study offers valuable insights for enhancing CFAA concrete's performance in harsh environments, encouraging future research into damage models and long-term performance under varying conditions.

Teymouri et al. (Contribution 3) experimentally investigate the durability of alkali-activated slag concrete (AASC) in a hydrochloric acid (HCl) environment, aiming to assess its potential as an eco-friendly alternative to conventional concrete. Thirteen distinct mix designs were explored, considering various parameters, including the type of alkaline solution, molarity, weight ratios, and alkali solution to slag ratios. After subjecting the samples to a six-month exposure in a HCl solution with pH = 3, visual inspections revealed that AASC had remarkable durability, contrasting with the substantial deterioration observed in ordinary Portland cement (OPC) concrete. The results underscored the superiority of AASC over OPC concrete in terms of HCl acid resistance, with AASC exhibiting minimal strength reduction and weight loss (one-tenth and one-fifth, respectively). Notably, samples activated with potassium hydroxide showed more pronounced strength reduction and weight loss than those with sodium hydroxide. The lower molarity of sodium hydroxide contributed to AASC's enhanced performance against acid attacks. Moreover, AASC samples with a $NaOH/Na_2SiO_3$ ratio of 1 demonstrated favorable durability compared to those with a ratio of three. The higher ratios of alkali solution to slag (0.6 compared to 0.4) resulted in increased strength reduction and weight loss. This study sheds light on essential mix design parameters for AASC in acidic environments, showcasing its potential as a durable and environmentally friendly alternative to traditional concrete.

Juárez-Alvarado et al. (Contribution 4) conducted a study on sustainable cementitious composites incorporating plant fibers and bio-aggregates to reduce environmental impacts. Four composites were examined: lechuguilla, flax-fiber-reinforced matrices, and wood and hemp shavings as bio-aggregates. The research encompassed nineteen mixtures, including control specimens, and evaluated the composites' hygrothermal, mechanical, and durability properties. Key findings include the adverse effect of fiber treatment on flexural strength in the lechuguilla and flax fiber composites, whereas untreated fibers and those with accelerated deterioration exhibited improved mechanical behavior. The presence of fibers increased porosity and reduced density, with flax fiber composites showing higher porosity due to the uneven fiber distribution. In addition, micrographs demonstrated no fiber embrittlement, with failure modes varying between untreated and treated fibers, impacting flexural and compressive behavior. Wood and hemp shavings' microstructures significantly affected the composites' physical and mechanical properties, with their high porosity reducing the overall compressive strength and bulk density. However, incorporating shavings improved hygrothermal properties, notably reducing thermal conductivity and achieving classification as moderate for moisture regulation efficiency.

The work of Ahmadi et al. (Contribution 5) explores the potential of using water treatment sludge (WTS) as a partial replacement for clay in fired clay brick production, aiming to address environmental concerns associated with traditional brick manufacturing processes. Various mixtures with WTS replacing clay in ratios ranging from 0% to 80% were investigated. The study revealed that an increase in WTS content led to improved plasticity and water absorption in the mixtures but resulted in decreased dry density and strength, particularly in unconfined compression strength. Higher WTS ratios also correlated with increased porosity and reduced bulk density. However, bricks containing up to 20% WTS and fired at 1000 °C exhibited enhanced properties suitable for moderate and severe exposure conditions. Efflorescence was primarily influenced by WTS content, and surprisingly, the firing temperature had minimal impact. Despite some reductions in strength, the research suggests that finding the right balance between WTS content and firing temperature holds promise for producing environmentally friendly bricks.

Khoshtina's review article (Contribution 6) explores the use of Sporosarcina pasteurii in sustainable construction through bio-cementation, inspired by natural calcium carbonate precipitation. While laboratory experiments demonstrate this process's potential, its practical application faces challenges due to process complexity and variable sensitivity. The review highlights bio-cementation's benefits in improving mechanical properties, reducing energy consumption, and lowering carbon footprint. Factors influencing bio-cementation include urease activity, environmental conditions, and application methods. Sporosarcina pasteurii shows robust performance, with enzymatic-induced calcium carbonate precipitation (EICP) proving superior to microbially induced precipitation (MICP) in some areas. However, MICP remains cost-effective. Particle granularity impacts bio-cementation outcomes. Promising research directions include repairing existing structures, interdisciplinary approaches, and employing numerical simulation models to advance this eco-friendly construction technique for a more durable and environmentally friendly industry.

The work of Mahat et al. (Contribution 7) explores a novel approach to tackling the environmental issue of waste tires by repurposing them in concrete columns. Recycled rubber tires are used as external reinforcements and formworks during concrete column construction, serving confinement functions. The experimental results show that these columns exhibit greater resistance to axial loads compared to traditional ones. Additionally, they display unique strain-softening behavior post-concrete failure. This innovative technique has the potential to enhance the structural performance of columns, making it a sustainable solution for retrofitting existing structures and constructing new ones.

Shi et al. (Contribution 8) aimed to assess and enhance the durability of concrete structures with fractures by investigating their seepage characteristics and the deterioration of their mechanical properties. Triaxial tests pre- and post-water penetration revealed significant impacts of fracture–seepage coupling on permeability and strength, with up to 32.8% decreased strength with increased fracture angle. Fracture–seepage–stress coupling further reduces strength, doubling the rate of decrease compared to intact concrete. Different fracture angles alter fracture expansion and damage modes, exacerbating concrete fragmentation. According to the research, enhancing concrete's seepage and fracture resistance is critical for extending structure lifespan.

Gil-Martín et al. (Contribution 9) discuss the limitations of the new Eurocode 2's guidance on concrete cover for reinforcement protection against carbonation-induced corrosion. While Eurocode 2 offers insights for 50- and 100-year service life designs, this article introduces a novel "square root of time" relationship, allowing for long-term structural designs and assessments. The newly proposed function adjusts carbonation front evolution to align with Eurocode 2 standards, enhancing the ability to design exceptionally durable structures and assess existing ones. The key revelation lies in the approximation of carbonation front evolution and a structure's expected service life using Eurocode 2 while disregarding resistance class against carbonation-induced corrosion. Instead, this method relies solely on concrete strength resistance and exposure class. The article's conclusions highlight a continuous formulation for the carbonation front based on the "square root of time" principle and in accordance with prEN 1992's minimum cover recommendations. This formulation, related to concrete compressive strength, aids in determining the requisite cover for carbonation protection, even considering indicative strength classes against carbonation-induced corrosion. Ultimately, this innovative expression is a valuable tool for designing exceptionally durable structures.

Fu et al. (Contribution 10) focus on the application of Polyvinyl-Alcohol-engineered cementitious composites (PVA-ECCs) in reinforced beams subjected to bending loads. A nonlinear finite element model is developed to simulate the behavior of PVA-ECC beams, incorporating constitutive models for PVA-ECCs, bond–slip behavior with reinforcing bars, and various design parameters. The study demonstrates that the numerical model accurately predicts the load–deflection curves and failure modes of PVA-ECC beams, offering a more precise estimation of ultimate loads and deformation capacities. Parameter analysis reveals the significant influence of the reinforcement ratio on the stiffness and ultimate load, while considering the tensile strength of PVA-ECCs in the tension zone yields more accurate results than existing design methods. Overall, the research contributes valuable insights into the behavior and design of reinforced PVA-ECC beams, offering potential advancements in structural engineering practices.

Author Contributions: M.K. and V.P. contributed to every part of the research described in this paper. All authors have read and agreed to the published version of the manuscript.

Acknowledgments: This Special Issue has been made possible by the contributions of several authors, reviewers, and editorial team members. Their efforts are greatly appreciated, and the Guest Editors would like to express their sincere thanks to all of them.

Conflicts of Interest: The authors declare no conflicts of interest.

List of Contributions

1. Azimi, Z.; Toufigh, V. Influence of Blast Furnace Slag on Pore Structure and Transport Characteristics in Low-Calcium Fly-Ash-Based Geopolymer Concrete. *Sustainability* **2023**, *15*, 13348. https://doi.org/10.3390/su151813348.
2. Peng, X.; Wang, Q.; Wu, J. Effect of Nanosilica on the Strength and Durability of Cold-Bonded Fly Ash Aggregate Concrete. *Sustainability* **2023**, *15*, 15413. https://doi.org/10.3390/su152115413.
3. Teymouri, M.; Behfarnia, K.; Shabani, A. Mix Design Effects on the Durability of Alkali-Activated Slag Concrete in a Hydrochloric Acid Environment. *Sustainability* **2021**, *13*, 8096. https://doi.org/10.3390/su13148096.
4. Juárez-Alvarado, C.A.; Magniont, C.; Escadeillas, G.; Terán-Torres, B.T.; Rosas-Diaz, F.; Valdez-Tamez, P.L. Sustainable Proposal for Plant-Based Cementitious Composites, Evaluation of Their Mechanical, Durability and Comfort Properties. *Sustainability* **2022**, *14*, 14397. https://doi.org/10.3390/su142114397.
5. Ahmadi, M.; Hakimi, B.; Mazaheri, A.; Kioumarsi, M. Potential Use of Water Treatment Sludge as Partial Replacement for Clay in Eco-Friendly Fired Clay Bricks. *Sustainability* **2023**, *15*, 9389. https://doi.org/10.3390/su15129389.
6. Khoshtinat, S. Advancements in Exploiting Sporosarcina pasteurii as Sustainable Construction Material: A Review. *Sustainability* **2023**, *15*, 13869. https://doi.org/10.3390/su151813869.
7. Mahat, M.; Acharya, M.; Acharya, M.; Mashal, M. Use of Waste Tires as Transverse Reinforcement and External Confinement in Concrete Columns Subjected to Axial Loads. *Sustainability* **2023**, *15*, 11620. https://doi.org/10.3390/su151511620.
8. Shi, Z.; Wu, C.; Wang, F.; Li, X.; Shan, C.; Xu, Y. Impact of Fracture–Seepage–Stress Coupling on the Sustainability and Durability of Concrete: A Triaxial Seepage and Mechanical Strength Analysis. *Sustainability* **2024**, *16*, 1187. https://doi.org/10.3390/su16031187.
9. Gil-Martín, L.M.; Hdz-Gil, L.; Molero, E.; Hernández-Montes, E. The Relationship between Concrete Strength and Classes of Resistance against Corrosion Induced by Carbonation: A Proposal for the Design of Extremely Durable Structures in Accordance with Eurocode 2. *Sustainability* **2023**, *15*, 7976. https://doi.org/10.3390/su15107976.
10. Fu, Q.-L.; Kang, S.-B.; Wang, D.-D. Numerical Modeling and Design Method for Reinforced Polyvinyl-Alcohol-Engineered Cementitious Composite Beams in Bending. *Sustainability* **2023**, *15*, 10130. https://doi.org/10.3390/su151310130.

References

1. Lu, X.; Plevris, V.; Tsiatas, G.; De Domenico, D. Editorial: Artificial Intelligence-Powered Methodologies and Applications in Earthquake and Structural Engineering. *Front. Built Environ.* **2022**, *8*, 876077. [CrossRef]
2. Tsiatas, G.C.; Plevris, V. Editorial: Innovative Approaches in Computational Structural Engineering. *Front. Built Environ.* **2020**, *6*, 39. [CrossRef]
3. de Brito, J.; Kurda, R. The past and future of sustainable concrete: A critical review and new strategies on cement-based materials. *J. Clean. Prod.* **2021**, *281*, 123558. [CrossRef]
4. Farahzadi, L.; Kioumarsi, M. Application of machine learning initiatives and intelligent perspectives for CO_2 emissions reduction in construction. *J. Clean. Prod.* **2023**, *384*, 135504. [CrossRef]
5. Shekarchi, M.; Ahmadi, B.; Azarhomayun, F.; Shafei, B.; Kioumarsi, M. Natural zeolite as a supplementary cementitious material—A holistic review of main properties and applications. *Constr. Build. Mater.* **2023**, *409*, 133766. [CrossRef]
6. Vishwakarma, V.; Uthaman, S. 9-Environmental impact of sustainable green concrete. In *Smart Nanoconcretes and Cement-Based Materials*; Liew, M.S., Nuguyen-Tri, P., Nguyen, T.A., Kakooei, S., Eds.; Elsevier: Amsterdam, The Netherlands, 2020; pp. 241–255. [CrossRef]
7. Guo, H.; Shi, C.; Guan, X.; Zhu, J.; Ding, Y.; Ling, T.-C.; Zhang, H.; Wang, Y. Durability of recycled aggregate concrete—A review. *Cem. Concr. Compos.* **2018**, *89*, 251–259. [CrossRef]
8. Tam, V.W.Y.; Soomro, M.; Evangelista, A.C.J. A review of recycled aggregate in concrete applications (2000–2017). *Constr. Build. Mater.* **2018**, *172*, 272–292. [CrossRef]
9. Yoo, D.-Y.; Banthia, N.; You, I.; Lee, S.-J. Recent advances in cementless ultra-high-performance concrete using alkali-activated materials and industrial byproducts: A review. *Cem. Concr. Compos.* **2024**, 105470. [CrossRef]

sustainability

Article

Influence of Blast Furnace Slag on Pore Structure and Transport Characteristics in Low-Calcium Fly-Ash-Based Geopolymer Concrete

Zahir Azimi * and Vahab Toufigh *

Department of Civil Engineering, Sharif University of Technology, Tehran 11155-9161, Iran
* Correspondence: zahirazimi5@gmail.com (Z.A.); toufigh@sharif.edu (V.T.); Tel.: +98-(21)-66164231 (V.T.); Fax: +98-(21)-66164828 (V.T.)

Abstract: Alkali-activated fly ash slag (AAFS) has emerged as a novel and environmentally sustainable construction material, garnering substantial attention due to its commendable mechanical attributes and minimal ecological footprint. This investigation delves into the influence of slag incorporation on the strength, pore structure, and transport characteristics of AAFS, encompassing various levels of fly ash replacement with slag. To assess the mechanical properties of AAFS concrete, unconfined compression and ultrasonic pulse velocity tests were conducted. Meanwhile, microstructural and mineralogical alterations were scrutinized through porosity, N_2-adsorption/desorption, and SEM/EDX assessments. In addition, transport properties were gauged using electrical surface resistivity, water permeability, and water vapor permeability tests. According to the results, a remarkable refinement in the pore volume was found by increasing the slag content. The volume of the gel pores and surface area increased significantly associated with the increase in tortuosity. Accordingly, Ca inclusion in the cross-linked sodium aluminosilicate hydrate gel remarkably reduced the transport properties.

Keywords: alkali-activated concrete; transport properties; pore structure; water permeability; electrical surface resistivity

check for updates

Citation: Azimi, Z.; Toufigh, V. Influence of Blast Furnace Slag on Pore Structure and Transport Characteristics in Low-Calcium Fly-Ash-Based Geopolymer Concrete. *Sustainability* **2023**, *15*, 13348. https://doi.org/10.3390/su151813348

Academic Editors: Mahdi Kioumarsi and Vagelis Plevris

Received: 8 August 2023
Revised: 30 August 2023
Accepted: 3 September 2023
Published: 6 September 2023

1. Introduction

In recent decades, global warming has become one of the most challenging issues throughout the world. Approximately 40% of global warming is due to the consumption of construction materials [1]. In particular, concrete as the second most widely used material in the world causes 5–8% of global carbon dioxide (CO_2) emissions [2]. Using by-product materials such as slag (SG), fly ash (FA), and silica fume as supplementary cementitious material can help to alleviate the adverse effects of cement [3,4]. Among new methods of green cement production, geopolymer concrete has demonstrated a potential of more than 70% reduction in global warming impact compared to traditional cement-based concrete [5–8].

In 1979, Davidovits represented a new cement-free concrete named geopolymer (GP). Geopolymer is a kind of inorganic polymer and a product of aluminosilicates reaction with alkali activators [9]. It has the potential to be utilized in a multitude of applications such as structural elements, soil stabilization, and immobilization of heavy metals [10–14]. The geopolymerization process includes three steps: (1) the alkali activator triggers the deconstruction of aluminosilicate source, (2) the alumina/silica-hydroxy polymerization generates geopolymeric gel, and (3) the fresh structure is stabilized [15–17].

Fly ash is a coal combustion by-products and is a great source of aluminosilicates. Generally, ambient-cured FA-based geopolymers cannot attain desirable strengths due to their low reactivity. However, heat curing can enhance cross-linked sodium aluminosilicate hydrate gel (N-A-S-H) [18]. For instance, at 75 °C for 24 h of heat curing, the compressive strength improved by about 50% compared to the ambient-cured sample [19]. To overcome the low reactivity of the FA particles in the alkali environment, adding high-CaO materials (such as SG, cement, CaO, and $Ca(OH)_2$) can effectively accelerates the reaction kinetics [20–23].

Blast furnace slag is a by-product of the iron- and steel-making industry. Owing to the high content of CaO, SG has been studied by many researchers as a partial replacement of the FA in geopolymers. In general, amorphous calcium aluminosilicate hydrate (C-A-S-H) has been identified as a dominant gel in the alkali-activated slag matrix. In FA/SG-based geopolymers with increasing SG content, C-A-S-H becomes a binding gel, and a hybrid gel of C-A-S-H and N-A-S-H (ascribed as N-C-A-S-H) is generated. Accordingly, Ca incorporation in the N-A-S-H structure leads to a well-compacted gel and a drastic increase in the strength of the geopolymer [24–26].

The FA/SG-based geopolymers' durability is highly dependent on the pore network and microstructural properties. The addition of SG to the FA-based geopolymers significantly decreases the porosity and refines the pore volume. The FA/SG-based geopolymers show high acid resistance. However, the greater content of Ca in the gel structure causes higher decalcification and more gypsum formation. It leads to expansion and, consequently, strength loss in the acid environment. Indeed, the high acid resistance of FA/SG-based geopolymers is due to its dense and tortuous pore solution, which causes relatively small penetration depth [27–30].

Among the concrete characteristics, mass transport through the pore solution is a parameter that depends on pore network topology, such as connectivity and tortuosity [31–33]. Water and vapor transport are the key parameters for the long-term durability of porous inorganic materials [34–36] since water is the main carrier of erosive ions and species (such as Cl^-, SO_4^{2-}, and CO_2) [37–40]. To measure the mass transport and pore structure along with direct transport measuring, non-destructive methods such as electrical resistivity and pulse velocity can be used to evaluate pore network characteristics and durability [41–45].

According to the available literature, numerous research studies have delved into the chemo-mechanical properties of geopolymers based on fly ash (FA) and slag (SG) [46–49]. However, there is a noticeable scarcity of publications addressing the transport characteristics of these geopolymers. The results of bulk diffusion tests revealed that all fly ash-based geopolymer concrete (GPC) specimens, regardless of their curing conditions, exhibited significantly high chloride content. Even at a depth of 25 mm beneath the surface, the chloride content exceeded 0.3% of the concrete mass. This observation suggests that low-calcium fly-ash-based GPCs display diminished resistance to chloride diffusion, likely attributable to reduced binding capacity, as well as well-connected pore structures and/or elevated levels of capillary pores [50]. In the case of rice-husk-ash-based geopolymer concrete incorporating a 15% ordinary Portland cement (OPC) replacement, the water absorption rate was notably low at 4.42%, a characteristic observed across all ages of the specimens [51]. This indicates that N-A-S-H possesses larger permeable voids compared to the denser and smaller pore system of C-A-S-H [52]. This distinction underscores the advantageous potential of porous geopolymers with a substantial proportion of interconnected pores for effectively removing heavy metals, which holds promise for their future applications in groundwater decontamination [53].

This study aims to investigate the pore structure and transport properties of FA/SG-based geopolymers, as there is currently no comprehensive research on this topic. The replacement ratios of FA to SG were 0, 10, 20, and 30%. The unconfined compression and ultrasonic pulse velocity tests were performed for evaluation of mechanical properties. Microstructural and mineralogical changes were evaluated by porosity, N_2-adsorption/desorption, and SEM/EDX tests. Along with this, transport properties were studied by electrical surface resistivity, water permeability, and water vapor permeability tests.

2. Materials and Preparation

Class F fly ash (FA) and ground granulated blast furnace slag (GGBS) were used as raw materials. Table 1 presents the chemical composition of the FA and GGBS. The X-ray diffraction (XRD) analysis of FA and GGBS is available in the previous study [54]. A combination of sodium hydroxide (NH) and sodium silicate (NS) was selected as an alkali activator. The chemical composition of NS was SiO_2 (=29.8 wt%), Na_2O (=14.19 wt%), and H_2O (=56 wt%), with a modulus (SiO_2/Na_2O) of 2.1. Ten-molar NH solution was prepared by dissolving 95–98% purified NH pallets in tap water one day before mixing. A polycarboxylate-based superplasticizer was utilized to enhance the concrete workability. Well-graded natural siliceous sand with a fineness modulus of 2.84 was used as fine aggregates. The coarse aggregates were crushed granite with a maximum nominal size of 12 mm. The particle size distribution of aggregates was in the range of ASTM C33-01 [55]. Fine and coarse aggregates were under the saturated surface dry (SSD) condition according to ASTM C128 and ASTM C127, respectively [56,57].

Table 1. Chemical composition (wt%) of the SG and FA.

Element	CaO	SiO_2	Al_2O_3	Fe_2O_3	MgO	SO_3	K_2O	Na_2O
GGBS	38.23	35.43	13.28	0.76	8.74	2.4	0.53	0.43
FA	4.3	57.3	29.5	11.7	-	0.2	-	-

In all mix designs, binder content, water/binder, alkali liquid/binder, coarse aggregate/fine aggregate, and NH solution molarity were 400 kg/m^3, 0.32, 0.4, 1.5, and 10 M, respectively. The concrete mix designs were based on the volumetric method. Based on a previous study, the optimum NS/NH solution was selected as 2.5 in this study [58]. Table 2 represents the mixture designs and the labels of the specimens.

Table 2. Mix design (kg/m^3).

Label	FA	GGBS	NH Solution	NS Solution	Sand	Gravel	Extra Water	Superplasticizer
GP0	400	0	45.71	114.29	722.76	1084.15	33.09	4
GP0	360	40	45.71	114.29	722.76	1084.15	33.09	4
GP0	320	80	45.71	114.29	722.76	1084.15	33.09	4
GP0	280	120	45.71	114.29	722.76	1084.15	33.09	4

Solid materials (binder and aggregate) were first mixed for three minutes to reach a homogenous mixture. NH and NS solutions were blended 30 min before adding. Alkali liquid and superplasticizer were then added to the dry mixture and mixed for another three minutes. Fresh concrete was placed in the 10 × 10 × 10 cm^3 cubic molds for unconfined compression and ultrasonic pulse velocity tests, 15 × 15 × 15 cm^3 cubic molds for water permeability, and cylindrical molds with a diameter of 10 cm and a height of 20 cm for electrical surface resistivity. Each specimen was prepared by pouring the mixture into three layers along with pounding 25 times. Although AAFS has good potential to cure in the ambient temperature, in this study, the specimens were cured at 70 °C for one day to reach a high degree of geopolymerization. They were then kept in the curing room at 20 ± 2 °C and 60 ± 5 RH for 28 days.

3. Testing Methods

3.1. Mechanical Properties

3.1.1. Unconfined Compression Test

The compressive strength was measured according to Figure 1 and BS EN 12390-3 standard [59] by the Dartec universal testing machine (UTM) with 1000 kN capacity.

Figure 1. Compressive strength apparatus.

3.1.2. Ultrasonic Pulse Velocity

The ultrasonic pulse velocity (UPV) test was carried out according to ASTM C597-09 [60]. The instrument consists of a transducer for wave generation, a receiving transducer, and a device for displaying the time it takes to pass the wave across the sample. The probes were lubricated to have a proper contact between the sample and the probes. The time taken by the wave to pass through the specimens was measured, and the ultrasonic pulse velocity (UPV) was calculated as follows:

$$\text{UPV} = \frac{L}{t} \tag{1}$$

where t is the time taken by the wave to transmit over the specimen, and L is the length of the specimen.

3.2. *Microstructural and Mineralogical Changes*

3.2.1. Porosity

The porosity was measured according to ASTM C642 [61]. The 10 cm × 5 cm discs were used to determine the porosity according to Equation (2) as follows:

$$\text{Porosity} = \frac{W_a - W_d}{W_a - W_w} \times 100 \tag{2}$$

where W_a is the weight of the saturated specimens in the air (g), W_d is the weight of the dried specimens (g), and W_w is the weight of saturated specimens in water (g).

3.2.2. Pore size Distribution and Specific Surface Area

Gas adsorption and desorption represent the prevailing techniques for assessing the surface area of powders and determining the size distribution of porous materials. To perform these measurements, we employed the Belsorp mini II surface area and pore size analyzer by Bel Company. Prior to subjecting the specimens to the N_2-adsorption/desorption test, they underwent vacuum drying and degassing. The experimental procedure was carried out at a temperature of −195.85 °C, involving the analysis of nitrogen adsorption/desorption isotherms for approximately 2 g of AAFS pastes. It is worth noting that the alkali liquid/binder and water/binder ratios of the AAFS paste samples were meticulously maintained to match those of the respective concrete specimens.

3.2.3. Scanning Electron Microscopy/Energy-Dispersive X-ray Spectroscopy

Scanning electron microscopy (SEM) and energy-dispersive X-ray spectroscopy (EDX) was used to investigate the microstructural and mineralogical characteristics of AAFS mixtures. The pieces were collected from the middle of the specimens. Before testing, the samples were solvent-vacuum dried, followed by polishing the surface and gold coating.

3.3. Transport Properties

3.3.1. Electrical Surface Resistivity

The electrical surface resistivity was conducted according to AASHTP TP 95-11 [62] at 28 days. The ESR was measured using four pin Wenner probe arrays, as shown in Figure 2. An alternative current potential difference was applied at the outer pins of the Wenner array, leading to current flow in the concrete. The two inner probes measured the resultant potential difference, and the resistivity was calculated using Equation (3) as follows:

$$\rho = R\frac{A}{L} = 2\pi aR = 2\pi a\frac{V}{I} \quad (3)$$

where ρ is resistivity in kΩ cm, R is resistance in kΩ, A is the surface area of the specimen in cm^2, L is the element length in cm, and a is the spacing between the probes in cm.

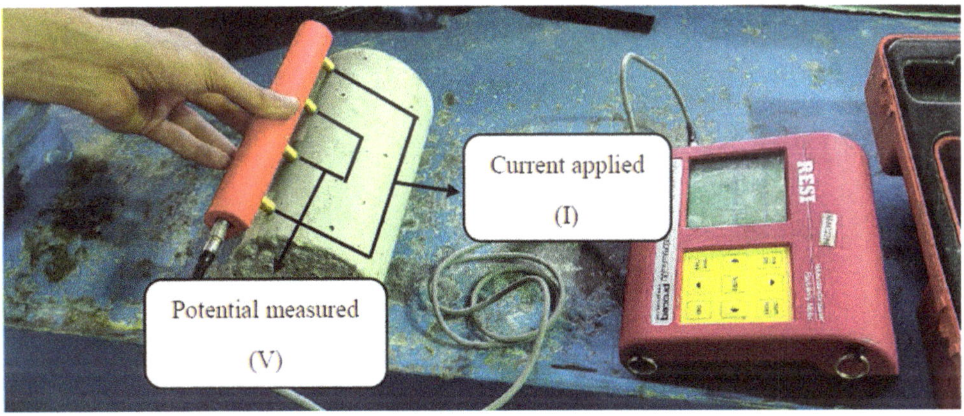

Figure 2. ESR test equipment.

3.3.2. Water Vapor Transmission

The water vapor transmission (WVT) was conducted according to ASTM E96/E96M-16 [63]. In this study, the dry cup method was performed. Figure 3 shows the WVT test apparatus. The silica gel as a desiccant agent filled the dish within 6 mm of the specimen. A 10 cm × 2.5 cm disc was attached to the dish. Furthermore, the specimen in contact with the container was sealed with wax to prevent the water vapor transport from this area. The apparatus was then placed in the humidity chamber. In this study, the humidity was kept at 95 ± 2% to apply a high water vapor gradient. Water vapor transmission was calculated according to Equation (4) as follows:

$$WVT = \frac{G}{tA} \quad (4)$$

where G is the weight change (g), t is time (hour), G/t is the slope of the straight line ($\frac{\text{g}}{\text{h}}$), and A is the test area (m^2).

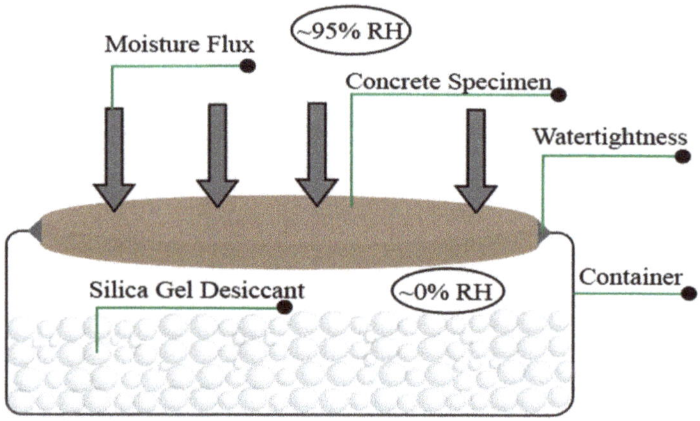

Figure 3. Water vapor transmission test apparatus.

3.3.3. Water Permeability

A water permeability device was used to measure the permeability coefficient of concrete, as can be seen in Figure 4a. The permeability test was carried out on 15 × 15 × 15 cm^3 cubic specimens after 28 days of curing. The sides of the specimen were coated and waterproofed to prevent water leakage from sides, as shown in Figure 4b. Water pressure was controlled using an air compressor with an 8-bar capacity. The water pressure was applied to the specimen surface at a 1 bar pressure and gradually increased to 5 bar after 5 min. The water level in the graduated capillary tube was recorded at different time intervals until the flow rate reached a constant value. The flow rate (cm^3/h) is determined by dividing the water volume into the specimen over the time interval between the records. The permeability coefficient is calculated using the Darcy law as follows:

$$K = \frac{M \times l}{A \times h} \quad (5)$$

where K is the coefficient of permeability (m/s), M is the flow rate (cm^3/s), l is the length of the specimen in the direction of water flow, A is the cross-section perpendicular to water flow, and h is the hydraulic head.

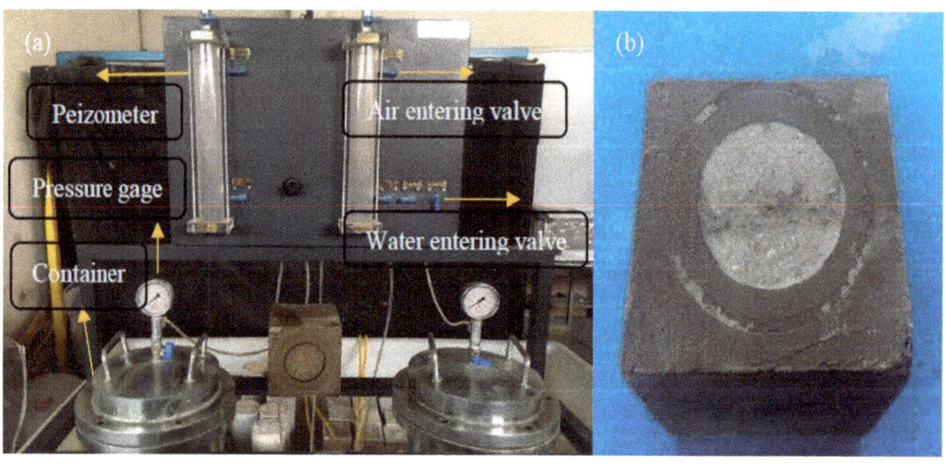

Figure 4. (**a**) Water permeability apparatus and (**b**) coated sample for water permeability test.

4. Results and Discussions

4.1. Mechanical Properties

4.1.1. Unconfined Compressive Strength

Figure 5 shows the compressive strength of AAFS specimens with different levels of slag replacement. Incorporating SG into FA has a significant effect on the compressive strength. The compressive strength of GP10, GP20, and GP30 increased by 32.35, 82.35, and 94.12%, respectively, compared to the GP0. This increase is mainly ascribed to the formation of C-A-S-H gel along with N-A-S-H. Also, Ca ions of SG can be dissolved in the N-A-S-H gel and form a new hybrid N-C-A-S-H gel. These new products lead the AAFS geopolymer to have dense structures [16,54,55]. The most significant improvement in the compressive strength was related to the GP20. This result was in good agreement with previous research [64].

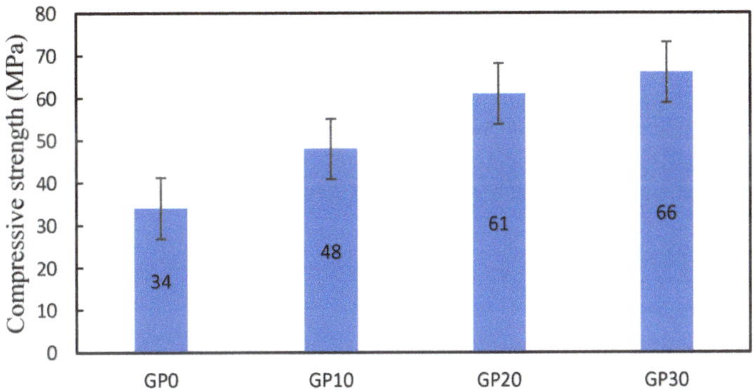

Figure 5. Compressive strength of the specimens (MPa).

4.1.2. Ultrasonic Pulse Velocity

The ultrasonic pulse velocity (UPV) is a measure of the speed of sound waves through a material, and it can be used to estimate the compressive strength of cement mortar and concrete. As shown in Figure 6, the UPV results for cement mortars blended with silica fume were in the range of 3774 and 4255 m/s. The UPV of GP10, GP20, and GP30 increased by 3.1, 5.2, and 12.8%, respectively, compared to the GP0. The UPV values increased with microstructure densification and increasing the compressive strength [65].

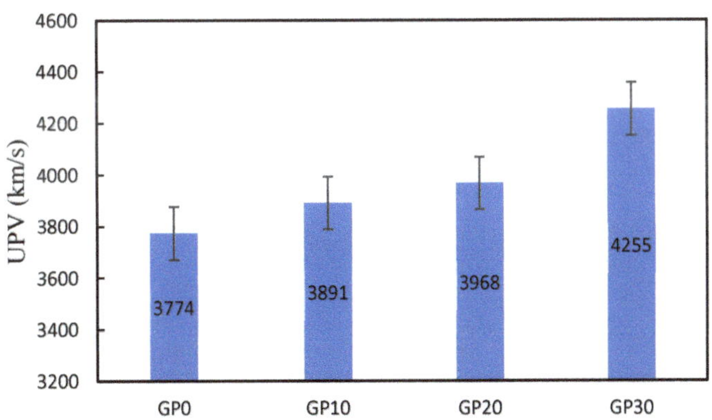

Figure 6. UPV of the specimens (Km/s).

To further investigate the relationship between UPV and compressive strength, the UPV results were plotted versus the compressive strength results, and the relationship between them was calculated. As shown in Figure 7, a good correlation was observed between UPV and compressive strength. The correlation between compressive strength and UPV is obtained by adopting the procedure presented by the ACI228. Variation in the tested cores' mixture ratios resulted in obtaining different strength–UPV correlations [66,67].

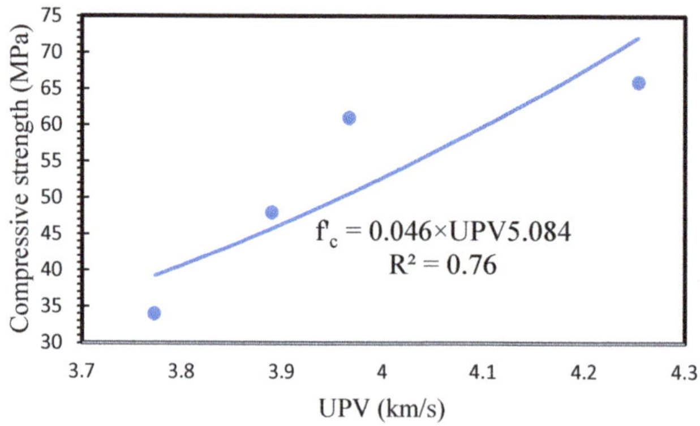

Figure 7. Relationship between UPV and compressive strength.

The correlation of the compressive strength and UPV in relation to the effect of the inclusion Ca ions in the N-A-S-H structure is complexly affected by the change in microstructure caused by further densification of the geopolymeric reaction [68]. Therefore, the relationship developed in the Figure 7 is case-specific due to the UPV and the compressive strength of geopolymer concrete. A linear relationship was predicted between the UPV and compressive strength of specimens with varying GGBS content.

4.2. Microstructural and Mineralogical Changes

4.2.1. Porosity

The porosity of the specimens and the relationship between porosity and compressive strength are shown in Figure 8. In comparison with GP0, the porosity of GP10, GP20, and GP30 decreased by 20.87%, 29.71%, and 36.95%, respectively. The reduction is related to higher FA dissolution in the presence of SG, which led to higher reaction products, followed by a denser matrix.

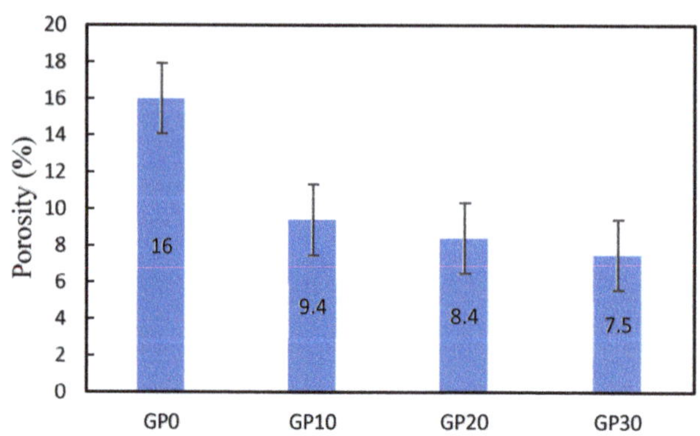

Figure 8. The porosity of the samples (%).

It is generally accepted that porosity has a direct relationship with compressive strength. A high value of the correlation was obtained between porosity and compressive strength (R^2 = 0.95). Figure 9 shows the relationship between compressive strength and porosity.

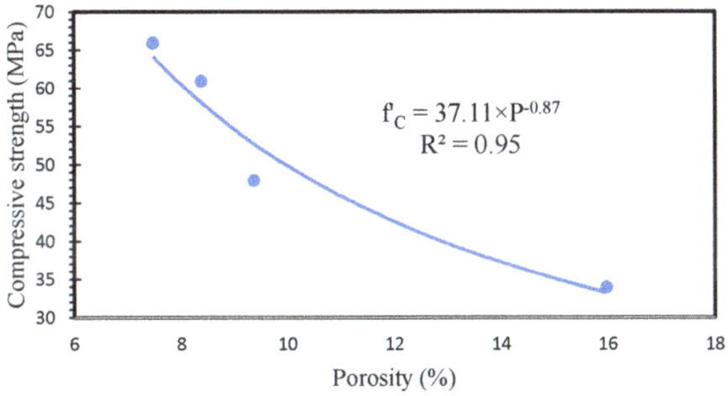

Figure 9. Relationship between porosity and compressive strength.

4.2.2. Pore Size Distribution and Surface Area

Figure 10 illustrates the pore size distribution of AAFS specimens. Barett, Joyner, and Halenda's (BJH) method was used to determine the pore size distribution in the mesopore and small macropore range [69]. This method is based on Kelvin's equation (see Equation (6)). It describes a relationship between the vapor pressure of a liquid and surface curvature. It can be concluded from this equation that the vapor pressure of the liquid on the convex surface is higher than that of the flat surface. Although the equation assumes the cylindrical pores, it can be useful in the porosimetry of the cement base materials.

$$\ln\left(\frac{P}{P_0}\right) = \frac{2\gamma V_M}{rRT}\cos\theta \quad (6)$$

where P is the vapor pressure, P_0 is the saturated pressure of the liquid in the plane surface, γ is the liquid surface tension, V_M is the molar volume of liquid, θ is the contact angle between the liquid and pore wall, R is the universal gas contact, and T is the thermodynamic temperature.

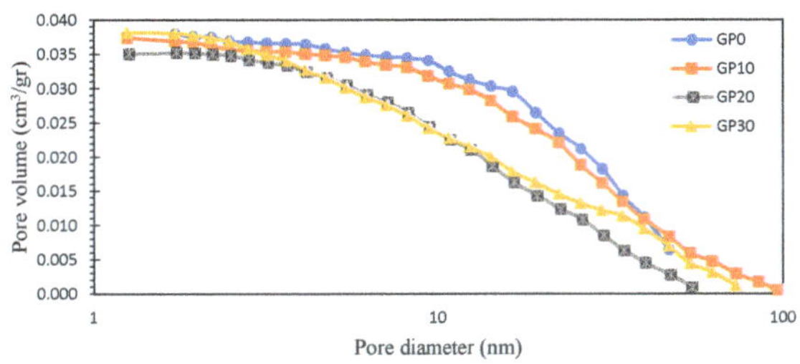

Figure 10. Pore size distribution of samples.

Aligizaki and Hanif et al. presented a pore size classification for cement-based materials [70]. Specifically, they defined pore sizes ranging from 2.5 to 10 nm as small capillary pores, those within the range of 10 to 50 nm as medium capillary pores, and pores sized between 50 to 10 µm as large capillary pores. In Table 3, the volume fractions of gel pores and capillary pores in the specimens are shown. As illustrated in Table 3 and Figure 10, the overall pore volume remained relatively stable with the inclusion of slag. However, a significant increase in gel pores was notably observed, which is indicative of enhanced pore network refinement. Specifically, the volume fraction of gel pores for GP10, GP20, and GP30 increased by 33.3%, 175%, and 216.7%, respectively, compared to GP0. These findings indicate a transformation of large N-A-S-H capillary pores into finer N-C-A-S-H gel pores. Furthermore, the most pronounced increase in the gel pore volume fraction was observed in GP20, aligning with the compressive strength results, where GP20 exhibited the most substantial strength enhancement.

Table 3. Volume fraction of gel and capillary pores (%).

Specimen	Gel Pores	Capillary Pores
GP0	12	88
GP10	16	84
GP20	33	67
GP30	38	62

The specific surface area was determined from the BET adsorption equation (see Equation (7)) suggested by Brunauer et al. [71]:

$$\frac{P}{V(P_0 - P)} = \frac{1}{V_m C} + \frac{C-1}{V_m C} \times \frac{P}{P_0} \quad (7)$$

where V is the volume of gas adsorbed at standard temperature and pressure (STP) (273.15 k and atmospheric pressure), P_0 is saturated pressure of N_2 gas, P is the partial vapor of N_2 gas in equilibrium with the surface at 77.4 k, V_m is the volume of N_2 adsorbed at STP to make a monolayer on the sample, and C is the dimensionless constant that is related to the enthalpy of adsorption of the N_2 gas on the specimen.

The value of the V was measured at four relative pressures (P/P_0), approximately ranging from 0.05 to 0.3. The BET equation can be evaluated and plotted by linear regression with $\frac{P}{V(P_0-P)}$ on the y-axis and $\frac{P}{P_0}$ on the x-axis and a y-intercept of $\frac{1}{V_m C}$. From the plotted line, the slope was obtained and V_m calculated for monolayer adsorption. Finally, the specific surface is calculated by Equation (8) as follows:

$$S_{BET} = \frac{V_m N_A s_m}{V} \quad (8)$$

where N_A is Avogadro's constant (6.022×10^{231}/mol), S_m is the effective cross-sectional area of the N_2 molecule (0.162 mm^2), and V is the molar volume of the N_2 molecule.

Odler [72] demonstrated a linear relationship between the volume of the hydration products and surface area values. He explained that, in the early ages of hydration, hydration products are more porous. As the hydration proceeds, and the free water in the pore solution is consumed, the pore volume becomes finer, and the pore solution becomes more interwoven. As a result of the hydration progress, more regions are developed in the microstructure, narrowing openings.

Figure 11 shows the specific surface area of AAFS specimens. The surface area of GP10, GP20, and GP30 increased by 60.92, 75.22, and 118.58%, respectively, in comparison with the GP0. Based on the SEM observations and pore structure analysis, adding slag to the GP0 led to more geopolymeric products, which have finer pores, connecting geopolymeric products [68]. Consequently, more regions developed. In the case of the AAFS materials, increasing the SG content in the AAF materials significantly increases the surface area of the AAFS materials. Figure 12 illustrates the relationship between the surface area and the compressive strength of the AAFS specimens. A high correlation was observed between surface area and compressive strength.

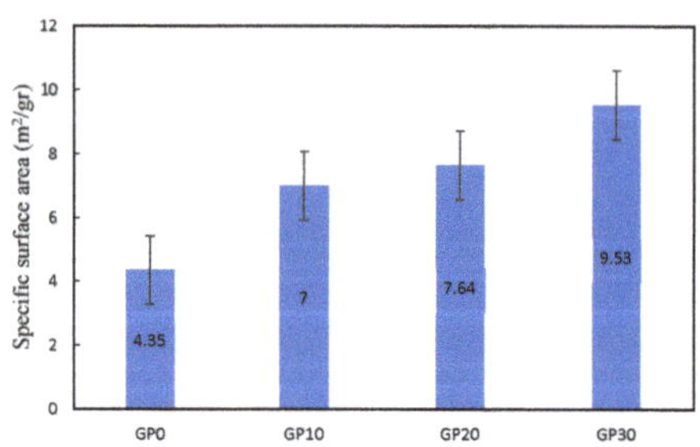

Figure 11. Specific surface area of the specimens (m^2/gr).

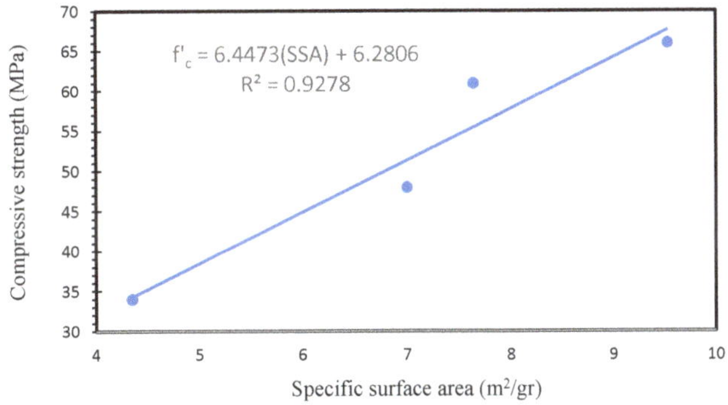

Figure 12. Relationship between specific surface area and compressive strength.

4.2.3. Microstructural Analysis (SEM/EDX)

Figure 13 indicates the SEM images of AAFS specimens. The most conspicuous observation was filling the matrix voids by gel progress and gel densification by Ca inclusion. The specimens gel locations were selected from the high-reactive areas and mostly observed as amorphous phases.

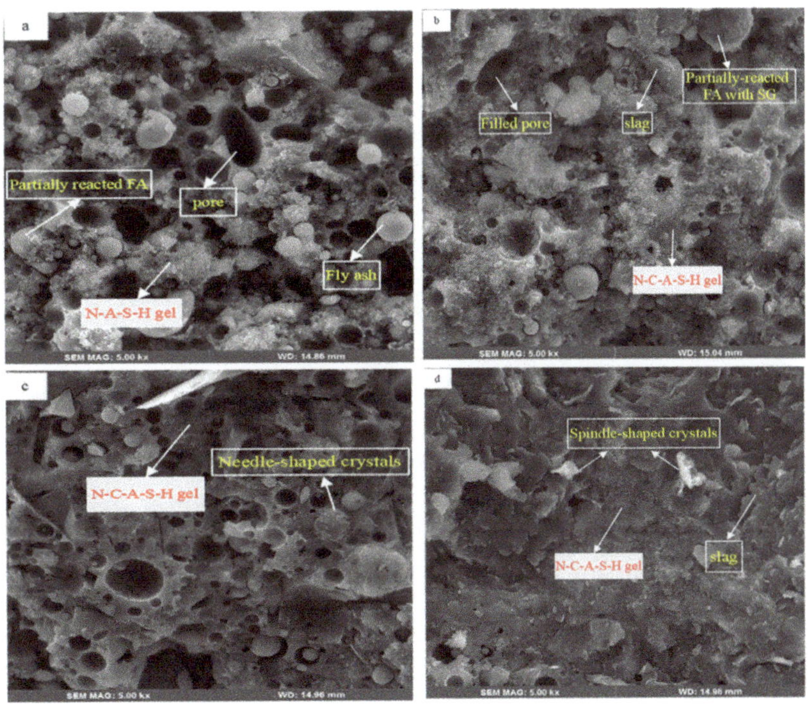

Figure 13. SEM images of samples having (**a**) 0% of SG, (**b**) 10% of SG, (**c**) 20% of SG, and (**d**) 30% of SG.

Figure 13a represents a high proportion of un-reacted FA spheres and voids in the matrix. These un-reacted FA can cause low mechanical properties and high transport through the matrix. From Figure 13b–d, irregular-shaped SG components are obvious. When comparing Figure 13a–d, it can be deduced that the un-reacted FA particles were either completely resolved or partially reacted owing to the SG incorporation associated with the high-alkalinity atmosphere, which led to high-geopolymerization products. Also, some filled pores in the GP10 confirmed the SG densification effect in the low contents of SG.

In Figure 13c, many needle-shaped crystals are displayed on the surface of FA spheres, as they have covered FA particles. These needle-shaped crystals have also been reported in previous studies claiming that due to their atomic characteristics, they are types of crystalline aluminosilicate products [73,74]. As shown in, Figure 13d, FA particles were rarely detected, which indicates that they mostly dissolved in the gel. Additionally, some spindle-shaped crystals were found in the GP30, as shown in the previous research [73]. This suggested that these crystals are likely calcium carbonates ($CaCO_3$) and observed in high contents of the SG.

Table 4 represents the mean ratio of the Ca/Si in the 10 points near the specific points shown in Figure 13a–d. The enhancement in the Ca/Si ratio by increasing the SG content confirmed that densification of the geopolymeric gel (N-A-S-H) was due to the Ca inclusion in the atomic structure of the gel.

Table 4. Ca/Si ratio analyzed via EDX analysis.

Specimens	Ca/Si
GP0	0.08
GP10	0.14
GP20	0.27
GP30	0.41

4.3. Transport Properties

4.3.1. Electrical Surface Resistivity

Figure 14 represents the electrical surface resistivity (ESR) of AAFS specimens. The ESR values ranged between 3.7 and 7.4 kΩ·cm. The results indicated that ESR values were enhanced by increasing the replacement level of the SG. The electrical resistance of GP10, GP20, and GP30 increased by 21.62, 70.27, and 100%, respectively, compared to the GP0. High ESR values indicate that the pore network has low connectivity. In other words, ESR is a beneficial parameter for evaluating material tortuosity.

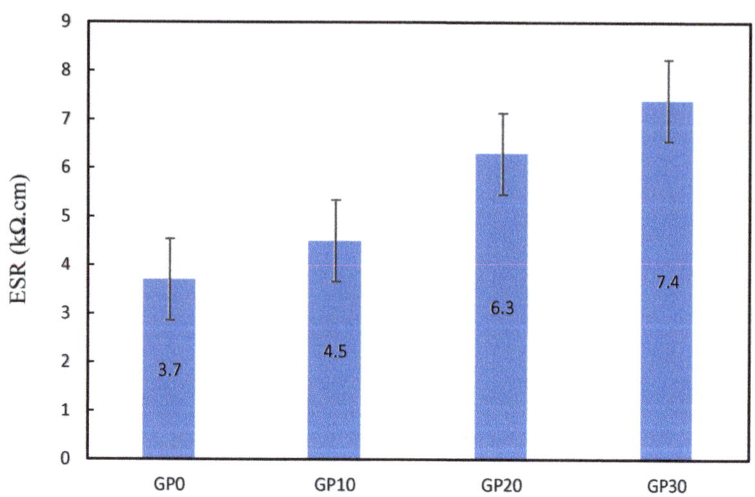

Figure 14. ESR values of the specimens.

Regarding the effect of Ca inclusion in the structure of the N-A-S-H gel, the pore structure and tortuosity of the AAFS specimens are shown schematically in Figure 15b. Comparing Figure 15a with Figure 15b exhibits the pore volume refinement and increase in the tortuosity by slag addition.

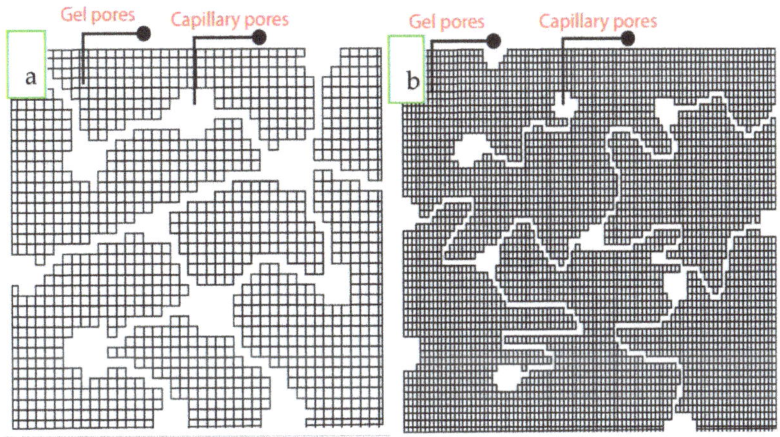

Figure 15. Schematically pore structure of (**a**) AAF and (**b**) AAFS.

Additionally, ESR has emerged as a valuable method for assessing chloride ion ingress and rebar corrosion in cement-based materials. A noteworthy reference point is the ESR value of 12 kΩ·cm, which has been established as the corrosion threshold for internal steel reinforcing bars. However, the observed low ESR values in AAFS specimens may be attributed to the presence of ions such as sodium in the pore solution [19]. Therefore, it is recommended to conduct further investigations into ESR and rebar corrosion in alkali-activated materials. Furthermore, Lee et al. demonstrated that the addition of slag resulted in a substantial reduction in chloride ion penetration depth, attributed to the denser and more convoluted matrix of AAFS compared to AAF [28]. The ESR results were plotted versus compressive strengths, and the correlation between them was calculated. As shown in Figure 16, a strong correlation was observed between ESR and compressive strength.

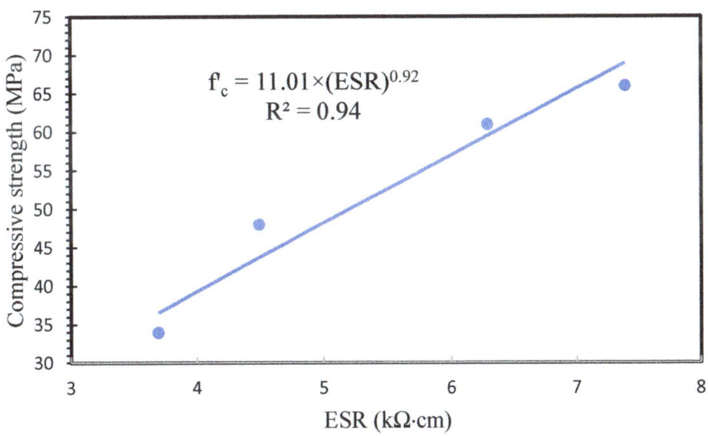

Figure 16. Relationship between ESR and compressive strength values.

4.3.2. Water Vapor Permeability

Water vapor permeability (WVP) values of the specimens are given in Figure 17. The WVP of GP10, GP20, and GP30 decreased by 20.79, 42.97, and 50%, respectively, compared to the GP0. The WVP values are good qualitative concurrence with compressive strength, water permeability, surface area, and porosity. The decrease in WVP with increasing SG is related to the refinement of pore size distribution to a denser and more tortuous matrix. As can be seen in the section of BET, electrical resistivity, and porosity, higher amounts of Ca in the N-A-S-H gel lead to a finer pore structure and more tortuous structure that does not permit higher water vapor transport through the gel.

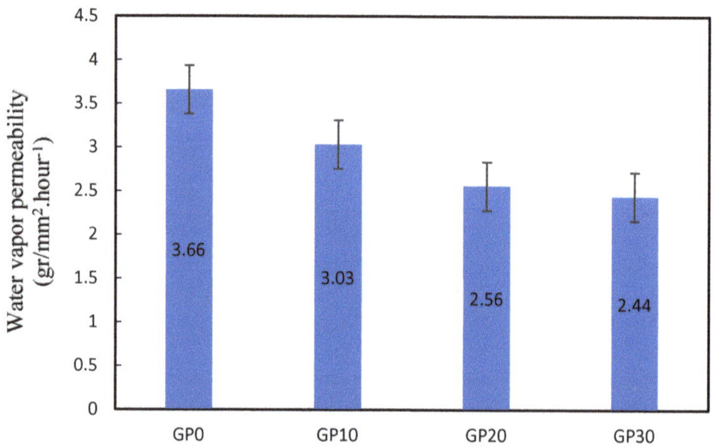

Figure 17. Water vapor permeability (gr/mm^2·hour^{-1}).

4.3.3. Water Permeability Coefficient

Figure 18 illustrates the water permeability coefficient (WPC) values of the AAFS specimens. Water permeability tests displayed that SG has a considerable influence on the permeability of AAF. The WPC of GP10, GP20, and GP30 decreased by 49.83, 69.89, and 95.88%, respectively, as compared to the GP0. Figure 19 shows the correlation between compressive strength/porosity and WPC. A strong linear correlation was observed between compressive strength–WPC and porosity–WPC. These results can be ascribed to the decrease in porosity and pore connectivity of the AAFS by increasing the SG content.

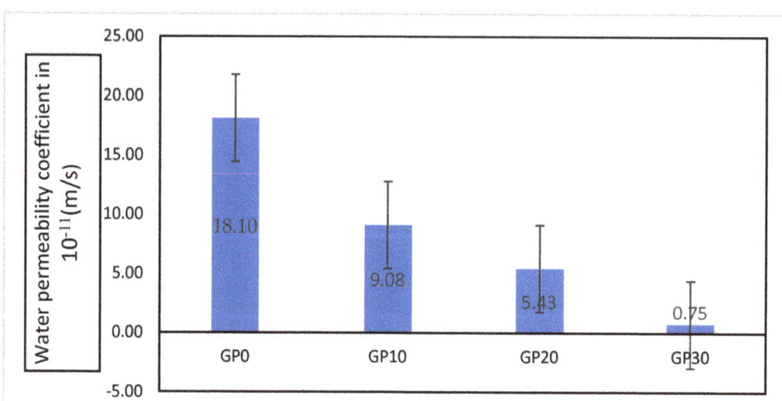

Figure 18. Water permeability coefficient (m/s).

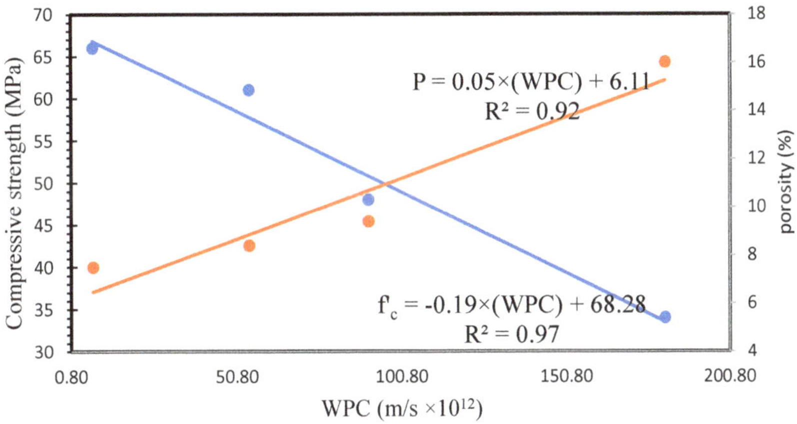

Figure 19. Relationship between compressive strength/porosity and WPC.

Rattanachu et al. [75] studied the relationship between compressive strength and permeability of the conventional concretes, and the results are shown in Figure 20. Comparing our results with conventional concretes revealed that the WPC of conventional concretes was quite higher in the low- and medium-strength concretes than in the AAFS concrete. However, in the high strengths (over 55 MPa), the WPC of conventional concrete tended to a constant value; however, in AAFS, it decreased sharply.

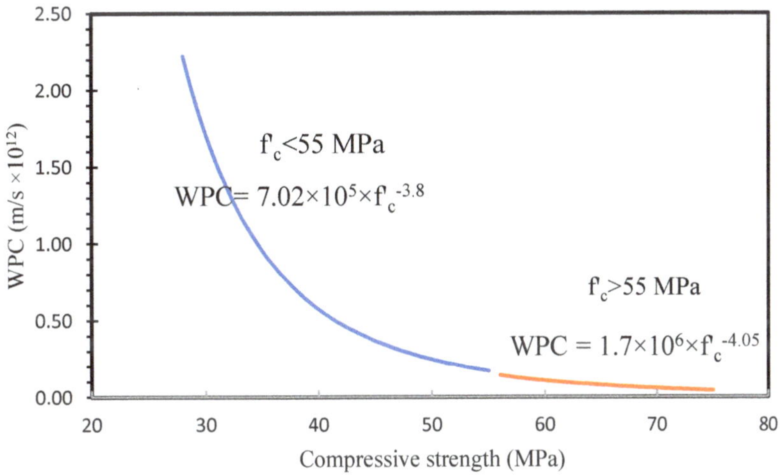

Figure 20. Relationship proposed for conventional concretes.

As a comparison, the WPC of AAFS is 170.2 and 102.5 times higher than that of conventional concretes at the compressive strength of 34 MPa and 66 MPa, respectively. The WPC of conventional concretes at the compressive strength of 34 MPa is 14.6 times higher than that of 66 MPa. Meanwhile, the WPC of AAFS concretes is 24.3 times higher. These findings confirmed that the N-A-S-H gel is more porous than CSH, and utilization of hybrid N-C-A-S-H gel can be an effective way to modify N-A-S-H gel and, consequently, reduce the water permeability.

5. Conclusions

Based on the experimental results, the following conclusions can be expressed:

- The compressive strength and UPV values of the AAFS enhanced with the increase in the slag content. The highest increase in the compressive strength appeared in 20% replacement of slag;
- By slag inclusion, a remarkable refinement in the pore volume was found, and the volume of the gel pores increased significantly;
- The increase in the surface area of the AAFS gel indicated that AAFS had quite more geopolymeric products than AAF gel. Additionally, in concurrence with electrical surface resistivity, the surface area showed that the microstructure of the AAFS gel is more tortuous than AAS gel;
- The SEM/EDX revealed that the increasing Ca ions in the N-A-S-H structure resulted in the highly compacted N-C-A-S-H gel;
- An increase in slag content up to 30% significantly decreased transport properties, which is related to the microstructural changes, i.e., pore volume refinement;
- The connection between UPV and compressive strength was characterized by an exponential relationship, signifying that minor UPV adjustments could result in significant alterations in compressive strength;
- SEM images clearly demonstrated that most of the unreacted FA particles were effectively eliminated due to the inclusion of GGBS, which was a result of the gel densification and highly alkaline environment;
- In comparison to concrete with a compressive strength of 34 MPa, conventional concrete with a strength of 66 MPa exhibited 14.6 times lower water permeability. However, at equivalent strength levels, AAFS showed 24.3 times lower water permeability, revealing the effect of GGBS on the alteration of pore structure.

Author Contributions: Conceptualization, Z.A. and V.T.; methodology, Z.A. and V.T.; formal analysis, Z.A.; investigation, Z.A. and V.T.; resources, Z.A. and V.T.; data curation, Z.A. and V.T.; writing—original draft preparation, Z.A.; writing—review and editing, Z.A. and V.T.; visualization, Z.A.; supervision, V.T.; project administration, V.T. All authors have read and agreed to the published version of the manuscript.

Funding: There is no external funding for this research.

Institutional Review Board Statement: No applicable.

Informed Consent Statement: No applicable.

Data Availability Statement: No new data were created or analyzed in this study. Data sharing is not applicable to this article.

Conflicts of Interest: The authors declare no conflict of interest.

References

1. Gursel, A.P.; Masanet, E.; Horvath, A.; Stadel, A. Life-cycle inventory analysis of concrete production: A critical review. *Cem. Concr. Compos.* **2014**, *51*, 38–48. [CrossRef]
2. Shariatmadari, N.; Hasanzadehshooiili, H.; Ghadir, P.; Saeidi, F.; Moharami, F. Compressive Strength of Sandy Soils Stabilized with Alkali-Activated Volcanic Ash and Slag. *J. Mater. Civ. Eng.* **2021**, *33*, 04021295. [CrossRef]
3. Ahmadi, M.; Hakimi, B.; Mazaheri, A.; Kioumarsi, M. Potential Use of Water Treatment Sludge as Partial Replacement for Clay in Eco-Friendly Fired Clay Bricks. *Sustainability* **2023**, *15*, 9389. [CrossRef]
4. Sattarifard, A.R.; Ahmadi, M.; Dalvand, A.; Sattarifard, A.R. Fresh and hardened-state properties of hybrid fiber–reinforced high-strength self-compacting cementitious composites. *Constr. Build. Mater.* **2022**, *318*, 125874. [CrossRef]
5. Tang, W.; Pignatta, G.; Sepasgozar, S.M.E. Life-cycle assessment of fly ash and cenosphere-based geopolymer material. *Sustainability* **2021**, *13*, 11167. [CrossRef]
6. Rabie, M.; Irshidat, M.R.; Al-Nuaimi, N. Ambient and Heat-Cured Geopolymer Composites: Mix Design Optimization and Life Cycle Assessment. *Sustainability* **2022**, *14*, 4942. [CrossRef]

7. Imtiaz, L.; Kashif-Ur-rehman, S.; Alaloul, W.S.; Nazir, K.; Javed, M.F.; Aslam, F.; Musarat, M.A. Life cycle impact assessment of recycled aggregate concrete, geopolymer concrete, and recycled aggregate-based geopolymer concrete. *Sustainability* **2021**, *13*, 13515. [CrossRef]
8. Bazli, M.; Ashrafi, H.; Rajabipour, A.; Kutay, C. 3D printing for remote housing: Benefits and challenges. *Autom. Constr.* **2023**, *148*, 104772. [CrossRef]
9. Davidovits, J. Geopolymers: Inorganic polymeric new materials. *J. Therm. Anal.* **1991**, *37*, 1633–1656. [CrossRef]
10. Razeghi, H.R.; Ghadir, P.; Javadi, A.A. Mechanical Strength of Saline Sandy Soils Stabilized with Alkali-Activated Cements. *Sustainability* **2022**, *14*, 13669. [CrossRef]
11. Nouri, H.; Ghadir, P.; Fatehi, H.; Shariatmadari, N.; Saberian, M. Effects of Protein-Based Biopolymer on Geotechnical Properties of Salt-Affected Sandy Soil. *Geotech. Geol. Eng.* **2022**, *40*, 5739–5753. [CrossRef]
12. Komaei, A.; Noorzad, A.; Ghadir, P. Stabilization and solidification of arsenic contaminated silty sand using alkaline activated slag. *J. Environ. Manag.* **2023**, *344*, 118395. [CrossRef] [PubMed]
13. Ghadir, P.; Ranjbar, N. Clayey soil stabilization using geopolymer and Portland cement. *Constr. Build. Mater.* **2018**, *188*, 361–371. [CrossRef]
14. Miraki, H.; Shariatmadari, N.; Ghadir, P.; Jahandari, S.; Tao, Z.; Siddique, R. Clayey soil stabilization using alkali-activated volcanic ash and slag. *J. Rock Mech. Geotech. Eng.* **2022**, *14*, 576–591. [CrossRef]
15. Yao, X.; Zhang, Z.; Zhu, H.; Chen, Y. Geopolymerization process of alkali-metakaolinite characterized by isothermal calorimetry. *Thermochim. Acta* **2009**, *493*, 49–54. [CrossRef]
16. Zhu, X.; Yan, D.; Fang, H.; Chen, S.; Ye, H. Early-stage geopolymerization revealed by 27Al and 29Si nuclear magnetic resonance spectroscopy based on vacuum dehydration. *Constr. Build. Mater.* **2021**, *266*, 121114. [CrossRef]
17. Park, S.; Pour-Ghaz, M. What is the role of water in the geopolymerization of metakaolin? *Constr. Build. Mater.* **2018**, *182*, 360–370. [CrossRef]
18. Atiş, C.D.; Görür, E.B.; Karahan, O.; Bilim, C.; Ilkentapar, S.; Luga, E. Very high strength (120 MPa) class F fly ash geopolymer mortar activated at different NaOH amount, heat curing temperature and heat curing duration. *Constr. Build. Mater.* **2015**, *96*, 673–678. [CrossRef]
19. Noushini, A.; Castel, A. The effect of heat-curing on transport properties of low-calcium fly ash-based geopolymer concrete. *Constr. Build. Mater.* **2016**, *112*, 464–477. [CrossRef]
20. Puligilla, S.; Mondal, P. Role of slag in microstructural development and hardening of fly ash-slag geopolymer. *Cem. Concr. Res.* **2013**, *43*, 70–80. [CrossRef]
21. Singh, B.; Rahman, M.R.; Paswan, R.; Bhattacharyya, S.K. Effect of activator concentration on the strength, ITZ and drying shrinkage of fly ash/slag geopolymer concrete. *Constr. Build. Mater.* **2016**, *118*, 171–179. [CrossRef]
22. Nath, P.; Sarker, P.K. Effect of GGBFS on setting, workability and early strength properties of fly ash geopolymer concrete cured in ambient condition. *Constr. Build. Mater.* **2014**, *66*, 163–171. [CrossRef]
23. Aliabdo, A.A.; Abd Elmoaty, A.E.M.; Salem, H.A. Effect of cement addition, solution resting time and curing characteristics on fly ash based geopolymer concrete performance. *Constr. Build. Mater.* **2016**, *123*, 581–593. [CrossRef]
24. Rafeet, A.; Vinai, R.; Soutsos, M.; Sha, W. Effects of slag substitution on physical and mechanical properties of fly ash-based alkali activated binders (AABs). *Cem. Concr. Res.* **2019**, *122*, 118–135. [CrossRef]
25. Ismail, I.; Bernal, S.A.; Provis, J.L.; San Nicolas, R.; Hamdan, S.; Van Deventer, J.S.J. Modification of phase evolution in alkali-activated blast furnace slag by the incorporation of fly ash. *Cem. Concr. Compos.* **2014**, *45*, 125–135. [CrossRef]
26. Karimpour, H.; Mazloom, M. Pseudo-strain hardening and mechanical properties of green cementitious composites containing polypropylene fibers. *Struct. Eng. Mech.* **2022**, *81*, 575–589.
27. Zhang, W.; Yao, X.; Yang, T.; Zhang, Z. The degradation mechanisms of alkali-activated fly ash/slag blend cements exposed to sulphuric acid. *Constr. Build. Mater.* **2018**, *186*, 1177–1187. [CrossRef]
28. Lee, N.K.; Lee, H.K. Influence of the slag content on the chloride and sulfuric acid resistances of alkali-activated fly ash/slag paste. *Cem. Concr. Compos.* **2016**, *72*, 168–179. [CrossRef]
29. Ren, J.; Zhang, L.; Nicolas, R.S. Degradation process of alkali-activated slag/fly ash and Portland cement-based pastes exposed to phosphoric acid. *Constr. Build. Mater.* **2020**, *232*, 117209. [CrossRef]
30. Komljenović, M.; Baščarević, Z.; Marjanović, N.; Nikolić, V. External sulfate attack on alkali-activated slag. *Constr. Build. Mater.* **2013**, *49*, 31–39. [CrossRef]
31. Yio, M.H.N.; Wong, H.S.; Buenfeld, N.R. 3D pore structure and mass transport properties of blended cementitious materials. *Cem. Concr. Res.* **2019**, *117*, 23–37. [CrossRef]
32. Shafikhani, M.; Chidiac, S.E. A holistic model for cement paste and concrete chloride diffusion coefficient. *Cem. Concr. Res.* **2020**, *133*, 106049. [CrossRef]
33. Wilson, W.; Gonthier, J.N.; Georget, F.; Scrivener, K. Tortuosity as a Key Parameter of Chloride Diffusion in LC3 Systems. In *Calcined Clays for Sustainable Concrete*; RILEM Bookseries; Springer: Singapore, 2020; Volume 25.
34. Addassi, M.; Johannesson, B. Reactive mass transport in concrete including for gaseous constituents using a two-phase moisture transport approach. *Constr. Build. Mater.* **2020**, *232*, 117148. [CrossRef]
35. Bai, Y.; Wang, Y.; Xi, Y. Modeling the effect of temperature gradient on moisture and ionic transport in concrete. *Cem. Concr. Compos.* **2020**, *106*, 103454. [CrossRef]

36. Casnedi, L.; Cappai, M.; Cincotti, A.; Delogu, F.; Pia, G. Porosity effects on water vapour permeability in earthen materials: Experimental evidence and modelling description. *J. Build. Eng.* **2020**, *27*, 100987. [CrossRef]
37. Zhang, P.; Cong, Y.; Vogel, M.; Liu, Z.; Müller, H.S.; Zhu, Y.; Zhao, T. Steel reinforcement corrosion in concrete under combined actions: The role of freeze-thaw cycles, chloride ingress, and surface impregnation. *Constr. Build. Mater.* **2017**, *148*, 113–121. [CrossRef]
38. Sola, E.; Ožbolt, J.; Balabanić, G.; Mir, Z.M. Experimental and numerical study of accelerated corrosion of steel reinforcement in concrete: Transport of corrosion products. *Cem. Concr. Res.* **2019**, *120*, 119–131. [CrossRef]
39. Samadi, P.; Ghodrati, A.; Ghadir, P.; Javadi, A.A. Effect of Seawater on the Mechanical Strength of Geopolymer/Cement Stabilized Sandy Soils. In *Proceedings of the TMIC 2022 Slope Stability Conference (TMIC 2022)*; Atlantis Press: Amsterdam, The Netherlands, 2023.
40. Kazemi, H.; Yekrangnia, M.; Shakiba, M.; Bazli, M.; Vatani Oskouei, A. Bond-slip behaviour between GFRP/steel bars and seawater concrete after exposure to environmental conditions. *Eng. Struct.* **2022**, *268*, 114796. [CrossRef]
41. Cardoso, R. Porosity and tortuosity influence on geophysical properties of an artificially cemented sand. *Eng. Geol.* **2016**, *211*, 198–207. [CrossRef]
42. Ghasemalizadeh, S.; Toufigh, V. Durability of Rammed Earth Materials. *Int. J. Geomech.* **2020**, *20*, 04020201. [CrossRef]
43. Ghassemi, P.; Toufigh, V. Durability of epoxy polymer and ordinary cement concrete in aggressive environments. *Constr. Build. Mater.* **2020**, *234*, 117887. [CrossRef]
44. Sepehrinezhad, A.; Toufigh, V. The evaluation of distributed damage in concrete based on sinusoidal modeling of the ultrasonic response. *Ultrasonics* **2018**, *89*, 195–205. [CrossRef] [PubMed]
45. Heidarnezhad, F.; Toufigh, V.; Ghaemian, M. Analyzing and predicting permeability coefficient of roller-compacted concrete (RCC). *J. Test. Eval.* **2021**, *49*, 20180718. [CrossRef]
46. Nofalah, M.-H.; Ghadir, P.; Hasanzadehshooiili, H.; Aminpour, M.; Javadi, A.A.; Nazem, M. Effects of Binder Proportion and Curing Condition on the Mechanical Characteristics of Volcanic Ash- and Slag-Based Geopolymer Mortars; Machine Learning Integrated Experimental Study. *Constr. Build. Mater.* **2023**, *395*, 132330. Available online: https://linkinghub.elsevier.com/retrieve/pii/S0950061823020469 (accessed on 1 September 2023). [CrossRef]
47. Ghadir, P.; Razeghi, H.R. Effects of sodium chloride on the mechanical strength of alkali activated volcanic ash and slag pastes under room and elevated temperatures. *Constr. Build. Mater.* **2022**, *344*, 128113. [CrossRef]
48. Matalkah, F.; Ababneh, A.; Aqel, R. Effects of nanomaterials on mechanical properties, durability characteristics and microstructural features of alkali-activated binders: A comprehensive review. *Constr. Build. Mater.* **2022**, *336*, 127545. [CrossRef]
49. Razeghi, H.R.; Geranghadr, A.; Safaee, F.; Ghadir, P.; Javadi, A.A. Effect of CO_2 exposure on the mechanical strength of geopolymer-stabilized sandy soils. *J. Rock Mech. Geotech. Eng.* 2023, in press. [CrossRef]
50. Noushini, A.; Castel, A.; Aldred, J.; Rawal, A. Chloride diffusion resistance and chloride binding capacity of fly ash-based geopolymer concrete. *Cem. Concr. Compos.* **2020**, *105*, 103290. [CrossRef]
51. Saloni; Parveen; Lim, Y.Y.; Pham, T.M. Influence of Portland cement on performance of fine rice husk ash geopolymer concrete: Strength and permeability properties. *Constr. Build. Mater.* **2021**, *300*, 124321. [CrossRef]
52. Mohamed, O.A.; Al Khattab, R. Fresh Properties and Sulfuric Acid Resistance of Sustainable Mortar Using Alkali-Activated GGBS/Fly Ash Binder. *Polymers* **2022**, *14*, 591. [CrossRef]
53. Zhang, X.; Zhang, X.; Li, X.; Liu, Y.; Yu, H.; Ma, M. Porous geopolymer with controllable interconnected pores—A viable permeable reactive barrier filler for lead pollutant removal. *Chemosphere* **2022**, *307*, 136128. [CrossRef] [PubMed]
54. Toufigh, V.; Karamian, M.H.; Ghasemalizadeh, S. Study of stress–strain and volume change behavior of fly ash-GBFS based geopolymer rammed earth. *Bull. Eng. Geol. Environ.* **2021**, *80*, 6749–6767. [CrossRef]
55. *ASTM C33/C33M-18*; Standard Specification for Concrete Aggregates. ASTM International: West Conshohocken, PA, USA, 2020.
56. *ASTM C128*; Standard Test Method for Specific Gravity and Absorption of Fine Aggregate. ASTM International: West Conshohocken, PA, USA, 1998.
57. *ASTM C127*; Standard Test Method for Density, Relative Density (Specific Gravity), and Absorption. ASTM International: West Conshohocken, PA, USA, 2015.
58. Ding, Y.; Dai, J.G.; Shi, C.J. Mechanical properties of alkali-activated concrete: A state-of-the-art review. *Constr. Build. Mater.* **2016**, *127*, 68–79. [CrossRef]
59. *BS EN 12390-3*; Testing Hardened Concrete, Part 3: Compressive Strength of Test Specimens. BSI: London, UK, 2019.
60. *ASTM C597*; Standard Test Method for Pulse Velocity Through Concrete. The American Society for Testing and Materials: West Conshohocken, PA, USA, 2016.
61. *ASTM C642*; Standard Test Method for Density, Absorption, and Voids in Hardened Concrete. ASTM International: West Conshohocken, PA, USA, 2006.
62. *AASHTO T 358*; Standard Method of Test for Surface Resistivity Indication of Concrete's Ability to Resist Chloride Ion Penetration. The American Association of State Highway and Transportation Officials: Washington, DC, USA, 2019.
63. *ASTM E96/E96M-16*; Standard Test Methods for Water Vapor Transmission of Materials, ASTM International. ASTM International: West Conshohocken, PA, USA, 2016.
64. Fang, G.; Ho, W.K.; Tu, W.; Zhang, M. Workability and mechanical properties of alkali-activated fly ash-slag concrete cured at ambient temperature. *Constr. Build. Mater.* **2018**, *172*, 476–487. [CrossRef]

65. Rahmati, M.; Toufigh, V. Evaluation of geopolymer concrete at high temperatures: An experimental study using machine learning. *J. Clean. Prod.* **2022**, *372*, 133608. [CrossRef]
66. Mata, R.; Ruiz, R.O.; Nuñez, E. Correlation between compressive strength of concrete and ultrasonic pulse velocity: A case of study and a new correlation method. *Constr. Build. Mater.* **2023**, *369*, 130569. [CrossRef]
67. Rahmati, M.; Toufigh, V.; Keyvan, K. Monitoring of crack healing in geopolymer concrete using a nonlinear ultrasound approach in phase-space domain. *Ultrasonics* **2023**, *134*, 107095. [CrossRef]
68. Mehta, A.; Siddique, R.; Ozbakkaloglu, T.; Shaikh, F.U.A.; Belarbi, R. Fly ash and ground granulated blast furnace slag-based alkali-activated concrete: Mechanical, transport and microstructural properties. *Constr. Build. Mater.* **2020**, *257*, 119548. [CrossRef]
69. Barrett, E.P.; Joyner, L.G.; Halenda, P.P. The Determination of Pore Volume and Area Distributions in Porous Substances. I. Computations from Nitrogen Isotherms. *J. Am. Chem. Soc.* **1951**, *73*, 373–380. [CrossRef]
70. Hanif, A.; Parthasarathy, P.; Ma, H.; Fan, T.; Li, Z. Properties improvement of fly ash cenosphere modified cement pastes using nano silica. *Cem. Concr. Compos.* **2017**, *81*, 35–48. [CrossRef]
71. Brunauer, S.; Emmett, P.H.; Teller, E. Adsorption of Gases in Multimolecular Layers. *J. Am. Chem. Soc.* **1938**, *60*, 309–319. [CrossRef]
72. Odler, I. Reply to the discussion by H.M. Jennings and J.J. Thomas of the paper "The BET-specific surface area of hydrated Portland cement and related materials.". *Cem. Concr. Res.* **2004**, *34*, 1961. [CrossRef]
73. Yazdi, M.A.; Liebscher, M.; Hempel, S.; Yang, J.; Mechtcherine, V. Correlation of microstructural and mechanical properties of geopolymers produced from fly ash and slag at room temperature. *Constr. Build. Mater.* **2018**, *191*, 330–341. [CrossRef]
74. Jang, J.G.; Lee, H.K. Effect of fly ash characteristics on delayed high-strength development of geopolymers. *Constr. Build. Mater.* **2016**, *102*, 260–269. [CrossRef]
75. Rattanachu, P.; Tangchirapat, W.; Jaturapitakkul, C. Water Permeability and Sulfate Resistance of Eco-Friendly High-Strength Concrete Composed of Ground Bagasse Ash and Recycled Concrete Aggregate. *J. Mater. Civ. Eng.* **2019**, *31*. [CrossRef]

sustainability

Article

Effect of Nanosilica on the Strength and Durability of Cold-Bonded Fly Ash Aggregate Concrete

Xiuzhong Peng [1,2], Qinghua Wang [2,3] and Jing Wu [2,4,*]

1 School of Civil Engineering, Southeast University, Nanjing 211189, China; pengxiuzhong163@163.com
2 College of Construction Engineering, Jilin University, Changchun 130021, China
3 Shanghai Geological Engineering Exploration (Group) Co., Ltd., Shanghai 200072, China
4 Beijing Institute of Architectural Design, Beijing 100045, China
* Correspondence: wujing3@biad.com.cn

Abstract: Cold-bonded Fly Ash Aggregate (CFAA), as an alternative to natural coarse aggregates, can prepare more lightweight, economical, and sustainable concrete. However, CFAA concrete has insufficient durability, which hinders its application in a salt-corrosion environment. Nanosilica (NS) has an advantage of high activity and is generally used as an efficient mineral admixture in engineering. This study aims to improve the strength and durability of CFAA concrete by incorporating NS. To this end, compression tests, splitting tensile tests, and microscopic analyses were performed to investigate the mechanical properties of the concrete containing different NS dosages. Subsequently, the dry–wet and freeze–thaw durability tests were conducted to evaluate the salt-corrosion resistance and the frost resistance in the water, Na_2SO_4 solution, and Na_2CO_3 solution. The results show the compressive and splitting tensile strength peak at 2 wt% NS dosage. In this instance, the concrete has an optimum microstructure and exhibits desirable salt-corrosion resistance in the late stage of dry–wet cycles. During freeze–thaw cycles, NS could improve the frost resistance of the concrete but scarcely diminished internal damage under sulfate attack. The study explores the long-term performance of NS-modified CFAA concrete, providing a simple and effective method to mitigate the concrete deterioration in a harsh environment.

Keywords: cold-bonded; fly ash aggregate concrete; nanosilica; mechanical properties; durability; microstructure

Citation: Peng, X.; Wang, Q.; Wu, J. Effect of Nanosilica on the Strength and Durability of Cold-Bonded Fly Ash Aggregate Concrete. *Sustainability* **2023**, *15*, 15413. https://doi.org/10.3390/su152115413

Academic Editors: Mahdi Kioumarsi and Vagelis Plevris

Received: 15 September 2023
Revised: 12 October 2023
Accepted: 27 October 2023
Published: 30 October 2023

1. Introduction

Cold-bonded Fly Ash Aggregate (CFAA) enables the disposal of plentiful industrial wastes to diminish environmental pollution and achieves the production process with less energy consumption than the sintered method [1–3]. In engineering, this aggregate is gradually becoming an alternative to natural aggregates to produce lightweight concrete [4–6]. CFAA concrete possesses the potential to decrease dead loads, reduce component sizes, and save transportation costs, having wide application prospects in building materials and structural design [1,7].

CFAA concrete is characterized by an economical and ecofriendly material [8–10]. The production process of CFAA includes pelletization and curing, achieving the recycling of fly ash, which is a by-product of thermal power plants [11–13]. Pelletization means that water is sprayed on a mixture of fly ash and cement in a rotating disc to form almost spherical particles. Curing aims to allow these particles to reach the desired strength through hydration reaction and pozzolanic reaction at ambient temperature. The manufacture of CFAA is simple and efficient, and only requires a small amount of energy during the curing process [14–16]. The superiority in cost makes CFAA concrete more attractive than other lightweight aggregate concrete, such as vermiculite concrete [17], perlite concrete [18], and pumice concrete [19]. Since aggregate makes up at least 60% of concrete by volume, the use of CFAA in concrete would consume fly ash rapidly [20,21]. The decrease in the large amount of the industrial waste contributes to avoiding the adverse effect on human health and nature [22].

Despite these advantages, CFAA concrete generally has a lower strength, owing to aggregate crushing and matrix cracking [23]. Ibrahim et al. [24] found CFAA concrete had weaker bond strength compared to ordinary concrete through reinforcement steel bar pull-out tests. Some studies focused on strengthening aggregates to improve concrete performance [25,26]. Patel et al. [27] took styrene butadiene rubber latex as a binder to manufacture fly ash aggregates, making the concrete obtain desirable compressive strength and transport properties. Yıldırım and Özturan [28] added polypropylene fibers to the aggregates so that the impact resistance of the concrete was enhanced. Moreover, some researchers pointed out that mineral admixtures can optimize the microstructure of a cementitious matrix and thus enhance the mechanical properties of CFAA concrete [29,30]. Joseph and Ramamurthy [31] advised that concrete could obtain a higher strength through replacing sand rather than cement with fly ash. Güneyisi et al. [32] reported that Silica Fume (SF) increased the compressive strength of self-compacting CFAA concrete, whereas fly ash might reduce the strength. Due to its high specific surface area, Nanosilica (NS) has better physical effects and chemical properties than conventional mineral admixtures [33,34]. Recent studies [35,36] discovered that NS with smaller particle sizes could accelerate cement hydration and have the higher pozzolanic activity to improve the mortar performance. Son et al. [37] revealed that the addition of NS decreased the generation rate of metastable hydrates, maintaining the long-term strength of calcium aluminate cement. With the development of nanotechnology, the use of NS in concrete has grown in recent years [38,39]. Wang et al. [40] demonstrated that NS-modified underwater concrete had higher compressive strength and lower permeability, contributing to resisting seawater attacks. Mukharjee and Barai [41] applied NS to recycled aggregate concrete to mitigate performance degradation due to the inherent defects of the aggregates. Farajzadehha et al. [42] found that the strength of plastic concrete increased with the growth of NS dosage through the uniaxial and triaxial compression tests. Murad et al. [43] studied the cyclic behavior of beam—column joints made with NS-modified concrete and discovered that the failure mode of the specimens was changed due to the incorporation of NS. Mashshay et al. [44] illustrated that NS could alleviate the postfire mechanical degradation of lightweight concrete.

Furthermore, the durability of CFAA concrete is a matter of concern, especially in a harsh environment. In saline soil areas, concrete suffers salt attacks and undergoes dry–wet and freeze–thaw cycles as precipitation and temperatures vary seasonally. As a result, the surface spallation and microcracks of the concrete can continually arise, leading to a dramatic reduction in performance. In terms of natural aggregate concrete, some advances in durability have been reported. Wang et al. [45] performed uniaxial compression tests to explore the concrete deterioration under the coupling of loads and wet–dry cycles. Liu et al. [46] stated that the physical and mechanical properties increased first and then decreased when the concrete was subjected to sulfate attacks and dry-wet cycles. Guo et al. [47] and Wang et al. [48] concluded that low dry–wet time ratio exacerbated the chemical erosion of sulfates and decreased the crystallization pressure. Zhou et al. [49] showed that salt attacks and initial load damage caused more strength loss during freeze–thaw cycles. Wang et al. [50] and Chen et al. [51] established freeze–thaw damage models to predict the damage degree of the concrete exposed to a saline environment. Yang et al. [52] believed that concrete deterioration would greatly accelerate under the coupling action of sulfate attacks, dry–wet cycles, and freeze–thaw cycles. However, the research on the durability of CFAA concrete is still in its infancy.

Present studies [45–52] are mostly related to the durability of natural aggregate concrete. It is difficult to directly apply these results to the durability improvement of lightweight concrete due to the differences in aggregates. Consequently, it is a considerable challenge to solve the deterioration of CFAA concrete in a salt-corrosion environment. To this end, it is necessary to find a simple and effective method to strengthen the matrix performance, helping the concrete resist salt erosion. In this study, partial cement was substituted with NS to improve the long-term performance of CFAA concrete subjected to sulfate and carbonate attacks.

This study aimed to explore the mechanical properties and durability of NS-modified CFAA concrete. Compression tests and splitting tensile tests were conducted to find the optimum NS dosage. The failure morphologies and microstructure images of the specimens were observed to analyze the influence mechanism of NS. Subsequently, the dry–wet and freeze–thaw cycle tests were performed to evaluate the salt-corrosion resistance and the frost resistance in the water, Na_2SO_4 solution, and Na_2CO_3 solution.

The highlights of this work are summarized below:

1. CFAA concrete had the optimum mechanical properties at 2% NS dosage.
2. The porosity of CFAA concrete was reduced obviously by the moderate use of NS.
3. NS improved the salt-corrosion resistance of CFAA concrete during dry–wet cycles.
4. The desirable freeze–thaw resistance under salt attacks could be achieved by incorporating NS in CFAA concrete.

2. Materials and Methods

2.1. Raw Materials

2.1.1. Cement

Ordinary Portland cement (Jilin Yatai Dinglu Cement Co., Ltd., Changchun, China) with a strength grade of 42.5 MPa was used in this experiment, conforming to the Chinese standard GB 175-2007 [53]. The properties and chemical composition of cement are listed in Table 1.

Table 1. Properties and chemical composition of cement, Silica Fume (SF), and Nanosilica (NS).

Items	Cement	SF	NS
Properties			
Average particle size (um)	–	0.1–0.3	0.02
Specific gravity	3.12	2.18	–
Specific surface area (m^2/kg)	358	18,000	200,000
Loss on ignition (wt%)	1.6	2.0	2.0
Chemical composition (wt%)			
SiO_2	20.62	93.30	>99.50
Al_2O_3	5.14	0.73	–
CaO	65.07	0.85	–
MgO	0.87	1.21	–
Fe_2O_3	3.96	0.49	–

2.1.2. Mineral Admixture

The added mineral admixtures included NS (Jiangsu Tianxing New Material Co., Ltd., Huaian, China) and SF (Changchun Siao Technology Co., Ltd., Changchun, China). NS, a white powder with high dispersion, has PH values ranging from 7 to 9 and a loose bulk density of 100 kg/m^3. SF can be combined with NS to optimize the particle gradation of cementitious materials [54]. Table 1 also gives the properties and chemical composition of both SF and NS.

2.1.3. Coarse Aggregate

CFAA (Gongyi Lanke Water Purification Material Co., Ltd., Zhengzhou, China) with particle sizes of 3–5 mm was chosen as the coarse aggregate, as shown in Figure 1a. In Figure 1b, the Scanning Electron Microscope (SEM) image of the aggregate indicated CFAA was full of Calcium-Silicate-Hydrate (C-S-H) gels. Figure 2a shows that the grading curve of the CFAA was within the boundaries of the nominal size grade (5–10 mm) according to the Chinese standard GB/T 17431.1-2010 [55]. It indicated that the particle gradation of the CFAA met the grading requirements of a lightweight coarse aggregate. The physical properties of CFAA are tabulated in Table 2.

(a) (b)

Figure 1. Characteristics of Cold-bonded Fly Ash Aggregate (CFAA): (**a**) the CFAA with particle sizes of 3–5 mm; (**b**) the Scanning Electron Microscope (SEM) image of the aggregate.

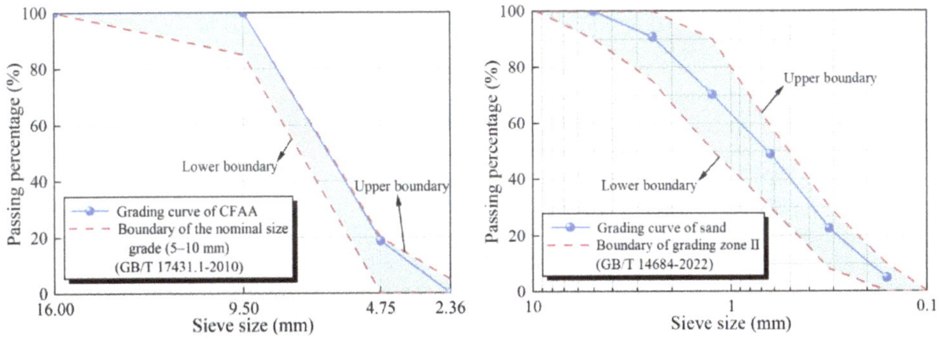

(a) (b)

Figure 2. Sieve test results of CFAA and sand: (**a**) grading curve of CFAA; (**b**) grading curve of sand.

Table 2. Physical properties of Cold-bonded Fly Ash Aggregate (CFAA) and sand.

Items	CFAA	Sand
Bulk density (kg/m^3)	920	1670
Apparent density (kg/m^3)	1582	2663
Water absorption (wt%)		
1 h	19	–
24 h	20	–

2.1.4. Sand

The fine aggregate was river sand with a fineness modulus of 2.58, and its sieve test results are illustrated in Figure 2b. The results showed that the gradation curve of the sand was in the range of the gradation zone II specified by the Chinese standard GB/T 14684-2022 [56]. Table 2 also presents the physical properties of sand.

2.1.5. Water

Tap water (Changchun Water Co., Ltd., Changchun, China) was used to prepare fresh concrete and cure specimens.

2.1.6. Superplasticizer

Polycarboxylate superplasticizer, a transparent liquid with a water reduction rate of 25%, was applied to attain the desired workability of the fresh concrete.

2.2. Mix Design

The mix design of CFAA concrete was performed by the loose volume method based on the Chinese standard JGJ/T 12-2019 [57]. The target compressive strength of the concrete was set to 40 MPa. During mix design, the total cementitious materials, the water-to-binder ratio, and the loose volume rate of sand were taken as 435 kg/m^3, 0.35, and 35 vol%, respectively. For CFAA concrete without NS, the mix proportion was labeled as N0. The cementitious materials of N0 contained 93 wt% cement and 7 wt% SF. The previous studies [58–62] demonstrated that the mechanical properties of concrete can be improved significantly when the NS dosage is within the range from 1 wt% to 4 wt%. In this study, NS partially replaced cement and SF to prepare CFAA concrete with different NS dosages (1 wt%, 2 wt%, 3 wt%, and 4 wt%). The resulting mix proportions were labeled N1, N2, N3, and N4, respectively. The dosage of superplasticizer was determined as 1.5 wt% of the total cementitious materials. Table 3 lists the mix proportions of CFAA concrete.

Table 3. Mix proportions of CFAA concrete.

Mix No.	NS Dosage (wt%)	Weight per Cubic Meter (kg/m^3)						
		Cement	SF	NS	CFAA	Sand	Water	Superplasticizer
N0	0	404.55	30.45	0	717.6	701.4	152.25	6.525
N1	1	400.50	30.15	4.35	717.6	701.4	152.25	6.525
N2	2	396.46	29.84	8.70	717.6	701.4	152.25	6.525
N3	3	392.41	29.54	13.05	717.6	701.4	152.25	6.525
N4	4	388.37	29.23	17.40	717.6	701.4	152.25	6.525

2.3. Specimen Preparation

The preparation process of specimens is depicted in Figure 3. CFAAs must be soaked in water for an hour and then dried with a towel. In a concrete mixer, these aggregates were mixed with cementitious materials and sand for 30 s. After adding water and superplasticizer, the mixture continued to be stirred for 150 s. The resulting fresh concrete was poured into molds to prepare the cubic specimens (100 mm × 100 mm ×100 mm) and the prismatic specimens (100 mm × 100 mm × 400 mm). In order to prevent the aggregates from floating in the mortar, the rodding method was adopted to compact the concrete mixture. After 24 h, all specimens were demolded and marked with identification codes. The specimens would be moved into a water storage tank within 30 min and cured at ambient temperature until the required age.

2.4. Saline Solution Preparation

In the western Jilin Province of China, there is a mass of sulfate ions and carbonate ions in the soil. The concentration of these ions ranges from 2.4 wt% to 5.1 wt% [50]. In this experiment, Na_2SO_4 solution and Na_2CO_3 solution were prepared with distilled water. The molar concentration of the solutions was set to 0.35 mol/L to ensure that the experimental conditions could simulate the actual situation in saline soil areas.

2.5. Test Methods

The strength tests and microscopic tests of CFAA concrete were conducted to determine the optimum NS dosage. The concrete with the optimum NS dosage was denoted as NOP. Subsequently, the durability in a salt-corrosion environment was studied through dry–wet and freeze–thaw cycle tests. The test procedure of the specimens is plotted in Figure 4.

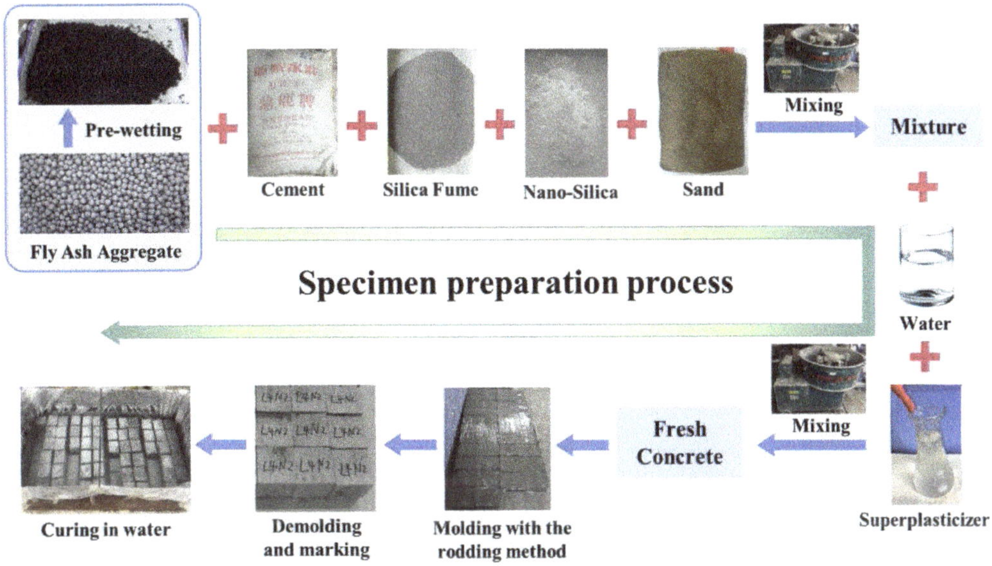

Figure 3. Preparation process of CFAA concrete specimens (the non-English characters appearing above the word "Cement" represent the brand of cement).

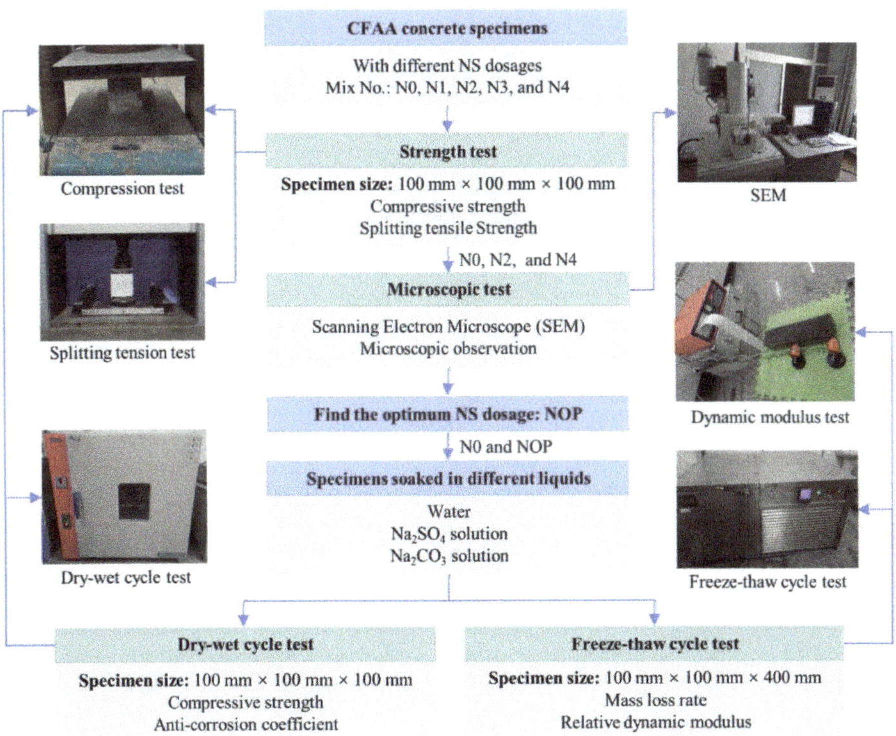

Figure 4. Test procedure of CFAA concrete specimens.

2.5.1. Compressive Strength Testing

Based on the Chinese standard GB/T 50081-2019 [63], the cubic specimens (100 mm × 100 mm × 100 mm) were utilized to perform the compressive strength tests. The concrete (N0, N1, N2, N3, and N4) contained the specimens at a series of ages (i.e., 3 days, 7 days, 14 days, 28 days, and 56 days). The YAW-2000 compression testing machine (Changchun New Testing Machine Co., Ltd., Changchun, China) was used in all compression tests. The loading rate was set to 0.5 MPa/s.

2.5.2. Splitting Tensile Strength Testing

The splitting tensile tests of the cubic specimens (100 mm × 100 mm × 100 mm) were carried on according to the standard GB/T 50081-2019 [63]. After 28 days of curing, the concrete (N0, N1, N2, N3, and N4) was tested using the DNS300 electronic universal testing machine (Changchun Research Institute for Mechanical Science Co., Ltd., Changchun, China) at a loading rate of 0.05 MPa/s.

2.5.3. Microscopic Testing

After the strength tests, three specimens, N0, N2, and N4, were made into SEM samples to observe the microstructure of the cementitious matrix.

2.5.4. Dry–Wet Cycle Testing

The cubic specimens (100 mm × 100 mm × 100 mm) of the concrete (N0 and NOP) were prepared to conduct dry–wet cycle tests. Based on the Chinese standard GB/T 50082-2009 [64], the process of the tests was the following. The 26-day specimens were taken out from the water storage tanks and wiped dry. The 101-3A electric blast drying oven (Shanghai Yetuo Technology Co., Ltd., Shanghai, China) was utilized to dry these specimens for 48 h at 80 °C (±5°C). After the specimens were cooled to the ambient temperature, the first dry–wet cycle was performed. To begin with, the cubes (N0 and NOP) were fully immersed in the distilled water, the Na_2SO_4 solution, and the Na_2CO_3 solution, respectively. After a soak of 15 h, the cubes were lifted out of the liquids and air-dried for an hour. Subsequently, the specimens were dried for 6 h at 80 °C (±5°C) using the oven and then cooled off for 2 h to 25–30 °C. In the end, these cubes were soaked in the liquids again to continue with the next dry–wet cycle. The steps mentioned above made up a 24 h dry-wet cycle. There were 60 dry–wet cycles in this experiment. Every 10 cycles, the cubes were removed from the liquids to test the compressive strength.

The anticorrosion coefficient K_f evaluates the salt-corrosion resistance of the specimens and is expressed as follows:

$$K_f = \frac{f_{cns}}{f_{cnw}} \tag{1}$$

where f_{cns} and f_{cnw} denote the compressive strength of the specimens subjected to n dry–wet cycles in saline solution and water, respectively.

2.5.5. Freeze–Thaw Cycle Testing

The rapid freezing method was adopted in freeze–thaw cycle tests in accordance with the standard GB/T 50082-2009 [64]. For the concretes N0 and NOP at 28 days, the prismatic specimens (100 mm × 100 mm × 400 mm) were moved into the TDRF-1 concrete rapid freezing and thawing test machine (Hangzhou Civit Instrument Equipment Co., Ltd., Hangzhou, China). Before the freeze–thaw cycles started, these specimens needed to be immersed in water, Na_2SO_4 solution, and Na_2CO_3 solution. Every liquid contained three N0 specimens and three NOP specimens. This experiment included 150 freeze–thaw cycles in total. Every 10 cycles, all specimens were taken out to acquire their mass (YP3001 electronic scale) and transverse fundamental frequency (DT-20 dynamic modulus tester). After the measurement, these specimens were put back to continue the cycle tests.

The mass loss rate ΔW_n represents the erosion degree of the specimen surface and is given as follows:

$$\Delta W_n = \frac{W_0 - W_n}{W_0} \times 100\% \quad (2)$$

where W_0 and W_n are the initial mass of specimens and the mass after n freeze–thaw cycles, respectively.

The relative dynamic modulus P estimates the internal damage of the specimens and is defined as follows:

$$P = \left(\frac{v_n}{v_0}\right)^2 \times 100\% \quad (3)$$

where v_0 and v_n denote the initial transverse fundamental frequency of specimens and the transverse fundamental frequency after n freeze–thaw cycles, respectively.

3. Results and Discussion

3.1. Mechanical Properties

3.1.1. Compressive Strength

Figure 5 presents the compressive strength and strength growth rate of CFAA concrete containing different NS dosages. The compressive strength of the concrete rises with increasing curing time. For the concrete at the same age, the strength growth rates peak at 2 wt% NS dosage. In the concrete, NS not only reacts with Calcium Hydroxide (CH) to form extra C-S-H gels but also fills nanoscale pores in the cementitious matrix [54]. The compressive strength would increase due to the strengthened matrix when the NS dosage varies from 0 to 2 wt%. Note that replacing cement partially with NS can save cement but lead to the less production of the CH resulting from cement hydration. When the CH is used up, the excess NS scarcely continues to produce C-S-H gels, reducing the total amount of the C-S-H gel. Meanwhile, the excess NS particles might agglomerate together to form weak-zones in the concrete [65]. Therefore, the compressive strength would decrease when the NS dosage exceeds 2 wt%. This result shows the optimum NS dosage of the concrete is 2 wt%, i.e., N2 is the NOP in Section 2.5.

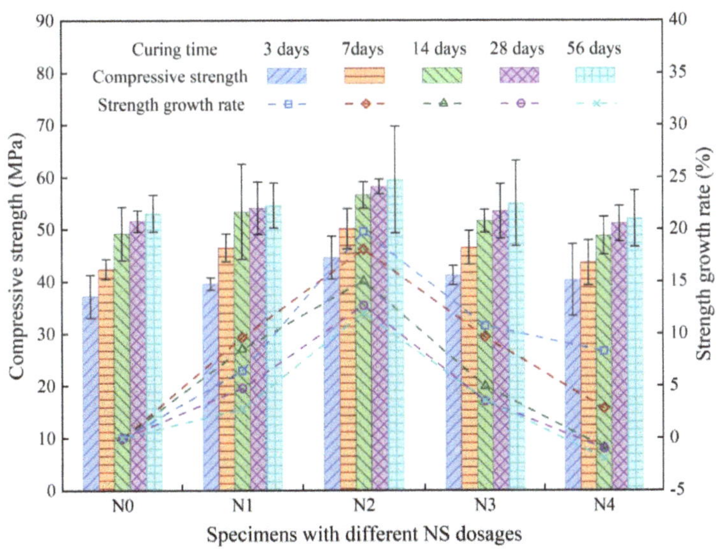

Figure 5. Compressive strength and strength growth rate of CFAA concrete containing different NS dosages.

As seen from Figure 5, the compressive strength of the 3-day concrete (N1, N2, N3, and N4) increased by 6.5%, 19.8%, 10.8%, and 8.3%, respectively, compared to N0. At 7 days, the strength growth rates were 9.7%, 18.1%, 9.7%, and 2.8%, respectively. After 14 days of curing, the growth of the strength started to slow down significantly, and the corresponding growth rates were 8.5%, 15.0%, 5.0%, and −0.9%, respectively. At 28 days, the strength growth rates of N1, N2, N3, and N4 further dropped to 4.8%, 12.7%, 3.6%, and −1.0%, respectively. The 56-day strength growth rates were 2.8%, 12.0%, 3.6%, and −1.9% for N1, N2, N3, and N4, respectively, and approximated to the 28-day strength growth rates. It can be found that the early-age strength growth rate was higher than the late-age strength growth rate for NS-modified CFAA concrete. The 3-day strength growth rate is the highest and up to 19.8% at 2 wt% NS dosage. This is because the NS particle act as a crystal nucleus of C-S-H gels in the early hydration stage [54]. At the same time, the pozzolanic reaction of NS consumes the CH and thus promotes cement hydration. These two reasons would accelerate the production of the C-S-H gels in the early stage and improve significantly the early compressive strength of the concrete.

Moreover, the strength growth rates of the concrete at 14 days, 28 days, and 56 days are less than zero when the NS dosage is 4 wt%. Under this condition, the concrete without NS had higher compressive strength, indicating that the excess NS is harmful to the concrete properties. It is recommended that the NS dosage should be determined by experiment when NS is used as a mineral admixture.

3.1.2. Splitting Tensile Strength

Figure 6 illustrates the splitting tensile strength and strength growth rate of CFAA concrete containing different NS dosages. Compared with N0 concrete, the splitting tensile strength of N1, N2, N3, and N4 increases by 7.8%, 11.2%, 4.1%, and −0.8%, respectively. This result shows that the splitting tensile strength is the highest at 2 wt% NS dosage and decreases when the NS dosage varies from 2 wt% to 4 wt%. This is because the NS with a moderate dosage produces more C-S-H gels through the pozzolanic reaction to fill pores in the cementitious matrix. The splitting tensile strength of the concrete improves as the matrix porosity decreases. On the contrary, the excess NS increases the number of pores and weak-zones in the matrix, reducing the splitting tensile strength.

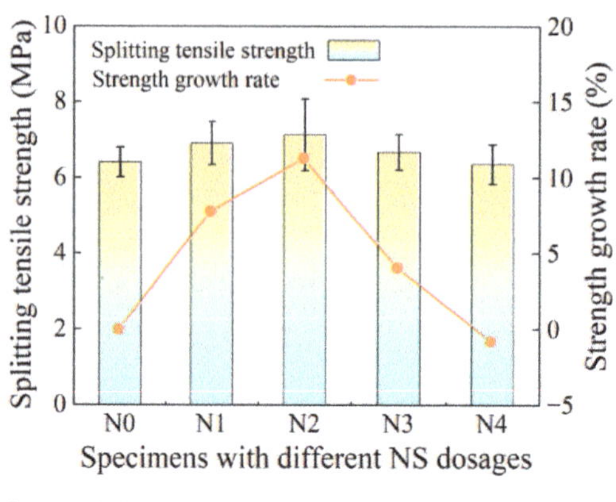

Figure 6. Splitting tensile strength and strength growth rate of CFAA concrete containing different NS dosages.

Based on the results above and the investigations [5,6], the relationship between splitting tensile strength and compressive strength for 28-day CFAA concrete can be obtained, as shown in Figure 7. Considering that the experiments had different specimen sizes, all data had been converted to the strength values of 100 mm cubes according to the standard

GB/T 50081-2019 [63]. Through regression analysis, the equation of the fitting line can be expressed as follows:

$$y = -0.809 + 0.14x \quad (4)$$

where x and y represent the compressive strength and the splitting tensile strength, respectively. The coefficient of determination, R^2, is 0.976, indicating that the equation has high goodness of fit. It can be found that the splitting tensile strength increases with the growth of the compressive strength, and their relationship is approximately linear. Note that the relationship is applicable to the cubic specimens (100 mm × 100 mm × 100 mm) with the nominal aggregate size grade of 5–10 mm. When the specimen size and aggregate gradation change, this relationship must be investigated further.

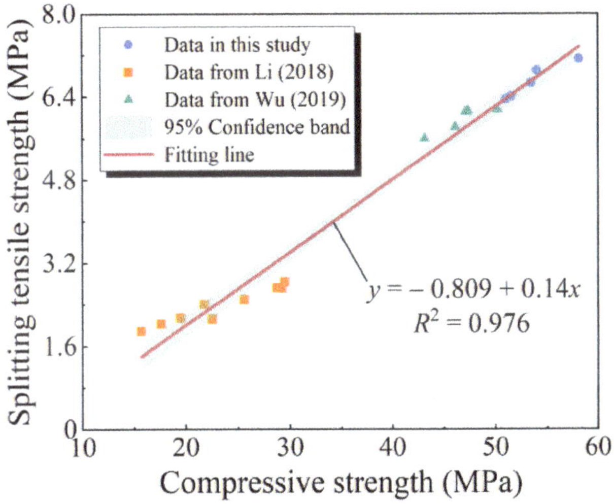

Figure 7. Relationship between splitting tensile strength and compressive strength for 28-day CFAA concrete (some data were sourced from Li [5] and Wu [6]).

3.1.3. Failure Morphology and SEM Analysis

The failure morphologies of CFAA concrete specimens in strength tests are shown in Figure 8. The CFAA concrete made with or without NS had similar failure morphology. During the compression tests, the top and bottom surfaces of the cubes are confined by friction. The initial cracks would occur at every corner of the cubes due to stress concentration. These cracks develop diagonally and converge in the middle of the cubes, resulting in severe concrete spallation from the lateral sides. The shape of the remaining concrete is close to the combination of two cones with opposing cone tops, as presented in Figure 8a. In the splitting tensile tests, the cubes are directly divided into two parts, and the fracture surface is approximately parallel to the surface of the cubes, as illustrated in Figure 8b.

Figure 8c depicts the details of the failed specimen after the compression testing. The result shows that the CFAAs are uniformly distributed on the fracture surface. It can also be observed that some CFAAs remain intact, and the interfacial debonding occurs between the aggregates and the matrix. The reason is that the aggregates have a spherical appearance and a smooth surface, leading to low interfacial strength. However, other CFAAs are crushed, and the damaged aggregates mainly concentrate in moist areas, as shown in Figure 8c. This may be because the moist areas resulting from the water penetration contain adequate water to ensure full cement hydration. The bonding between the aggregates and the matrix would be enhanced. Furthermore, the aggregates consist of C-S-H gels and have a similar strength to the matrix. Consequently, the fracture surface might penetrate the aggregates when the interfacial strength is enhanced. In conclusion, the compressive strength of the concrete is related to the aggregate strength, the matrix strength, and the bonding strength between the aggregates and the matrix. These factors also influence the splitting tensile strength of lightweight concrete [18,66,67]. This can explain why the strong correlation exists between the splitting tensile strength and the compressive strength of CFAA concrete in Section 3.1.2.

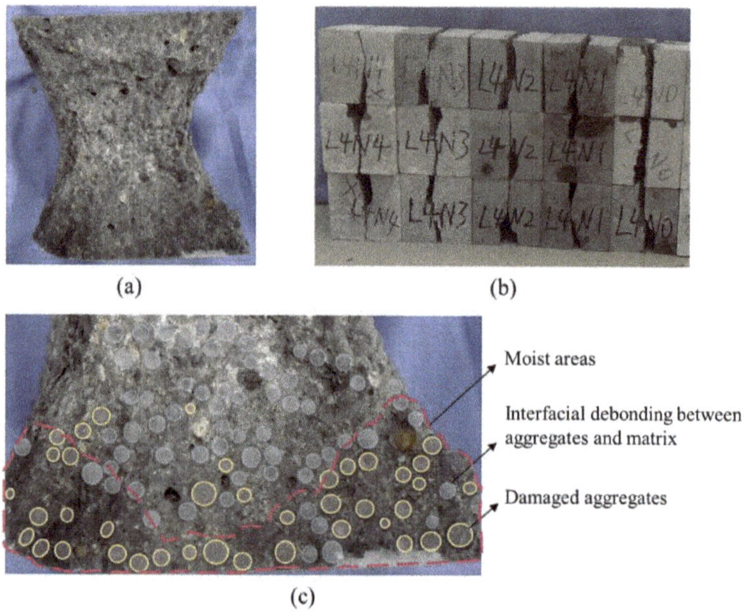

Figure 8. Failure morphologies of CFAA concrete specimens in strength tests: (**a**) after compression testing; (**b**) after splitting tensile testing; (**c**) details of the failed specimen after the compression testing.

After 28 days of curing, the three most representative cubes, N0, N2, and N4, were made into SEM samples to observe their microstructure. The SEM images of the CFAA concrete (N0, N2, and N4) are presented in Figure 9. There are many pores among the C-S-H gels in the concrete without NS, as illustrated in Figure 9a. With the addition of 2 wt% NS, the number and size of pores decrease greatly, and the C-S-H gels become more homogeneous, as shown in Figure 9b. The reason is that NS reacting with CH produces the additional C-S-H gels to fill the big holes so that the matrix is enhanced. At the same time, NS can fill the nanoscale pores in the matrix to make the microstructure much denser. However, when the NS dosage increases to 4 wt%, pores and microcracks would arise in the matrix so that the microstructure gets loose, as depicted in Figure 9c. Compared with N0, larger pores are observed in N4. Furthermore, the initiation of microcracks might relate to the agglomeration of NS particles because the drying shrinkage of concrete can contribute to forming the microcracks around the agglomerated NS particles [37]. Overall, the moderate NS significantly optimizes the microstructure of the matrix, whereas the excess NS causes more microscale defects in the matrix.

3.2. Durability Properties

The strength tests in Section 3.1 show that the optimum NS dosage is N2. During the durability tests, the specimens, N0 and N2, were soaked in different liquids (water, Na_2SO_4 solution, and Na_2CO_3 solution). The dry–wet and freeze–thaw cycle tests were conducted to explore the durability of CFAA concrete.

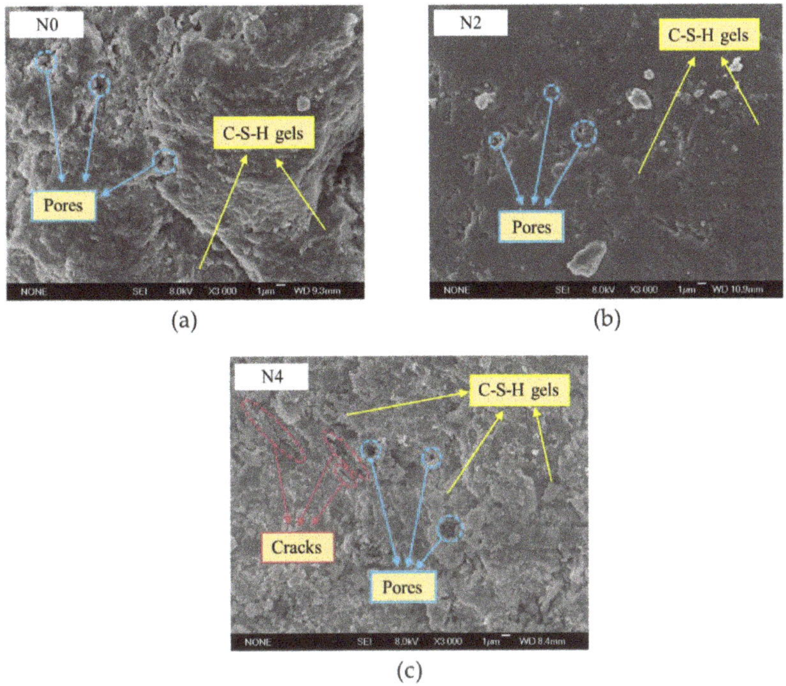

Figure 9. SEM images of 28-day CFAA concrete containing different NS dosages: (**a**) N0; (**b**) N2; (**c**) N4.

3.2.1. Results of the Dry–Wet Cycle Testing

The compressive strength of N0 and N2 subjected to the dry–wet cycles in water are listed in Table 4. The compressive strength slightly rises as the number of dry–wet cycles increases. This is because, in the wet period of each cycle, the cement continues to react with water so that the concrete is enhanced. To eliminate the influence of cement hydration on the strength, the results in Table 4 were used to calculate the anticorrosion coefficient through Equation (1).

Table 4. Compressive strength of CFAA concrete subjected to dry–wet cycles in water.

Dry–Wet Cycles		0	10	20	30	40	50	60
N0	Mean (MPa)	51.93	52.40	52.40	52.50	52.97	53.37	53.57
	COV [1] (%)	5.90	4.58	8.51	9.86	5.32	7.16	8.60
N2	Mean (MPa)	56.03	56.57	57.03	57.07	57.20	57.80	57.83
	COV [1] (%)	6.52	2.37	5.10	4.44	4.85	5.50	1.44

[1] COV denotes the coefficient of variation.

Figure 10 shows the compressive strength and the anticorrosion coefficient of the CFAA concrete subjected to the dry–wet cycles in the Na_2SO_4 and Na_2CO_3 solutions. It can be found that the compressive strength of N2 is always higher than that of N0 due to the pozzolanic reaction and filler effects of NS. Moreover, the standard deviation of these strength values varied from 0.21 to 4.35, and the maximum value was about 20 times the minimum value. The main reason for such a big difference might be the use of the rodding method during the casting of concrete. The rodding method is manually operated and has difficulty in ensuring that the concrete specimens have a similar degree of compaction. The standard deviation of the compressive strength values is generally between 1 and 5 in this experiment. In rare cases, the degree of concrete compaction in the specimens might be close to each other. This would cause the possibility that the standard deviation could decrease and be significantly less than 1.

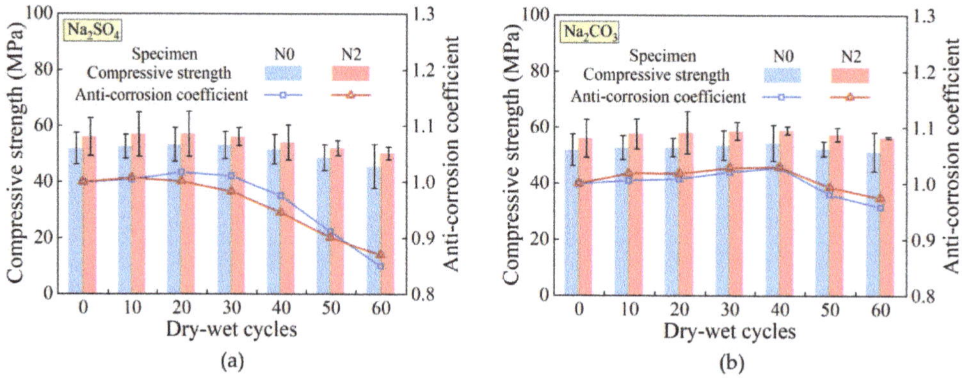

Figure 10. Compressive strength and anticorrosion coefficient of CFAA concrete subjected to dry–wet cycles in different saline solutions: (**a**) Na_2SO_4 solution; (**b**) Na_2CO_3 solution.

As observed from Figure 10, the anticorrosion coefficient slightly rises in the early stage of dry–wet cycles and then significantly decreases. The increase in the anticorrosion coefficient is attributed to the filler effects of the reaction products and the salt crystals. However, when the pore size hinders the growth of the crystals, crystallization pressure arises and causes the formation of microcracks around the pores. Meanwhile, sulfates or carbonates consume the CH and thus destroy the alkaline environment inside the concrete, decreasing the matrix strength. Therefore, crystallization pressure and matrix degradation lead to a reduction in the anticorrosion coefficient. In the Na_2SO_4 solution, the anticorrosion coefficient of N0 and N2 reaches the peak at cycles 10 and 20, respectively. In the Na_2CO_3 solution, the highest anticorrosion coefficient in N0 and N2 occurs at cycle 40. This indicates that CFAA concrete is more susceptible to erosion in the Na_2SO_4 solution than in the Na_2CO_3 solution. This is because sulfate ions can react with CH and Calcium-Aluminate-Hydrate (C-A-H) gels to form Ettringite (Et). The volume of the Et is more than 2.5 times that of the original C-A-H gels, resulting in the generation of expansive stress within the concrete. Hence, in the Na_2SO_4 solution, the concrete is simultaneously subjected to the crystallization pressure and the internal expansive stress, accelerating the performance degradation.

In Figure 10a, for the anticorrosion coefficient, the peak of N0 occurs later than that of N2 in the Na_2SO_4 solution. The reason may be that N0 has more voids than N2. The voids in N0 would take more dry–wet cycles to be filled with the reaction products and crystals. After 40 cycles, the anticorrosion coefficient of N2 falls slower than that of N0 in the Na_2SO_4 solution. This is because N2 has a higher matrix strength than N0, contributing to resisting the internal expansive stress and the crystallization pressure. Thus, NS can improve the corrosion resistance of the concrete to sulfates in the late stage of the dry–wet cycle. In Figure 10b, the anticorrosion coefficient of N2 is higher than that of N0 in the Na_2CO_3 solution. During cycles 40 to 60, the anticorrosion coefficient decreases by 6.86% and 5.35% in N0 and N2, respectively. This shows that N2 has a stronger corrosion resistance than N0 in the Na_2CO_3 solution, especially in the late dry–wet cycles. Consequently, NS can enhance the corrosion resistance of CFAA concrete to carbonates.

3.2.2. Results of the Freeze–Thaw Cycle Testing

Figure 11 presents the mass loss rate of the specimens (N0 and N2) under freeze–thaw cycles. During the freeze–thaw cycles, the pore solutions are converted into solids due to low temperature, causing the frost-heaving force around the pores. When the frost-heaving force exceeds the tensile strength of the matrix, the concrete spallation would arise and gradually get severe. Hence, the mass loss rate rises with the increasing freeze–thaw cycles. In Figure 11, the mass loss rate in saline solutions exceeds that in water after 20 cycles, because in the thaw period of the cycles, some reaction products and salts are taken out of the concrete as the pore water moves, increasing the porosity of surfaces. The concrete in saline solutions would have much severer surface spallation than in water during the freeze–thaw cycles. Meanwhile, the mass loss rate in the Na_2SO_4 solution is higher than that in the Na_2CO_3 in the late freeze–thaw cycles. Under sulfate attacks, C-A-H gels are consumed to generate the expansive Et, dramatically reducing the matrix strength near surfaces. For this reason, concrete spallation would be severer in the Na_2SO_4 solution than in the Na_2CO_3 solution with the growing freeze–thaw cycles.

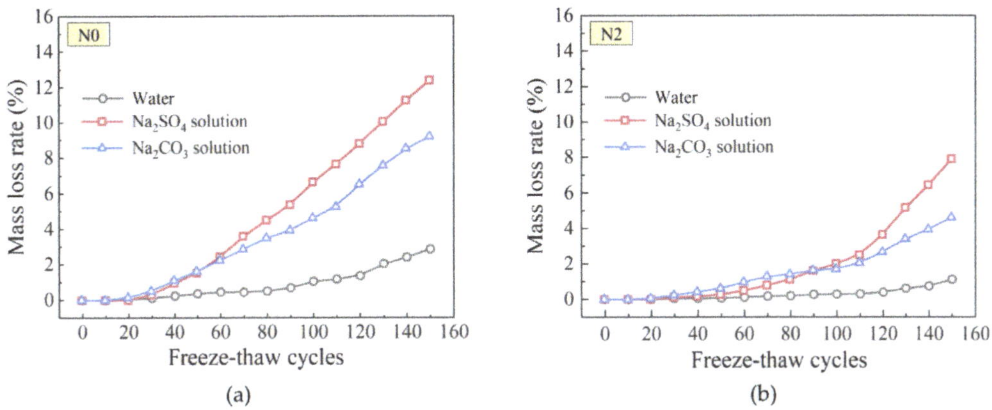

Figure 11. Mass loss rate of the specimens under freeze–thaw cycles: (**a**) N0; (**b**) N2.

In Figure 11a,b, at cycle 150, the mass loss rate of N2 in water, Na_2SO_4 solution, and Na_2CO_3 solution, compared with N0, decreases by 61.2%, 36.1%, and 49.7%, respectively. This result indicates that the CFAA concrete has stronger surface spallation resistance after incorporating NS. The reason is that the pozzolanic effect of NS enhances the matrix strength, contributing to resisting the frost-heaving force. Therefore, NS can improve the surface spallation resistance of CFAA concrete during freeze–thaw cycles.

Figure 12 illustrates the relative dynamic modulus of the specimens (N0 and N2) under freeze–thaw cycles. In addition to the frost-heaving force, the osmotic pressure resulting from the concentration difference of pore solutions would arise in the freezing stage of the cycles. The osmotic pressure and the frost-heaving force cause the continued growth of microcracks around the pores, increasing the internal damage of the concrete. For this reason, the relative dynamic modulus decreases as the number of freeze–thaw cycles grows. In Figure 12, the concrete in saline solutions has a significantly higher relative dynamic modulus than that in water in the late stage of the freeze–thaw cycles. This is because the saline solution has a lower freezing point than water. At the same temperature, the frost-heaving force in saline solution is weaker compared with water, resulting in less internal damage. Moreover, the relative dynamic modulus in the Na_2SO_4 solution is gradually lower than that in the Na_2CO_3 solution after 30 cycles. The reason is that sulfate attacks can destroy the C-A-H gels and generate the expansive Et. In consequence, the decreased matrix strength and the increased internal stress accelerate the internal damage of the concrete.

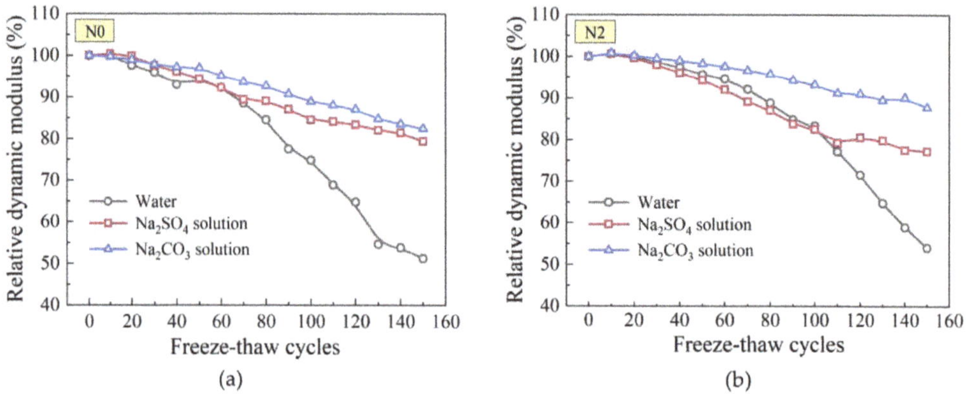

Figure 12. Relative dynamic modulus of the specimens under freeze–thaw cycles: (**a**) N0; (**b**) N2.

In Figure 12a,b, at cycle 150, compared with N0, the relative dynamic modulus of N2 in water, Na_2SO_4 solution, and Na_2CO_3 solution increases by 5.4%, −2.7%, and 6.5%, respectively. The matrix has fewer pores and higher strength with the addition of NS, contributing to resisting the osmotic pressure and the frost-heaving force. Thus, NS can reduce the internal damage of CFAA concrete during freeze–thaw cycles in the water and the Na_2CO_3 solution. However, N2 has a lower relative dynamic modulus than N0 in the Na_2SO_4 solution. This exhibits the fact that NS can hardly improve the internal frost resistance of the concrete subjected to sulfate attacks.

4. Conclusions

To obtain satisfactory long-term performance, Nanosilica (NS)-modified Cold-bonded Fly Ash Aggregate (CFAA) concrete was prepared to conduct strength tests and microstructure analysis. The durability was evaluated by dry–wet and freeze–thaw cycle tests in different saline solutions. The main conclusions are drawn as follows:

1. After incorporating NS, the compressive and splitting tensile strength of CFAA concrete is enhanced, and the optimum NS dosage is 2 wt%. NS also plays a crucial role in improving the early-age strength of the concrete.
2. The strength of CFAA concrete is related to the aggregate strength, the matrix strength, and the bonding strength between the aggregate and the matrix. NS can reduce the porosity and improve the properties of the cementitious matrix, whereas excess NS would cause more microscale deficiencies in the matrix.
3. During dry–wet cycles, CFAA concrete has stronger corrosion resistance to carbonates than to sulfates. The addition of NS effectively alleviates the concrete deterioration in the late dry–wet cycles.
4. Under freeze–thaw conditions, NS-modified CFAA concrete exhibits desirable spallation resistance and internal frost resistance. However, NS could barely diminish the internal damage due to sulfate attacks.

This paper focuses on utilizing NS to enhance the strength and durability of CFAA concrete. The research results facilitate alleviating the concrete deterioration under salt attacks, offering a reference for the potential application of the concrete in a harsh environment. Further studies would benefit from performing the quantitative analysis of the existing data to establish damage models for CFAA concrete. It would also be valuable to investigate the long-term performance of the concrete subjected to the coupling action of sulfates and carbonates. Several factors could be considered in experiments, such as the NS dosage, the salt concentration, and the ratio of sulfates to carbonates.

Author Contributions: Conceptualization, Q.W.; investigation: J.W.; methodology, J.W.; supervision, Q.W.; visualization: X.P.; writing—original draft preparation, X.P.; writing—review and editing, X.P. All authors have read and agreed to the published version of the manuscript.

Funding: This research received no external funding.

Institutional Review Board Statement: Not applicable.

Informed Consent Statement: Not applicable.

Data Availability Statement: The authors confirm that the data supporting the findings of this study are available within the article.

Conflicts of Interest: The authors declare no conflict of interest.

References

1. Tajra, F.; Elrahman, M.A.; Stephan, D. The Production and Properties of Cold-Bonded Aggregate and Its Applications in Concrete: A Review. *Constr. Build. Mater.* **2019**, *225*, 29–43. [CrossRef]
2. Ren, P.; Ling, T.-C.; Mo, K.H. Recent Advances in Artificial Aggregate Production. *J. Clean Prod.* **2021**, *291*, 125215. [CrossRef]
3. Liu, J.; Li, Z.; Zhang, W.; Jin, H.; Xing, F.; Tang, L. The Impact of Cold-Bonded Artificial Lightweight Aggregates Produced by Municipal Solid Waste Incineration Bottom Ash (MSWIBA) Replace Natural Aggregates on the Mechanical, Microscopic and Environmental Properties, Durability of Sustainable Concrete. *J. Clean Prod.* **2022**, *337*, 130479. [CrossRef]
4. Shang, X.; Chang, J.; Yang, J.; Ke, X.; Duan, Z. Life Cycle Sustainable Assessment of Natural vs Artificial Lightweight Aggregates. *J. Clean Prod.* **2022**, *367*, 133064. [CrossRef]
5. Li, B. The Mechanical Properties of Cold-Formed Thin-Walled Steel Skeleton-Flyash Ceramsite Concrete Wallboard Study. Master's Thesis, Jilin University, Changchun, China, 2018.
6. Wu, J. Study on Mechanical Properties and Durability of Nano SiO_2 Modified Flyash Ceramsite Concrete. Master's Thesis, Jilin University, Changchun, China, 2019.
7. Their, J.M.; Özakça, M. Developing Geopolymer Concrete by Using Cold-Bonded Fly Ash Aggregate, Nano-Silica, and Steel Fiber. *Constr. Build. Mater.* **2018**, *180*, 12–22. [CrossRef]
8. Tang, P.; Brouwers, H.J.H. Integral Recycling of Municipal Solid Waste Incineration (MSWI) Bottom Ash Fines (0–2 mm) and Industrial Powder Wastes by Cold-Bonding Pelletization. *Waste Manage.* **2017**, *62*, 125–138. [CrossRef]
9. Thomas, J.; Harilal, B. Sustainability Evaluation of Cold-Bonded Aggregates Made from Waste Materials. *J. Clean Prod.* **2019**, *237*, 117788. [CrossRef]
10. Öz, H.Ö.; Yücel, H.E.; Güneş, M.; Köker, T.Ş. Fly-Ash-Based Geopolymer Composites Incorporating Cold-Bonded Lightweight Fly Ash Aggregates. *Constr. Build. Mater.* **2021**, *272*, 121963. [CrossRef]
11. Narattha, C.; Chaipanich, A. Phase Characterizations, Physical Properties and Strength of Environment-Friendly Cold-Bonded Fly Ash Lightweight Aggregates. *J. Clean Prod.* **2018**, *171*, 1094–1100. [CrossRef]
12. Dong, B.; Chen, C.; Wei, G.; Fang, G.; Wu, K.; Wang, Y. Fly Ash-Based Artificial Aggregates Synthesized through Alkali-Activated Cold-Bonded Pelletization Technology. *Constr. Build. Mater.* **2022**, *344*, 128268. [CrossRef]
13. Zorlu, K.; Binal, A. A Cold-Binding Aggregate Production Technique and Performance Evaluation under Ageing Tests. *J. Build. Eng.* **2022**, *45*, 103569. [CrossRef]
14. Joseph, G.; Ramamurthy, K. Workability and Strength Behaviour of Concrete with Cold-Bonded Fly Ash Aggregate. *Mater. Struct.* **2009**, *42*, 151–160. [CrossRef]
15. Geetha, S.; Ramamurthy, K. Environmental Friendly Technology of Cold-Bonded Bottom Ash Aggregate Manufacture through Chemical Acti-vation. *J. Clean Prod.* **2010**, *18*, 1563–1569. [CrossRef]
16. Patel, J.K.; Patil, H.; Patil, Y.; Vesmawala, G. Production and Performance of Alkali-Activated Cold-Bonded Lightweight Aggregate in Concrete. *J. Build. Eng.* **2018**, *20*, 616–623. [CrossRef]
17. Kumar, R.; Srivastava, A.; Lakhani, R. Industrial Wastes-Cum-Strength Enhancing Additives Incorporated Lightweight Aggregate Concrete (LWAC) for Energy Efficient Building: A Comprehensive Review. *Sustainability* **2022**, *14*, 331. [CrossRef]
18. Sahoo, S.; Selvaraju, A.K.; Suriya Prakash, S. Mechanical Characterization of Structural Lightweight Aggregate Concrete Made with Sintered Fly Ash Aggregates and Synthetic Fibres. *Cem. Concr. Compos.* **2020**, *113*, 103712. [CrossRef]
19. Gündüz, L.; Uğur, İ. The Effects of Different Fine and Coarse Pumice Aggregate/Cement Ratios on the Structural Concrete Properties without Using Any Admixtures. *Cem. Concr. Res.* **2005**, *35*, 1859–1864. [CrossRef]
20. Gesoğlu, M.; Güneyisi, E.; Öz, H.Ö. Properties of Lightweight Aggregates Produced with Cold-Bonding Pelletization of Fly Ash and Ground Granulated Blast Furnace Slag. *Mater. Struct.* **2012**, *45*, 1535–1546. [CrossRef]
21. Nadesan, M.S.; Dinakar, P. Structural Concrete Using Sintered Flyash Lightweight Aggregate: A Review. *Constr. Build. Mater.* **2017**, *154*, 928–944. [CrossRef]
22. Mohanta, N.R.; Murmu, M. Alternative Coarse Aggregate for Sustainable and Eco-Friendly Concrete—A Review. *J. Build. Eng.* **2022**, *59*, 105079. [CrossRef]

23. Güneyisi, E.; Gesoglu, M.; Azez, O.A.; Öz, H.Ö. Physico-Mechanical Properties of Self-Compacting Concrete Containing Treated Cold-Bonded Fly Ash Lightweight Aggregates and SiO_2 Nano-Particles. *Constr. Build. Mater.* **2015**, *101*, 1142–1153. [CrossRef]
24. Ibrahim, M.A.; Atmaca, N.; Abdullah, A.A.; Atmaca, A. Mechanical Properties of Concrete Produced by Light Cement-Based Aggregates. *Sustainability* **2022**, *14*, 15991. [CrossRef]
25. Güneyisi, E.; Gesoğlu, M.; Algın, Z.; Yazıcı, H. Effect of Surface Treatment Methods on the Properties of Self-Compacting Concrete with Recycled Aggregates. *Constr. Build. Mater.* **2014**, *64*, 172–183. [CrossRef]
26. Gomathi, P.; Sivakumar, A. Accelerated Curing Effects on the Mechanical Performance of Cold Bonded and Sintered Fly Ash Aggregate Concrete. *Constr. Build. Mater.* **2015**, *77*, 276–287. [CrossRef]
27. Patel, J.; Patil, H.; Patil, Y.; Vesmawala, G. Strength and Transport Properties of Concrete with Styrene Butadiene Rubber Latex Modified Lightweight Aggregate. *Constr. Build. Mater.* **2019**, *195*, 459–467. [CrossRef]
28. Yıldırım, H.; Özturan, T. Impact Resistance of Concrete Produced with Plain and Reinforced Cold-Bonded Fly Ash Aggregates. *J. Build. Eng.* **2021**, *42*, 102875. [CrossRef]
29. Güneyisi, E.; Gesoğlu, M.; Booya, E. Fresh Properties of Self-Compacting Cold Bonded Fly Ash Lightweight Aggregate Concrete with Different Mineral Admixtures. *Mater. Struct.* **2012**, *45*, 1849–1859. [CrossRef]
30. Gesoğlu, M.; Güneyisi, E.; Alzeebaree, R.; Mermerdaş, K. Effect of Silica Fume and Steel Fiber on the Mechanical Properties of the Concretes Produced with Cold Bonded Fly Ash Aggregates. *Constr. Build. Mater.* **2013**, *40*, 982–990. [CrossRef]
31. Joseph, G.; Ramamurthy, K. Influence of Fly Ash on Strength and Sorption Characteristics of Cold-Bonded Fly Ash Aggregate Concrete. *Constr. Build. Mater.* **2009**, *23*, 1862–1870. [CrossRef]
32. Güneyisi, E.; Gesoğlu, M.; Booya, E.; Mermerdaş, K. Strength and Permeability Properties of Self-Compacting Concrete with Cold Bonded Fly Ash Lightweight Aggregate. *Constr. Build. Mater.* **2015**, *74*, 17–24. [CrossRef]
33. Wang, Y.; Hughes, P.; Niu, H.; Fan, Y. A New Method to Improve the Properties of Recycled Aggregate Concrete: Composite Addition of Basalt Fiber and Nano-Silica. *J. Clean Prod.* **2019**, *236*, 117602. [CrossRef]
34. Federowicz, K.; Techman, M.; Sanytsky, M.; Sikora, P. Modification of Lightweight Aggregate Concretes with Silica Nanoparticles—A Review. *Materials* **2021**, *14*, 4242. [CrossRef] [PubMed]
35. Bai, S.; Guan, X.; Li, H.; Ou, J. Effect of the Specific Surface Area of Nano-Silica Particle on the Properties of Cement Paste. *Powder Technol.* **2021**, *392*, 680–689. [CrossRef]
36. Sun, H.; Zhang, X.; Zhao, P.; Liu, D. Effects of Nano-Silica Particle Size on Fresh State Properties of Cement Paste. *KSCE J. Civ. Eng.* **2021**, *25*, 2555–2566. [CrossRef]
37. Son, H.M.; Park, S.M.; Jang, J.G.; Lee, H.K. Effect of Nano-Silica on Hydration and Conversion of Calcium Aluminate Cement. *Constr. Build. Mater.* **2018**, *169*, 819–825. [CrossRef]
38. Paredes, J.A.; Gálvez, J.C.; Enfedaque, A.; Alberti, M.G. Matrix Optimization of Ultra High Performance Concrete for Improving Strength and Durability. *Materials* **2021**, *14*, 6944. [CrossRef]
39. Wang, Y.; Zhang, X.; Fang, J.; Wang, X. Mechanical Properties of Recycled Concrete Reinforced by Basalt Fiber and Nano-Silica. *KSCE J. Civ. Eng.* **2022**, *26*, 3471–3485. [CrossRef]
40. Wang, Y.; Gu, L.; Zhao, L. Beneficial Influence of Nanoparticles on the Strengths and Microstructural Properties of Non-Dispersible Underwater Concrete. *KSCE J. Civ. Eng.* **2021**, *25*, 4274–4284. [CrossRef]
41. Mukharjee, B.B.; Barai, S.V. Influence of Nano-Silica on the Properties of Recycled Aggregate Concrete. *Constr. Build. Mater.* **2014**, *55*, 29–37. [CrossRef]
42. Farajzadehha, S.; Ziaei Moayed, R.; Mahdikhani, M. Comparative Study on Uniaxial and Triaxial Strength of Plastic Concrete Containing Nano Silica. *Constr. Build. Mater.* **2020**, *244*, 118212. [CrossRef]
43. Murad, Y.Z.; Aljaafreh, A.J.; AlMashaqbeh, A.; Alfaouri, Q.T. Cyclic Behaviour of Heat-Damaged Beam−Column Joints Modified with Nano-Silica, Nano-Titanium, and Nano-Alumina. *Sustainability* **2022**, *14*, 10916. [CrossRef]
44. Mashshay, A.F.; Hashemi, S.K.; Tavakoli, H. Post-Fire Mechanical Degradation of Light-weight Concretes and Maintenance Strategies with Steel Fibers and Nano-Silica. *Sustainability* **2023**, *15*, 7463. [CrossRef]
45. Wang, L.; Gao, M.; Zhang, J. Effect of Continuous Loading Coupled with Wet–Dry Cycles on Strength Deterioration of Concrete. *Sustainability* **2022**, *14*, 13407. [CrossRef]
46. Liu, J.; Li, A.; Yang, Y.; Wang, X.; Yang, F. Dry–Wet Cyclic Sulfate Attack Mechanism of High-Volume Fly Ash Self-Compacting Concrete. *Sustainability* **2022**, *14*, 13052. [CrossRef]
47. Guo, J.-J.; Wang, K.; Guo, T.; Yang, Z.-Y.; Zhang, P. Effect of Dry–Wet Ratio on Properties of Concrete Under Sulfate Attack. *Materials* **2019**, *12*, 2755. [CrossRef] [PubMed]
48. Wang, K.; Guo, J.; Yang, L. Effect of Dry–Wet Ratio on Sulfate Transport-Reaction Mechanism in Concrete. *Constr. Build. Mater.* **2021**, *302*, 124418. [CrossRef]
49. Zhou, J.; Wang, G.; Liu, P.; Guo, X.; Xu, J. Concrete Durability after Load Damage and Salt Freeze–Thaw Cycles. *Materials* **2022**, *15*, 4380. [CrossRef]
50. Wang, B.; Pan, J.; Fang, R.; Wang, Q. Damage Model of Concrete Subjected to Coupling Chemical Attacks and Freeze-Thaw Cycles in Saline Soil Area. *Constr. Build. Mater.* **2020**, *242*, 118205. [CrossRef]
51. Chen, D.; Deng, Y.; Shen, J.; Sun, G.; Shi, J. Study on Damage Rules on Concrete under Corrosion of Freeze-Thaw and Saline Solution. *Constr. Build. Mater.* **2021**, *304*, 124617. [CrossRef]

52. Yang, S.; Han, M.; Chen, X.; Song, J.; Yang, J. Effect of Sulfate Crystallization on Uniaxial Compressive Behavior of Concrete Subjected to Combined Actions of Dry–Wet and Freeze–Thaw Cycles. *J. Cold Reg. Eng.* **2023**, *37*, 04022015. [CrossRef]
53. *GB 175-2007*; Common Portland Cement. Standards Press of China: Beijing, China, 2007. Available online: https://www.doc88.com/p-9049016034394.html (accessed on 9 November 2007).
54. Wang, Y.; Xu, Z.; Wang, J.; Zhou, Z.; Du, P.; Cheng, X. Synergistic Effect of Nano-Silica and Silica Fume on Hydration Properties of Cement-Based Materials. *J. Therm. Anal. Calorim.* **2020**, *140*, 2225–2235. [CrossRef]
55. *GB/T 17431.1-2010*; Lightweight Aggregates and Its Test Methods—Part 1: Lightweight Aggregates. Standards Press of China: Beijing, China, 2010. Available online: https://www.doc88.com/p-531467649446.html (accessed on 2 September 2010).
56. *GB/T 14684-2022*; Sand for Construction. Standards Press of China: Beijing, China, 2022. Available online: https://www.doc88.com/p-89959796011700.html (accessed on 15 April 2022).
57. *JGJ/T 12-2019*; Technical Standard for Application of Lightweight Aggregate Concrete. China Architecture & Building Press: Beijing, China, 2019. Available online: https://www.doc88.com/p-9592946311306.html (accessed on 30 July 2019).
58. Naji Givi, A.; Abdul Rashid, S.; Aziz, F.N.A.; Salleh, M.A.M. Experimental Investigation of the Size Effects of SiO_2 Nano-Particles on the Mechanical Properties of Binary Blended Concrete. *Compos. Pt. B-Eng.* **2010**, *41*, 673–677. [CrossRef]
59. Du, H.; Du, S.; Liu, X. Effect of Nano-Silica on the Mechanical and Transport Properties of Lightweight Concrete. *Constr. Build. Mater.* **2015**, *82*, 114–122. [CrossRef]
60. Atmaca, N.; Abbas, M.L.; Atmaca, A. Effects of Nano-Silica on the Gas Permeability, Durability and Mechanical Properties of High-Strength Lightweight Concrete. *Constr. Build. Mater.* **2017**, *147*, 17–26. [CrossRef]
61. Afzali Naniz, O.; Mazloom, M. Effects of Colloidal Nano-Silica on Fresh and Hardened Properties of Self-Compacting Lightweight Concrete. *J. Build. Eng.* **2018**, *20*, 400–410. [CrossRef]
62. Wang, X.F.; Huang, Y.J.; Wu, G.Y.; Fang, C.; Li, D.W.; Han, N.X.; Xing, F. Effect of Nano-SiO_2 on Strength, Shrinkage and Cracking Sensitivity of Lightweight Aggregate Concrete. *Constr. Build. Mater.* **2018**, *175*, 115–125. [CrossRef]
63. *GB/T 50081-2019*; Standard for Test Methods of Concrete Physical and Mechanical Properties. China Architecture & Building Press: Beijing, China, 2019. Available online: https://www.doc88.com/p-7884725357290.html (accessed on 19 June 2019).
64. *GB/T 50082-2009*; Standard for Test Methods of Long-Term Performance and Durability of Ordinary Concrete. China Architecture & Building Press: Beijing, China, 2009. Available online: https://www.doc88.com/p-3999737606121.html (accessed on 30 November 2009).
65. Abd Elrahman, M.; Chung, S.-Y.; Sikora, P.; Rucinska, T.; Stephan, D. Influence of Nanosilica on Mechanical Properties, Sorptivity, and Microstructure of Lightweight Concrete. *Materials* **2019**, *12*, 3078. [CrossRef]
66. Satpathy, H.P.; Patel, S.K.; Nayak, A.N. Development of Sustainable Lightweight Concrete Using Fly Ash Cenosphere and Sintered Fly Ash Aggregate. *Constr. Build. Mater.* **2019**, *202*, 636–655. [CrossRef]
67. Domagała, L. Durability of Structural Lightweight Concrete with Sintered Fly Ash Aggregate. *Materials* **2020**, *13*, 4565. [CrossRef]

sustainability

MDPI

Article

Mix Design Effects on the Durability of Alkali-Activated Slag Concrete in a Hydrochloric Acid Environment

Mohammad Teymouri [1], Kiachehr Behfarnia [2] and Amirhosein Shabani [3,*]

[1] Department of Civil and Environmental Engineering, University of Nebraska-Lincoln, Lincoln, NE 68588-0531, USA; mteymourimoogooee2@huskers.unl.edu
[2] Department of Civil Engineering, Isfahan University of Technology, Isfahan 84156-83111, Iran; kia@iut.ac.ir
[3] Department of Civil Engineering and Energy Technology, Oslo Metropolitan University, Pilestredet 35, 0166 Oslo, Norway
* Correspondence: amirhose@oslomet.no

Abstract: Because of its high strength, energy reduction, and low environmental impact, researchers have encouraged considering alkali-activated slag concrete (AASC) as a potential alternative to conventional concrete. In this study, the impact of mix design parameters on the durability of AASC, made with ground granulated blast furnace slag and activated with different alkaline solutions (NaOH, KOH, and Na_2SiO_3) immersed up to six months in a hydrochloric acid bath with pH = 3, has been investigated. A total of 13 mix designs were made in a way that, in addition to the type of alkaline solution, considered three other parameters, namely the molarity of alkaline solutions, the weight ratio of alkaline solutions to slag, and the weight ratio of alkaline solutions to sodium silicate. Visual inspections displayed that the AASC samples almost remained intact after exposure to an HCl acid solution with pH = 3 for up to 6 months, while the OPC sample experienced deleterious deterioration. The results clearly show that AASC outperformed OPC concrete when it comes to durability in an HCl acid solution. The strength reduction and weight loss of AASC compared with OPC concrete were approximately one-tenth and one-fifth, respectively. The AASC samples containing potassium hydroxide showed a higher strength reduction and weight loss in the HCl acid solution than the samples made with sodium hydroxide.

Keywords: alkali-activated concrete; slag; alkaline solution; hydrochloric acid

Citation: Teymouri, M.; Behfarnia, K.; Shabani, A. Mix Design Effects on the Durability of Alkali-Activated Slag Concrete in a Hydrochloric Acid Environment. *Sustainability* **2021**, *13*, 8096. https://doi.org/10.3390/su13148096

Academic Editor: Emilio Bastidas-Arteaga

Received: 28 May 2021
Accepted: 17 July 2021
Published: 20 July 2021

1. Introduction

Ordinary Portland cement (OPC) concrete has adverse impacts on the environment because of its production, the release of huge quantities of CO_2, and consuming a large amount of energy [1]. Moreover, its durability is also a significant drawback, especially in acidic environments [2,3]. Thus, researchers have always been searching to compensate for OPC's shortcomings and to reduce its environmental impact. An array of materials have been suggested, one of which has extensive popularity is alkali-activated concrete [4–6]. In general, alkali-activated concrete is a mixture of an aluminosilicate precursor activated with alkaline solutions [7,8]. Metakaolin, ground granulated blast furnace slag (GGBFS), and fly ash are common aluminosilicate sources. Sodium hydroxide (NaOH), potassium hydroxide (KOH), and sodium silicate (Na_2SiO_3) are the mainly used alkaline solutions, where sodium silicate is in solution with NaOH and KOH. Ground granulated blast furnace slag is a solid waste of steel production reported to have an acceptable performance in alkali-activated concrete as an aluminosilicate source [9]. Alkali-activated slag concrete (AASC) is an environmentally friendly product characterized by high percentages of alumina and silica activated with alkaline solutions, which provides a possibility for the production of cement-free concrete. AASC with relatively large amounts of silicon oxide and aluminum oxide is considered a potential alternative to OPC concrete for non-structural concrete and reinforced concrete structures [4,10,11].

It is reported that AASC is environmentally friendly because of its lower greenhouse emissions compared with OPC concrete, but it is also economically sustainable owing to its lower energy consumption [12–14]. Because of its high final strength, energy reduction, and low environmental impacts, researchers have encouraged considering AASC as a potential alternative for conventional concrete, especially in the precast concrete industry [15,16], while high shrinkage (3.3 times the OPC drying shrinkage as reported in [17]) and carbonation are two major imperfections of AASC compared with conventional concrete. AASC offers favorable properties, some of which include durability in aggressive and acidic environments [4,17]. Numerous studies have been performed to investigate the diverse characteristics of AASC, such as the mechanical properties [7,18–20], shrinkage [21,22], carbonation [23,24], and elevated temperature endurance [25]. The main hydration product of AASC is calcium aluminum silicate hydrate (C-A-S-H) gel, with a quite quick setting time and a low Ca/Si ratio compared with OPC concrete [13,20].

Cities in the developing world face outstanding economic and human losses caused by man-made or natural hazards, and the amount of loss is affected by the quality of the preventive measures and emergency management [26,27]. Concrete can be subjected to an acid attack in diverse circumstances. Conditions such as acid rain, acid river, sewerage, and chemical facility are examples of acidic environments that can bring about irreparable damage to concrete structures [28]. A few studies have been carried out to investigate the durability of AASC in acetic acid, sulfuric acid, nitric acid, and phosphoric acid environments [29–32]. Acid attack is a vital issue regarding concrete performance that adversely affects concrete durability. AASC is reported to have superior durability in acidic environments compared with conventional concrete. The main reasons for the high durability are the low permeability and less calcium in its composition compared with OPC concrete [29,32].

In an actual environment, the acid attack is not a pure acid, but often mixed acids. For example, in acid rain, the acid attack on concrete contains a mixture of sulfuric acid, nitric acid, and hydrochloric acid [33]. OPC concrete is alkaline and is susceptible to acidic environments [3]. The equilibrium of the cement matrix is quickly disturbed when the pH of the concrete pore solution drops. At pH values lower than 12.6, the first hydration product that dissolves is calcium hydroxide (portlandite), which converts to a calcium salt of the acid, hydrogels of silicium, aluminum, and ferric oxide [34]. The corrosion rate of concrete after an acid attack is a complex process governed by the combination of the dissolution, precipitation, and transport processes, as determined by the cement chemical composition, paste matrix reactivity, aggregate reactivity, aggregate grading, and concrete composition [35].

Hydrochloric (HCl) acid is a strong acid categorized as a mineral acid found in municipal sewage, acid rain, and industrial effluent. HCl acid is utilized in the metal industry as well as in the manufacture of fertilizers, dyes, and pigments, and the wastewater from these industries can be very aggressive for concrete structures [36]. When OPC paste comes in contact with HCl acid, calcium chloride ($CaCl_2$) is produced [28]. $CaCl_2$ is a highly soluble salt that can be leached out and causes concrete deterioration [37]. It is reported that the most important factor governing the corrosion rate in an acid attack on cement-based materials is calcium salt solubility, and the second factor is acid strength [38]. HCl acid is considered an aggressive acid because of its highly soluble calcium salts and dissociation ability [39]. In addition, the formation of Friedel's salt, which is an expansive product, accelerates concrete deterioration [34].

A few studies have investigated the durability of AASC in HCl acid attack [40,41]. It has been reported that AASC has a better durability compared with conventional concrete in aggressive environments [42]. Thunuguntla et al. investigated the effect of mix design parameters on the mechanical and durability properties of AASC in three different acidic environments, including hydrochloric acid solution, and NaOH was considered the only alkaline solution. Eight different mix designs were formulated for the tests. The results showed that the NaOH concentration was found to be the most influential parameter on the mechanical strength and durability characteristics of AASC [41]. Molarity and type of alkaline solutions are two vital elements in the mix design of AASC [43]. There is no denying that variability in the chemical composition of slag may impact its durability, and concretes prepared using various slags may have different resistance in aggressive media [44].

In the literature, a limited number of studies focused on a systematic approach for considering the effect of the mix design parameters of AASC on its performance and durability in acidic environments. Concrete durability is a crucial property, as concrete structures are required to sustain and endure throughout the entire service life acceptably. In the vast majority of studies, the acid resistance of alkali-activated concrete with a fixed mix design was studied. To fill this knowledge gap, an experimental study is required to investigate the effect of the diverse mix design parameters on the durability and mechanical properties of AASC exposed to an HCl acid solution. Thus, in this paper, the impact of mix design parameters on the durability of AASC samples, made with GGBFS and activated with different alkaline solutions (NaOH, KOH, and Na_2SiO_3) immersed up to six months in a hydrochloric acid bath with pH = 3, has been investigated. As well as the type of alkaline solution, three other parameters, including the molarity of alkaline solutions, the weight ratio of alkaline solutions to slag, the weight ratio of alkaline solutions to sodium silicate, were considered. OPC concrete samples were also made as a control mixture. The XRF test measured the chemical compositions of the slag and Portland cement. For all AASC and OPC concrete specimens, compressive strength reduction and weight loss were monitored at predetermined intervals.

2. Materials and Methods

In this study, slag as a silica-aluminate precursor was used according to ASTM C989M. The chemical compositions of GGBFS and Portland cement, based on the XRF test, are presented in Table 1. Additionally, GGBFS specific gravity and Blaine fineness were equal to 2.85 gr/cm^3 and 400 m^2/kg, respectively. The weight ratio of Al_2O_3/SiO_2 and CaO/SiO_2, basicity coefficient (Kb = (CaO + MgO)/(SiO_2 + Al_2O_3)), and hydration modulus ((CaO + MgO + Al_2O_3)/SiO_2) were equal to 0.447, 1.079, 0.916, and 1.77, respectively.

Table 1. Chemical compositions of GGBFS and Portland cement based on XRF test (wt.%).

	CaO	SiO_2	Al_2O_3	MgO	TiO_2	MnO	S	K_2O	Fe_2O_3	NaO_2	SO_3	L.O.I *
GGBFS	36.52	38.35	10.88	8.77	1.48	1.25	1.21	0.93	0.52	0.49	-	0.26
Cement	63.50	21.50	5.10	2.30	-	-	-	0.93	3.80	-	2.00	0.70

* Loss of Ignition.

Alkaline solutions were added to the binder in liquid form. The alkaline solutions were sodium hydroxide (NaOH), potassium hydroxide (KOH), and sodium silicate (Na_2SiO_3), which had a SiO_2/Na_2O ratio equal to 2.5 (Na_2O = 14%, SiO_2 = 35%, and water = 51%). The hydration process of alkali-activated slag is influenced by the sodium content and silica module [45]. It should be stated that water glass (NaOH) and potassium hydroxide (KOH) flakes were dissolved in water to make the solutions with a needed molarity based on the mix design parameters. The mixing process was carried out using a mixer with 60 L volume. Note that the preparation of the alkaline solution released heat, however, these solutions were not added to the mixture right away after preparation. After preparation, it took at least 5 min before being added to the final mixture. As a result, the temperature almost reached the ambient temperature.

The physical properties of aggregates are presented in Table 2. The fine aggregates were crushed sand. Sand equality and water absorption were measured in accordance with ASTM D2419 and ASTM C128, respectively. Gravel (Crushed stone) with a maximum aggregate size (MSA) of 19.5 mm was used as the coarse aggregates. The saturated surface dry specific gravity and water absorption were measured as per ASTM C127. It should be noted that limestone aggregates were used because they are widely used in a majority of projects in Iran. The grading of aggregates met the requirements of ASTM C33.

Table 2. Physical properties of aggregates.

Type of Aggregate	Fineness Module	Sand Equality	SSD Specific Gravity (gr/cm^3)	Water Absorption (%)
Fine	2.99	77	2.47	2.06
Course	-	-	2.59	0.76

The AASC mix proportions were selected based on previous studies [14,15]. As mentioned before, the selected parameters included the alkaline solution type, the molarity of alkaline solutions, the weight ratio of alkaline solutions to slag, and the weight ratio of alkaline solutions to sodium silicate. A total of 13 mix designs were made in a way that four selected parameters are considered, as shown in Table 3. It should be noted that when a parameter (for example, NaOH molarity) was changed, the other mix design parameters (such as NaOH/Na2SiO3, alkaline solution/slag, and alkaline solution type) remained constant.

Table 3. AASC mix designs.

Mix Code	Alkaline Solution (Kg/m^3)	Slag (Kg/m^3)	The Weight Ratio of NaOH (KOH) to Na_2SiO_3	The Weight Ratio of Alkaline Solution to Slag	Molarity	Type of Alkaline Solution
N6041	158	394	1	0.4	6	NaOH
N6043	158	394	3	0.4	6	NaOH
N10041	158	394	1	0.4	10	NaOH
N10043	158	394	3	0.4	10	NaOH
N14041	158	394	1	0.4	14	NaOH
N14043	158	394	3	0.4	14	NaOH
N60404	158	394	0.4	0.4	6	NaOH
N100404	158	394	0.4	0.4	10	NaOH
N6061	207	345	1	0.6	6	NaOH
N10063	207	345	3	0.6	10	NaOH
K6041	158	394	1	0.4	6	KOH
K6043	158	394	3	0.4	6	KOH
K10043	158	394	3	0.4	10	KOH

To evaluate the acid resistance of the AASC mixes, samples were exposed to an HCl acid bath with pH = 3. The pH was selected based on previous research [29,30,44,46]. The ratio of water to solid materials was 0.50, and the weight percentage of the aggregates in the mixture was 77%. To keep the workability of all of the AASC samples the same, a naphthalene-based superplasticizer was used [14,15].

At first, the aggregates and slag were mixed for 3 min, then the alkaline solutions were added. In the end, water and the admixture were added, and the mixture thoroughly mixed for 5 min, followed by a rest for 1 min, and remixing for 3 min. Cube molds ($100 \times 100 \times 100$ mm) were used to prepare the AASC samples. The prepared samples were covered with plastic sheets to minimize water evaporation. They were kept at ambient temperature (23 ± 2 Celsius) for 24 h, and were then taken out of molds and placed in water baths for two weeks before immersion in acid solutions. The reason for selecting this time is that, based on the compressive results, a majority of AASC specimens gained nearly 80% of 180 days compressive strength in the first 14 days, as discussed in the Results and Discussion section. To have a constant pH level in acid baths, the pH of the container was measured continuously with a pH meter twice a week, and the HCl acid was added as needed. Three replicate cubes from each mix design at predetermined interval days (28, 90, 120, and 180 days) were used for the compressive strength tests and weight loss measurements.

The compressive strength of the OPC concrete cubic samples ($100 \times 100 \times 100$ mm), as a control mix, was 60 MPa at 28 days, and its water to cement ratio was 0.28. Furthermore, the same aggregates that were used in the AASC mixtures were also used in the OPC samples. For mixing, first aggregate and cement were mixed for 3 min and then water was gradually added to the mixture. Moreover, a polycarboxylate superplasticizer was used to enhance the workability. After casting, OPC concrete specimens were kept at ambient temperature (23 ± 2 Celsius) for 24 h, and then were taken out of molds and cured in water for up to 28 days. Then, they were transferred to the HCl acid bath.

Because of the deficiency of a standard procedure for evaluating acid attacks on concrete [29], the resistance to acid attacks on AASC and conventional concrete was tested by immersion in an HCl acid bath as presented. Deterioration of specimens was tested by compressive strength reduction and weight loss at predetermined intervals (28, 90, 120, and 180 days after immersion in HCl acid bath) based on the procedure used by [29,44,46]. A reduction in strength or weight loss was calculated using following equation ((A–B)/A); where A is the average compressive strength or weight of three cubic samples cured in water, and B is the average compressive strength or weight of three cubic samples cured in an HCl acid bath.

The process was as follows: after casting, the AASC samples were kept at ambient temperature (23 ± 2 Celsius) for 24 h and were taken out of the cube molds and cured in water for 14 days. Then, the specimens were taken out of the water, were kept at ambient temperature for three hours, and were sandpapered on all six sides before weighing. Afterward, they were immersed in HCl acid baths until the experiment time (90, 120, and 180 days). The samples were weighed again after 3 h and after being sandpapered on all six sides. The weight difference before and after immersion in the HCl acid solutions was calculated as the weight loss. The compressive strength of the cubic samples was measured according to EN 12390-3, and three cubes were tested at predetermined intervals. The companion specimens, cured in potable water, were also tested to find the strength reduction.

3. Results and Discussion

In the following paragraphs, firstly, the compressive strength development of the AASC cured in water is discussed. Then, the impacts of four mix design parameters on the acid resistance of AASC exposed to an HCl acid solution are exploited. Finally, a comparison between the results of this study and the reference OPC concrete is discussed.

3.1. Compressive Strength of AASC Samples before Acid Exposure

The compressive strength results of the OPC concrete and AASC specimens cured in water for up to 180 days are shown in Figure 1. After 14 days of curing in water, most AASC samples showed a high compressive strength, where more than 44 MPa is considerable. This agrees well with previous findings [47], as reposted by [48], that AASC has a rapid development of compressive strength at early ages [49]. Gruskovnjak et al. reported that a faster strength development of ASSC compared with OPC is owed to a thin protective layer that covers the unhydrated slag grain systems. In the OPC concrete, this layer is considerably thicker, causing slower compressive strength growth [50]. Moreover, different hydration products in the two systems can also be considered as another main reason for the different strength development rates [12,13].

The results show that the mix design contains 10 M KOH, the weight ratio of the slag to alkaline solution equals 0.4, and the weight ratio of KOH to Na_2SiO_3 equals 3 (Mix code: K10043), which gained 86% of its 180 days compressive strength after 14 days. All three mix designs made with KOH gained more than 60 MPa compressive strength after 14 days of curing in water. Reddy et al. observed the same results [51]. It can be argued that KOH is a stronger alkaline solution than NaOH, and when it comes in contact with slag, a very rapid reaction happens, leading to gaining a considerable percentage of compressive strength at early ages.

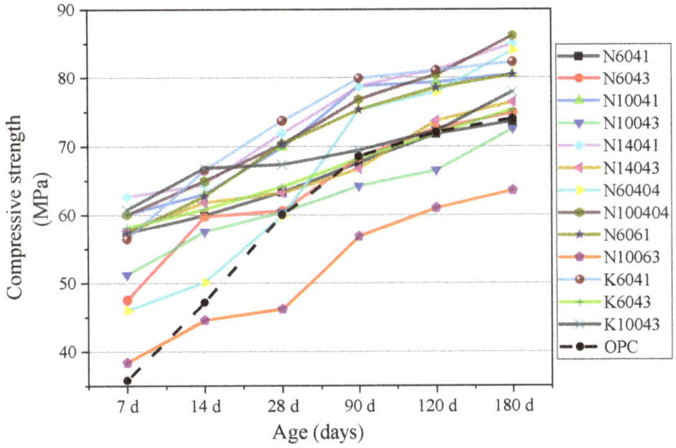

Figure 1. Compressive strength of AASC samples (100 × 100 × 100 mm) cured in water before exposure to HCl acid.

At 28 days, the lowest and highest compressive strength were 46.2 and 73.7 MPa related to N10063 and K6041, respectively. According to the literature, when the NaOH or KOH molarities are increased and the other parameters are the same, a higher compressive strength is achieved for all ages [52]. This can be related to the higher alkalinity of the AASC brought about by the increase in molarity, leading to the formation of more hydration products [49,53]. Fang et al. reported that the increase of NaOH from 10 to 12 led to a 23% increase in compressive strength [54]. The concentration of alkaline solutions increased the reaction rates and thus led to a higher compressive strength. Aliabdo et al. used 10 M, 12 M, and 14 M NaOH solutions, and reported an increase in molarity that significantly impacted the compressive strength at the age of 28 days [7].

3.2. *Effect of Mix Design Parameters on Performance in Acid Attack*

The HCl acid attack mechanism on AASC is a reaction between calcium-based compounds in the concrete paste and a solution of hydrogen chloride. This reaction initiates the formation of highly soluble calcium salt ($CaCl_2$). Visual inspections show that AASC samples were almost intact after exposure to HCl acid solution for up to 6 months, while the OPC sample experienced a deleterious deterioration, as shown in Figure 2. Although this comparison proved that AASC is superior to OPC in resisting HCl acid attack, other conducted tests also confirmed this observation, as discussed in the following sections.

(**a**) (**b**)

Figure 2. (**a**) OPC and (**b**) AASC samples (100 × 100 × 100 mm) appearance after six months of immersion in an HCl solution with pH = 3.

3.2.1. Type of Alkaline Activator

Numerous studies investigated AASC durability [29,31,44,55] and the correlation between the durability and microstructural characteristics [56] containing NaOH as an alkaline activator. Potassium hydroxide (KOH) has always been a potential alternative for sodium hydroxide (NaOH) in alkali-activated concrete [12]. In this study, two alkaline solutions were used, caustic soda (NaOH) and potassium hydroxide. KOH is a stronger alkaline solution than NaOH [57], and its price is at least two times greater than NaOH. It is reported that alkali-activated materials activated by KOH have acceptable mechanical and durability properties [46,58–60]. As shown in Figure 3, replacing KOH with caustic soda when other mix design parameters remained the same resulted in more strength reduction, especially after 120 and 180 days of exposure to the acid solution.

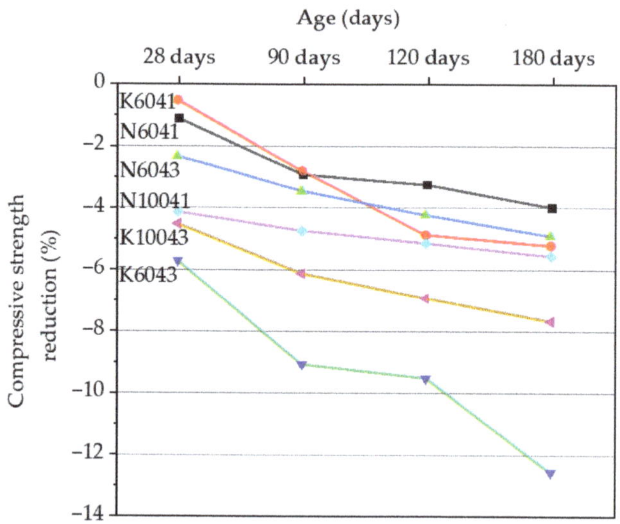

Figure 3. Compressive strength reduction of AASC with different alkaline activators immersed in an HCl acid solution.

Based on the results, the AASC samples made with KOH immersed in HCl acid for six months showed at least a 30% increase in strength reduction compared with the specimens made with NaOH. The obtained results showed that a progressive compressive strength reduction was observed with an increase in exposure time for all of the mixes. Figure 3 shows that N6043 experienced roughly a 5% strength reduction after 180 days of immersion in HCl acid. However, by replacing KOH (K6043), the strength reduction was about 13% at the same age. The cause of the unsatisfying performance lies in how KOH reacts to slag [61]. When potassium hydroxide reacts with slag, the reaction is rapid, creating a structure with countless tiny unfilled holes and a less homogeneous microstructure [61]. When this porous structure exposes aggressive environments such as an acid solution, the acid expands in the pores, causing concrete deterioration and strength reduction [62].

Thus, because of the undesirable compressive strength and costly prices of KOH compared with NaOH, its application does not have technical justification or financial benefits. A limited number of studies have been conducted on the effect of acid attack on alkali-activated slag concrete made with KOH. Bakharev reported that alkali-activated concreted made with fly ash and activated by KOH resulted in a larger pore diameter, which deleteriously impacted the concrete durability in the sulfuric acid solution [46]. The compressive strength reduction of the reference OPC concrete is reported in Section 4, and a comparison between the OPC and AASC results is also discussed.

Figure 4 shows the effect of the four selected mix design parameters on the weight loss of AASC samples immersed in HCl solutions for up to 180 days. Figure 4a indicates that when KOH replaced NaOH, weight loss increased approximately 67% after 180 days of immersion in an HCl acid bath. Hence, the replacement of KOH caused adverse effects on the durability of AASC. The results indicated that KOH replacement increased the weight loss of the AASC specimens by 100%, 123%, and 67% on days 90, 120, and 180, respectively. To explain this observation, as discussed before, it can be argued that KOH rapidly reacts with slag and creates a less homogeneous microstructure with countless tiny unfilled holes. This weak microstructure allows acid agents to penetrate the AASC structure and deteriorate bonds [61]. Hence, it seems unsatisfying to replace NaOH as a common activator with a reasonable price with KOH when the durability of AASC matters. The effect of other parameters (as shown in Figure 4) will be discussed in the corresponding section.

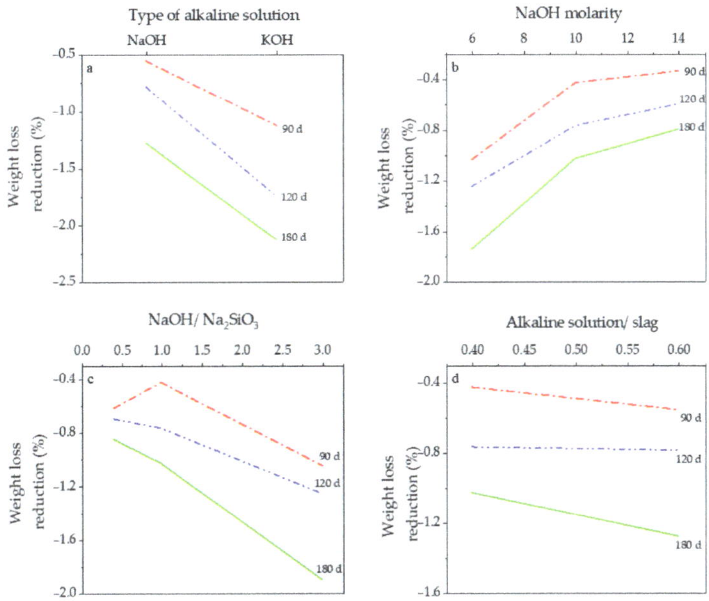

Figure 4. The effect of mix design parameters on the weight loss of AASC samples immersed in an HCl acid solution (**a**) Type of alkaline solution, (**b**) NaOH molarity, (**c**) $NaOH/Na_2SiO_3$, and (**d**) weight ratio of the alkaline solution to slag.

3.2.2. Sodium Hydroxide Molarity

One of the principal mix design parameters that can tremendously affect the alkali-activated slag concrete characteristics is the molarity of the alkaline solutions [41]. Figure 5 shows the compressive strength reduction of the AASC samples with different molarities immersed in an HCl acid solution for up to six months. In Figure 5a, the weight ratio of $NaOH/Na_2SiO_3$ equals 1, and in Figure 5b, the ratio is 3. Regardless of the NaOH to Na_2SiO_3 ratio, an increase in the NaOH concentration led to more strength reduction at all ages. Therefore, AASC samples containing less molarity of caustic soda had a superior performance in the HCl acid environment. Therefore, although a higher NaOH molarity led to a higher compressive strength for the AASC cured in water, the acid resistance of the AASC improved when the lower molarity of NaOH was used. Thunuguntla investigated the durability of AASC using 1 M and 8 M NaOH in contact with HCl, H_2SO_4, and HNO_3 for up to 56 days. The loss of compressive strength for the specimens immersed in HCl acid was about 1 to 4% that is very little. He concluded that AASC provides outstanding resistance to all of the acid solutions used here compared with OPC [41].

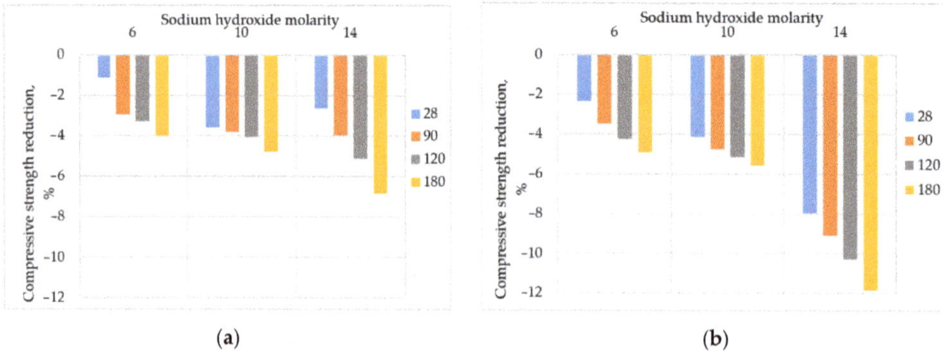

Figure 5. Compressive strength reduction of AASC with different NaOH molarities in an HCl acid solution for (**a**) $NaOH/Na_2SiO_3$ = 1 and (**b**) $NaOH/Na_2SiO_3$ = 3.

Figure 5a,b shows that the $NaOH/Na_2SiO_3$ ratio directly affects the influent of NaOH molarity. In Figure 5a, the AASC samples exposed to HCl acid for 180 days lost around 7% of their compressive strength, but the strength reduction was approximately 12% in Figure 5b. The underlying reason is that the sodium hydroxide amount was higher, and in the $NaOH/Na_2SiO_3$ ratio it equaled 3. The higher sodium hydroxide molarity increased the reaction rate between the slag and alkaline solutions, creating AASC with irregular microstructures and causing more deterioration in acid attacks. After 180 days of exposure to the acid, when the sodium hydroxide molarity changed from 6 to 14, the strength reduction increased by approximately 71% (Figure 5a) and 140% (Figure 5b).

Figure 4b shows the weight loss reduction of the AASC samples with NaOH concentrations of 6 M, 10 M, and 14 M. The results indicated that an increase in NaOH molarity led to a decrease in weight loss. The main reason for this is that when the NaOH molarity increased from 6 to 14, the alkalinity of the concrete increased. It should be noted that AASC samples made with 14 M NaOH had a dark green color compared with the specimens made with a lower molarity, which had a pale green color. The mix design, which had the highest concentration (14 M), experienced 55% less weight loss than the AASC samples with the lowest molarity (6 M) after six months of exposure to the HCl acid solution.

3.2.3. Weight Ratio of Sodium Hydroxide to Sodium Silicate

In a majority of investigations, two combined alkaline solutions were used. Sodium hydroxide and sodium silicate are the two most used alkaline activators in the literature [44,52,63]. Thus, the weight ratio of these two alkaline solutions can be an influential parameter when AASC mix design is to be selected. In this study, three ratios, 0.4, 1, and 3, were used. The effect of the ratio of $NaOH/Na_2SiO_3$ on the strength reduction was evaluated when the other parameters were the same. Figure 6a,b shows the strength reduction when the NaOH molarities were 6 and 10, respectively. Regardless of the concentration of the sodium hydroxide solution, the highest strength reduction occurred when the $NaOH/Na_2SiO_3$ ratio equaled 3 at all ages. In addition, when the ratio increased from 1 to 3, strength reduction at 180 days rose approximately 23% and 16% for 6 M and 10 M NaOH, respectively. To explain this phenomenon, it can be said that the lower $NaOH/Na_2SiO_3$ ratio means the more sodium silicate in the paste, which is a great source of Si, enhancing the production of the C-S-H (or C-A-S-H) gel and providing a denser AASC. Therefore, when the ratio was one, the AASC samples showed a better performance after acid attacks in the long run.

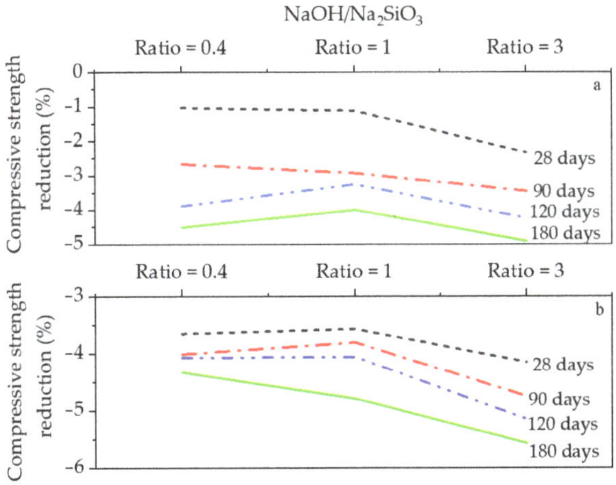

Figure 6. Compressive strength reduction of AASC with different $NaOH/Na_2SiO_3$ ratios for (**a**) 6 M NaOH and (**b**) 10 M NaOH.

It should be noted that when it comes to workability, the lowest ratio ($NaOH/Na_2SiO_3$ = 0.4) adversely affects the workability of the fresh concrete. Concrete workability is an indispensable part of concrete production and implementation. It should be taken into consideration when the mix design parameters are selected. The impact of the ratio of $NaOH/Na_2SiO_3$ on ASSC weight loss is shown in Figure 4c. It can be concluded from Figure 4c that an increase in the ratio of $NaOH/Na_2SiO_3$ from 0.4 to 3 caused the escalation of weight loss. After six months of immersion of AASC samples in the HCl acid solution, weight loss for $NaOH/Na_2SiO_3$ = 3 became more than two times greater than the specimens with $NaOH/Na_2SiO_3$ = 0.4. The principal reason is that the higher ratio of $NaOH/Si_2O_3$ decreased the sodium silicate content, which is an excellent source of Si in the concrete paste. Si is an essential constituent for the production of C-(A)-S-H gel in alkali-activated slag concrete.

3.2.4. Weight Ratio of Alkaline Solution to Slag

In the OPC concrete, the water to cement ratio is a critical parameter, especially when it comes to durability. The weight ratio of the alkaline solution to slag has a critical role in the properties and performance of the alkali-activated materials [41]. Limited studies have evaluated the impacts of this essential parameter on AASC durability. In most studies, a ratio of alkaline solution to slag was selected in a range of 0.4 to 0.6 [29,41,44,63,64]. When the ratio increased from 0.4 to 0.6, the quantity of alkaline solution in the concrete increased, and a lower amount of slag existed in the mixture as paste. In this study, two weight ratios of the alkaline solutions to slag were used, 0.4 and 0.6.

Figure 7 shows the impact of the weight ratio of the alkaline solution to slag on the strength reduction of AASC specimens exposed to HCl acid solutions for up to 6 months. Figure 7a shows that altering the weight ratio of the alkaline solution to slag from 0.4 to 0.6 led to nearly 10% more strength reduction for the AASC specimens subjected to HCl acid solution for six months. Likewise, for 10 M NaOH and $NaOH/Na_2SiO_3$ = 3 (Figure 7b), more alkaline solution caused nearly 29% more strength reduction. In Figure 7a,b, the AASC samples with a ratio of 0.4 had the lowest strength reduction at all ages. The fundamental reason is that the higher ratio of alkaline solution to slag reduced the adequate amount of slag and paste in the mixture. A definite amount of alkaline solutions is needed to activate the entire volume of slag in the mixture, thus up to that volume, the presence of an alkaline solution enhanced the mechanical properties of AASC [64]. The same results are observed in [41], where an increase in paste content led to an enhancement of acid attack for AASC mixes.

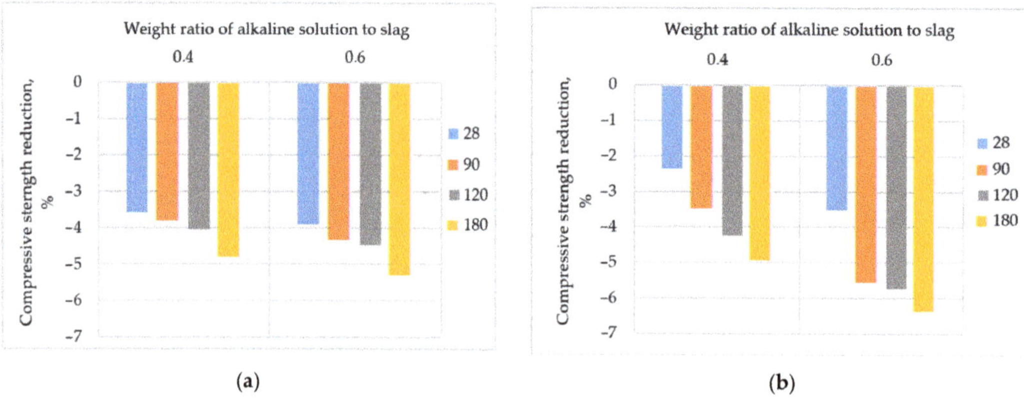

Figure 7. Compressive strength reduction of AASC with different alkaline solution to slag ratios for (**a**) NaOH molarity = 6, $NaOH/Na_2SiO_3$ = 1 and (**b**) NaOH molarity = 10, $NaOH/Na_2SiO_3$ = 3.

In addition, the alkaline solutions consisted of a considerable amount of water. Hence, when the ratio of the alkaline solution to slag increased, extra alkaline solutions created countless capillary pores, which could develop a poor structure, inferior performance, and deterioration in acid attacks [64]. The findings of this study pointed out that more alkaline solutions initially, with a ratio of 0.6, resulted in a lower resistance in acid attack, and a fewer alkaline solutions, with a ratio of 0.4, led to a better performance of AASC after exposure to the acid.

Figure 4d shows the weight loss percentage under the influence of the weight ratio of alkaline solutions to slag. The ratio did not significantly affect the weight loss after 90 and 120 days of immersion of the AASC samples in HCl acid. When the ratio of alkaline solutions to slag changed from 0.4 to 0.6 after 180 days, the weight loss reduction increased by about 20%. As discussed above, when the weight ratio of the alkaline solutions to slag increased, an excessive amount of alkaline solutions consisting of a substantial amount of water in the AASC mixture would be available, which could negatively affect the performance of AASC for a long time. As reported in [41], the weight loss of AASC tended to decrease when the solution to binder ratio increased. The main reason is that more paste in the structure reduced the voids and ingress of acid agents.

4. Comparison of OPC Concrete and AASC in HCl Acid Solution

In this section, the performance of OPC concrete is presented and compared with AASC in an HCl acid solution. As stated before, alkali-activated materials show acceptable mechanical properties and durability in acidic media [65,66]. Nevertheless, when acid attacks OPC concrete, calcium components, especially calcium hydroxide, hydrated calcium silicate, and hydrated calcium aluminate, convert into calcium salts, leading to deterioration and cracking [3].

As presented in Figure 8a, the strength reduction of OPC concrete after six months of immersion in HCl acid was approximately 38%. However, the minimum and maximum strength reductions of AASC samples with the same exposure time to HCl acid for all of the mix designs were 4% and 12.6%, respectively. In other words, the minimum and maximum AASC strength reductions were one-tenth and one-third of OPC concrete, respectively. This means a well-designed alkali-activated slag concrete has a superior performance when it comes to strength reduction. Accordingly, a concrete structure or element erected by OPC concrete that deteriorated after two years due to acid attacks would acceptably work for at least 20 years, if built by AASC with well-selected mix design parameters.

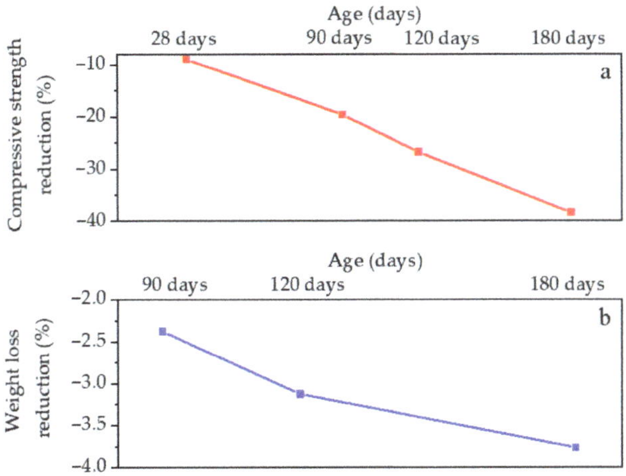

Figure 8. (**a**) Compressive strength and (**b**) weight loss reduction of OPC concrete in an HCl acid solution.

The weight loss of the OPC concrete was 3.76% after 180 days of immersion in an HCl acid solution, as shown in Figure 8b. Nonetheless, the minimum weight loss for the AASC samples was 0.79%. This means that the weight loss of AASC was nearly one-fifth of the OPC concrete. The better performance of AASC could be attributed to several factors. The main factor is the different structures between the two concretes. It is reported that AASC has a calcium (aluminate) silicate (C-(A)-S-H) hydrate structure similar to the OPC hydration products, but with a lower ratio of Ca/Si compared with OPC concrete [49]. The chemical and physical characteristics of slag (Na_2O content in NaOH and silicate module (SiO_2/Na_2O) in Na_2SiO_3) are other factors that can negatively or positively impact the durability of AASC [49]. In this study, the amount of CaO in the slag (Table 1) was about half of the CaO amount in the Portland cement. Hence, AASC samples have lower calcium components in their composition, leading to a better performance in acid attacks.

The obtained results of the weight loss reduction test are in agreement with the literature. Torgal et al. reported a 2.6% average weight loss for AASC specimens after exposure to three acid solutions (HCl, H_2SO_4, and HNO_3) for 28 days, which was less than half of the observed weight loss for OPC concrete [67]. Bakharev stated 1.15% and 12.43% weight loss for alkali active concrete made with fly ash activated by NaOH and Na_2SiO_3 after 60 days of exposure to acetic acid and sulfuric acid solutions, respectively [46]. Munn et al. investigated the weight loss for OPC and alkali-activated concrete made with fly ash and activated by NaOH and Na_2SiO_3. He reported a 40.9% weight loss for OPC concrete after a month of exposure to a 10% sulfuric acid solution, and 5.2% weight loss for alkali-activated concrete after two months of exposure to 10% sulfuric acid [68].

5. Conclusions

This study investigated the effects of four mix design parameters on the durability of alkali-activated slag concrete in an HCl acid solution. Having a clear understanding of the mix design parameters of alkali-activated slag concrete when there is no standard design procedure method can help researchers develop and design AASC efficiently. The results showed that AASC outperforms OPC concrete when it comes to durability in an HCl acid solution. The AASC samples contained potassium hydroxide as an alkaline activator, and showed a higher strength reduction and weight loss in the HCl acid solution than the samples made with sodium hydroxide. The lower molarity of the sodium hydroxide caused a superior performance of AASC in acid attacks. Moreover, the AASC samples with $NaOH/Na_2SiO_3 = 1$ compared with the samples with a ratio equal to three had an acceptable performance after acid attacks. While it might seem that the more alkaline solution would bring about a superior performance in an HCl solution, the results pointed out that when the weight ratio of the alkali solution to slag increased from 0.4 to 0.6, the strength reduction and weight loss increased. Furthermore, for a well-designed mixture, strength reduction and weight loss of AASC compared with OPC concrete were approximately one-tenth and one-fifth, respectively.

Author Contributions: Conceptualization, M.T. and K.B.; methodology, M.T. and K.B.; validation, M.T. and K.B.; formal analysis, M.T. and K.B.; investigation, M.T., K.B. and A.S.; resources, M.T. and K.B.; data curation, M.T., K.B. and A.S.; writing—original draft preparation, M.T. and A.S.; writing—review and editing, M.T., K.B. and A.S.; visualization, M.T., K.B. and A.S.; supervision, K.B.; project administration, K.B.; funding acquisition, A.S. All authors have read and agreed to the published version of the manuscript.

Funding: This research received no external funding.

Institutional Review Board Statement: Not applicable.

Informed Consent Statement: Not applicable.

Data Availability Statement: No new data were created or analyzed in this study. Data sharing is not applicable to this article.

Conflicts of Interest: The authors declare no conflict of interest.

References

1. Sonebi, M.; Ammar, Y.; Diederich, P. 15-Sustainability of cement, concrete and cement replacement materials in construction. In *Sustainability of Construction Materials*, 2nd ed.; Khatib, J.M., Ed.; Woodhead Publishing: Cambridge, UK, 2016; pp. 371–396. [CrossRef]
2. Raju, P.S.N.; Dayaratnam, P. Durability of concrete exposed to dilute sulphuric acid. *Build. Environ.* **1984**, *19*, 75–79. [CrossRef]
3. Zivica, V.; Bajza, A. Acidic attack of cement based materials—A review: Part 1. Principle of acidic attack. *Constr. Build. Mater.* **2001**, *15*, 331–340. [CrossRef]
4. Mohamed, O.A. A review of durability and strength characteristics of alkali-activated slag concrete. *Materials* **2019**, *12*, 1198. [CrossRef] [PubMed]
5. Li, X.; Ling, T.-C.; Mo, K.H. Functions and impacts of plastic/rubber wastes as eco-friendly aggregate in concrete—A review. *Constr. Build. Mater.* **2020**, *240*, 117869. [CrossRef]
6. Amran, M.; Debbarma, S.; Ozbakkaloglu, T. Fly ash-based eco-friendly geopolymer concrete: A critical review of the long-term durability properties. *Constr. Build. Mater.* **2021**, *270*, 121857. [CrossRef]
7. Aliabdo, A.A.; Abd Elmoaty, A.E.M.; Emam, M.A. Factors affecting the mechanical properties of alkali activated ground granulated blast furnace slag concrete. *Constr. Build. Mater.* **2019**, *197*, 339–355. [CrossRef]
8. Panda, B.; Ruan, S.; Unluer, C.; Tan, M.J. Investigation of the properties of alkali-activated slag mixes involving the use of nanoclay and nucleation seeds for 3D printing. *Compos. Part B Eng.* **2020**, *186*, 107826. [CrossRef]
9. Yi, H.; Xu, G.; Cheng, H.; Wang, J.; Wan, Y.; Chen, H. An Overview of Utilization of Steel Slag. *Procedia Environ. Sci.* **2012**, *16*, 791–801. [CrossRef]
10. Yang, K.-H.; Song, J.-K. Workability Loss and Compressive Strength Development of Cementless Mortars Activated by Combination of Sodium Silicate and Sodium Hydroxide. *J. Mater. Civ. Eng.* **2009**, *21*, 119–127. [CrossRef]
11. Adesina, A. Properties of Alkali Activated Slag Concrete Incorporating Waste Materials as Aggregate: A Review. *Mater. Sci. Forum* **2019**, *967*, 214–220. [CrossRef]
12. Shi, C.; Roy, D.; Krivenko, P. *Alkali-Activated Cements and Concretes*; CRC press: Boca Raton, FL, USA, 2003.
13. Pacheco-Torgal, F.; Labrincha, J.; Leonelli, C.; Palomo, A.; Chindaprasit, P. *Handbook of Alkali-Activated Cements, Mortars and Concretes*; Elsevier: Amsterdam, The Netherlands, 2014.
14. Shojaei, M.; Behfarnia, K.; Mohebi, R. Application of alkali-activated slag concrete in railway sleepers. *Mater. Des.* **2015**, *69*, 89–95. [CrossRef]
15. Mohebi, R.; Behfarnia, K.; Shojaei, M. Abrasion resistance of alkali-activated slag concrete designed by Taguchi method. *Constr. Build. Mater.* **2015**, *98*, 792–798. [CrossRef]
16. Surul, O.; Bilir, T.; Gholampour, A.; Sutcu, M.; Ozbakkaloglu, T.; Gencel, O. Recycle of ground granulated blast furnace slag and fly ash on eco-friendly brick production. *Eur. J. Environ. Civ. Eng.* **2020**, 1–19. [CrossRef]
17. Collins, F.; Sanjayan, J.G. Effect of pore size distribution on drying shrinking of alkali-activated slag concrete. *Cem. Concr. Res.* **2000**, *30*, 1401–1406. [CrossRef]
18. Awoyera, P.; Adesina, A. Durability Properties of Alkali Activated Slag Composites: Short Overview. *Silicon* **2020**, *12*, 987–996. [CrossRef]
19. Bondar, D.; Ma, Q.; Soutsos, M.; Basheer, M.; Provis, J.L.; Nanukuttan, S. Alkali activated slag concretes designed for a desired slump, strength and chloride diffusivity. *Constr. Build. Mater.* **2018**, *190*, 191–199. [CrossRef]

20. Puertas, F.; González-Fonteboa, B.; González-Taboada, I.; Alonso, M.M.; Torres-Carrasco, M.; Rojo, G.; Martínez-Abella, F. Alkali-activated slag concrete: Fresh and hardened behaviour. *Cem. Concr. Compos.* **2018**, *85*, 22–31. [CrossRef]
21. Ye, H.; Radlińska, A. Shrinkage mechanisms of alkali-activated slag. *Cem. Concr. Res.* **2016**, *88*, 126–135. [CrossRef]
22. Zhu, X.; Tang, D.; Yang, K.; Zhang, Z.; Li, Q.; Pan, Q.; Yang, C. Effect of $Ca(OH)_2$ on shrinkage characteristics and microstructures of alkali-activated slag concrete. *Constr. Build. Mater.* **2018**, *175*, 467–482. [CrossRef]
23. Li, N.; Farzadnia, N.; Shi, C. Microstructural changes in alkali-activated slag mortars induced by accelerated carbonation. *Cem. Concr. Res.* **2017**, *100*, 214–226. [CrossRef]
24. He, J.; Gao, Q.; Wu, Y.; He, J.; Pu, X. Study on improvement of carbonation resistance of alkali-activated slag concrete. *Constr. Build. Mater.* **2018**, *176*, 60–67. [CrossRef]
25. Mohabbi Yadollahi, M.; Dener, M. Investigation of elevated temperature on compressive strength and microstructure of alkali activated slag based cements. *Eur. J. Environ. Civ. Eng.* **2019**, 1–15. [CrossRef]
26. Shabani, A.; Kioumarsi, M.; Plevris, V.; Stamatopoulos, H. Structural Vulnerability Assessment of Heritage Timber Buildings: A Methodological Proposal. *Forests* **2020**, *11*, 881. [CrossRef]
27. Shabani, A.; Kioumarsi, M.; Zucconi, M. State of the art of simplified analytical methods for seismic vulnerability assessment of unreinforced masonry buildings. *Eng. Struct.* **2021**, *239*, 112280. [CrossRef]
28. Khitab, A.; Arshad, M.; Ffaisal, A.; Khan, I. Development of an Acid Resistant Concrete: A Review. *Int. J. Sustain. Constr. Eng. Technol.* **2013**, *4*, 33–38.
29. Bakharev, T.; Sanjayan, J.G.; Cheng, Y.B. Resistance of alkali-activated slag concrete to acid attack. *Cem. Concr. Res.* **2003**, *33*, 1607–1611. [CrossRef]
30. Komljenović, M.; Baščarević, Z.; Marjanović, N.; Nikolić, V. External sulfate attack on alkali-activated slag. *Constr. Build. Mater.* **2013**, *49*, 31–39. [CrossRef]
31. Madhuri, G.; Srinivasa Rao, K. Performance of alkali-activated slag concrete against sulphuric acid attack. *Asian J. Civ. Eng.* **2018**, *19*, 451–461. [CrossRef]
32. Ren, J.; Zhang, L.; San Nicolas, R. Degradation process of alkali-activated slag/fly ash and Portland cement-based pastes exposed to phosphoric acid. *Constr. Build. Mater.* **2020**, *232*, 117209. [CrossRef]
33. Miyamoto, S.; Minagawa, H.; Hisada, M. Deterioration rate of hardened cement caused by high concentrated mixed acid attack. *Constr. Build. Mater.* **2014**, *67*, 47–54. [CrossRef]
34. Allahverdi, A.; Škvára, F. Acidic corrosion of hydrated cement based materials. Part 1. Mechanism of the phenomenon. *Ceram. Silik.* **2000**, *44*, 114–120.
35. Beddoe, R.E.; Dorner, H.W. Modelling acid attack on concrete: Part I. The essential mechanisms. *Cem. Concr. Res.* **2005**, *35*, 2333–2339. [CrossRef]
36. Yun, T.; Kwak, S.-Y. Recovery of hydrochloric acid using positively-charged nanofiltration membrane with selective acid permeability and acid resistance. *J. Environ. Manag.* **2020**, *260*, 110001. [CrossRef]
37. Türkel, S.; Felekoğlu, B.; Dulluc, S. Influence of various acids on the physico-mechanical properties of pozzolanic cement mortars. *Sadhana* **2007**, *32*, 683–691. [CrossRef]
38. Allahverdi, A.; Škvára, F. Acidic corrosion of hydrated cement based materials—Part 2. Kinetics of the phenomenon and mathematical models. *Ceram.Silik.* **2000**, *44*, 152–160.
39. Pavlík, V. Corrosion of hardened cement paste by acetic and nitric acids part I: Calculation of corrosion depth. *Cem. Concr. Res.* **1994**, *24*, 551–562. [CrossRef]
40. Özcan, A.; Karakoç, M.B. The Resistance of Blast Furnace Slag- and Ferrochrome Slag-Based Geopolymer Concrete Against Acid Attack. *Int. J. Civ. Eng.* **2019**, *17*, 1571–1583. [CrossRef]
41. Thunuguntla, C.S.; Gunneswara Rao, T.D. Effect of mix design parameters on mechanical and durability properties of alkali activated slag concrete. *Constr. Build. Mater.* **2018**, *193*, 173–188. [CrossRef]
42. Matalkah, F.; Salem, T.; Soroushian, P. Acid resistance and corrosion protection potential of concrete prepared with alkali aluminosilicate cement. *J. Build. Eng.* **2018**, *20*, 705–711. [CrossRef]
43. Nagaraj, V.K.; Babu, D.L.V. Assessing the performance of molarity and alkaline activator ratio on engineering properties of self-compacting alkaline activated concrete at ambient temperature. *J. Build. Eng.* **2018**, *20*, 137–155. [CrossRef]
44. Bakharev, T.; Sanjayan, J.G.; Cheng, Y.B. Sulfate attack on alkali-activated slag concrete. *Cem. Concr. Res.* **2002**, *32*, 211–216. [CrossRef]
45. Chi, M. Effects of dosage of alkali-activated solution and curing conditions on the properties and durability of alkali-activated slag concrete. *Constr. Build. Mater.* **2012**, *35*, 240–245. [CrossRef]
46. Bakharev, T. Resistance of geopolymer materials to acid attack. *Cem. Concr. Res.* **2005**, *35*, 658–670. [CrossRef]
47. Collins, F.G.; Sanjayan, J.G. Workability and mechanical properties of alkali activated slag concrete. *Cem. Concr. Res.* **1999**, *29*, 455–458. [CrossRef]
48. Rajamane, N.P.; Nataraja, M.C.; Lakshmanan, N.; Dattatreya, J.K.; Sabitha, D. Sulphuric acid resistant ecofriendly concrete from geopolymerisation of blast furnace slag. *Indian J. Eng. Mater. Sci.* **2012**, *19*, 357–367.
49. Amer, I.; Kohail, M.; El-Feky, M.S.; Rashad, A.; Khalaf, M.A. A review on alkali-activated slag concrete. *Ain Shams Eng. J.* **2021**. [CrossRef]

50. Gruskovnjak, A.; Lothenbach, B.; Holzer, L.; Figi, R.; Winnefeld, F. Hydration of alkali-activated slag: Comparison with ordinary Portland cement. *Adv. Cem. Res.* **2006**, *18*, 119–128. [CrossRef]
51. Reddy, A.; Uppalapati, V. Performance of Alkali Activated Slag and Alkali Activated Slag + Fly Ash with various Alkali Activators. *Int. J. Eng. Technol. Res.* **2014**, *2*, 73–78.
52. Behfarnia, K.; Taghvayi Yazeli, H.; Khalili Khasraghi, M.B. The Effect of Alkaline Activator on Workability and Compressive Strength of Alkali-Activated Slag Concrete. *AUT J. Civ. Eng.* **2017**, *1*, 55–60. [CrossRef]
53. Manjunath, G.S.; Giridhar, C.; Jadhav, M. Compressive Strength Development in Ambient Cured Geo-Polymer Mortar. *Int. J. Earth Sci. Eng.* **2011**, *4*, 830–834.
54. Fang, G.; Ho, W.K.; Tu, W.; Zhang, M. Workability and mechanical properties of alkali-activated fly ash-slag concrete cured at ambient temperature. *Constr. Build. Mater.* **2018**, *172*, 476–487. [CrossRef]
55. Beltrame, N.A.M.; da Luz, C.A.; Perardt, M.; Hooton, R.D. Alkali activated cement made from blast furnace slag generated by charcoal: Resistance to attack by sodium and magnesium sulfates. *Constr. Build. Mater.* **2020**, *238*, 117710. [CrossRef]
56. Shariati, M.; Shariati, A.; Trung, N.T.; Shoaei, P.; Ameri, F.; Bahrami, N.; Zamanabadi, S.N. Alkali-activated slag (AAS) paste: Correlation between durability and microstructural characteristics. *Constr. Build. Mater.* **2021**, *267*, 120886. [CrossRef]
57. Wanger, V. *Manual of Chemical Technology*; Forgotten Books: London, UK, 2012.
58. Idir, R.; Cyr, M.; Pavoine, A. Investigations on the durability of alkali-activated recycled glass. *Constr. Build. Mater.* **2020**, *236*, 117477. [CrossRef]
59. Abdul Rahim, R.H.; Rahmiati, T.; Azizli, K.A.; Man, Z.; Nuruddin, M.F.; Ismail, L. Comparison of Using NaOH and KOH Activated Fly Ash-Based Geopolymer on the Mechanical Properties. *Mater. Sci. Forum* **2015**, *803*, 179–184. [CrossRef]
60. Sabitha, D.; Dattatreya, J.K.; Sakthivel, N.; Bhuvaneshwari, M.; Sathik, S.A.J. Reactivity, workability and strength of potassium versus sodium-activated high volume fly ash-based geopolymers. *Curr. Sci.* **2012**, *103*, 1320–1327.
61. Brough, A.R.; Atkinson, A. Sodium silicate-based, alkali-activated slag mortars: Part I. Strength, hydration and microstructure. *Cem. Concr. Res.* **2002**, *32*, 865–879. [CrossRef]
62. Bernal, S.A.; Rodríguez, E.D.; de Gutiérrez, R.M.; Provis, J.L. Performance of alkali-activated slag mortars exposed to acids. *J. Sustain. Cem. Based Mater.* **2012**, *1*, 138–151. [CrossRef]
63. Robayo-Salazar, R.A.; Aguirre-Guerrero, A.M.; Mejía de Gutiérrez, R. Carbonation-induced corrosion of alkali-activated binary concrete based on natural volcanic pozzolan. *Constr. Build. Mater.* **2020**, *232*, 117189. [CrossRef]
64. Behfarnia, K.; Rostami, M. The Effect of Alkaline Solution-to-Slag Ratio on Permeability of Alkali Activated Slag Concrete. *Int. J. Civ. Eng.* **2017**, 16. [CrossRef]
65. Wang, A.; Zheng, Y.; Zhang, Z.; Liu, K.; Li, Y.; Shi, L.; Sun, D. The Durability of Alkali-Activated Materials in Comparison with Ordinary Portland Cements and Concretes: A Review. *Engineering* **2020**, *6*, 695–706. [CrossRef]
66. Pacheco-Torgal, F.; Abdollahnejad, Z.; Camões, A.F.; Jamshidi, M.; Ding, Y. Durability of alkali-activated binders: A clear advantage over Portland cement or an unproven issue? *Constr. Build. Mater.* **2012**, *30*, 400–405. [CrossRef]
67. Fernando, P.-T.; Castro-Gomes, J.; Jalali, S. Durability and Environmental Performance of Alkali-Activated Tungsten Mine Waste Mud Mortars. *J. Mater. Civ. Eng.* **2010**, *22*, 897–904. [CrossRef]
68. Robert Munn, X.S.; Marosszeky, M. Experimental Study of Geopolymer Concrete Resistance to Sulphuric Acid Attack. In Proceedings of the 11DBMC International Conference on Durability of Building Materials and Components ISTANBUL, Istanbul, Turkey, 11 May 2008.

sustainability

MDPI

Article

Sustainable Proposal for Plant-Based Cementitious Composites, Evaluation of Their Mechanical, Durability and Comfort Properties

César A. Juárez-Alvarado [1], Camille Magniont [2], Gilles Escadeillas [2], Bernardo T. Terán-Torres [1], Felipe Rosas-Diaz [1] and Pedro L. Valdez-Tamez [1,*]

[1] Facultad de Ingeniería Civil, Universidad Autónoma de Nuevo León, C. Pedro de Alba s/n, San Nicolás de los Garza 66455, NL, Mexico
[2] LMDC, Université de Toulouse, INSAT, UPS, 31077 Toulouse, France
* Correspondence: pedro.valdeztz@uanl.edu.mx

Abstract: This research evaluates four sustainable cementitious composites with sustainable plant fibers and bio-aggregates: (1) cementitious matrix composite with lechuguilla fibers (LFC) and (2) with flax fibers (FFC); and (3) cementitious matrix composite with wood shavings (WSC) and (4) with hemp shavings (HSC). The fibers are for reinforcement and the shavings act as bio-aggregates as a total replacement for limestone aggregates. The lechuguilla (LF) and flax (FF) fibers were treated; wood (WS) and hemp (HS) bio-aggregates were also processed. Nineteen mixtures were manufactured, and five were used as controls, and the hygrothermal, mechanical, and durability properties were evaluated. The results for LFC and FFC showed that fiber treatment negatively affected flexural–compressive strength; untreated LFC with accelerated deterioration had better mechanical behavior, higher density, and lower porosity than FFC. Strength and density decreased, but porosity increased with increasing fiber volume (Vf). Regarding WSC and HSC, the microstructure of WS and HS had a significant effect on the physical and mechanical properties. The high porosity influenced the results obtained, since it decreased compressive strength and bulk density; however, thermal conductivity, hygroscopicity, and vapor resistance showed better behavior in most cases than the control specimens, i.e., without bio-aggregates.

Keywords: cementitious composite; sustainability; plant-based materials; hygroscopicity; durability; bio-aggregate

Citation: Juárez-Alvarado, C.A.; Magniont, C.; Escadeillas, G.; Terán-Torres, B.T.; Rosas-Diaz, F.; Valdez-Tamez, P.L. Sustainable Proposal for Plant-Based Cementitious Composites, Evaluation of Their Mechanical, Durability and Comfort Properties. *Sustainability* **2022**, *14*, 14397. https://doi.org/10.3390/su142114397

Academic Editors: Mahdi Kioumarsi and Vagelis Plevris

Received: 4 October 2022
Accepted: 26 October 2022
Published: 3 November 2022

1. Introduction

The increase in the world's population has led to the construction of larger and more expensive infrastructures that are designed for a specific minimum duration. The world population was 7942 million in 2022 according to the report of the United Nations Department of Economic and Social Affairs, and population growth is expected to continue increasing over time, mainly in cities in Africa, Asia, and Latin America, with projected growth reaching 9.7 billion in 2050 [1]. The need for affordable and sustainable housing is an inherently global problem, and numerous challenges remain in producing environmentally friendly construction products [2].

The rising problem of resource depletion and global pollution has challenged many researchers to look for new materials based on renewable resources. The construction sector struggles with four main environmental impacts: greenhouse gas emissions; energy consumption; natural resource consumption; and waste production. The alternative technologies currently adopted in buildings will have immediate consequences concerning energy consumption and emission patterns [3]. The consideration and application of sustainability practices in construction can generate a 35% reduction in CO_2 emissions, and water and energy consumption savings of 30% to 50%, respectively, and a reduction in solid waste disposal costs of up to 90% [4–7].

In this context, the concept of "sustainable construction" has been used to characterize construction that includes environmental criteria in the project conception, in the form of construction and maintenance, and, when the time comes, demolition of the infrastructure [8]. Most products came from renewable resources until the beginning of the 20th century; however, the enormous growth of the petrochemical industry slowed down the growth of biotechnological products [9]. The combination of plant materials with non-organic matrices to produce competitive composite materials is increasingly gaining attention based on the strategy of preventing the clearing of agroforestry resources as well as producing good economic returns for their cultivation, resulting in valuable benefits for the environment, favoring a comfortable habitat (humidity, thermal, and acoustic management) [10–12].

The agroindustry is a sector that mainly serves agriculture to meet the demand for food supplies. The unit operations implemented in the production chain of any raw material bring environmental problems, namely, the high production of associated residues [13]. An analysis of the use given to the residues obtained from different agro-industrial sectors shows, for instance, that the brewing industry uses 8% of the grain nutrients, while in the palm oil and cellulose industries, less than 9% and 30% are used, respectively, and in the coffee industry, only 9.5% of the weight of the fresh fruit is used for the preparation of the beverage, leaving 90.5% as residue [14]. Depending on the type of raw material, it can be treated to reduce the negative environmental impacts and to transform it into valuable products that can be used as raw materials for other production processes and generate additional economic income. Thus, one of the challenges of research in this area is to optimize the results and diversify the use of these raw materials. Hence, achieving greater waste utilization depends on creating a strong value proposition throughout the value chain [15]

The main environmental benefit of bio-based building materials is their carbon storage capacity, which could contribute to climate change limitations. For instance, the use of these by-products can bring enormous benefits to reducing the cost of most of the world's industries, such as biotechnology or construction, among others [13]. The widespread use of plant materials and their products is attributed to properties such as low density, high electrical resistance, specific modulus and stiffness, non-abrasive properties, and biodegradability, among others. On the other hand, plant materials are raw materials from renewable sources and are of high availability, although to use them correctly it is necessary to know their properties and limitations in more detail to include them in new projects. The use of these materials has allowed for the availability of cheap, biodegradable, and recyclable construction materials from renewable resources, and under certain conditions of preparation, the insulating effect is comparable to that of other synthetic materials widely used in the industry, such as polystyrene, but with greater variability in their mechanical properties, which allow diverse applications given the difficulty to recycle the material of synthetic origin [16]. Natural fibers and their use in construction materials will have a positive impact on the construction projects in which they are used, generating a reduction in greenhouse gas emissions, climate control energy savings (operational energy), and less waste generation [11,17]. On the other hand, the main disadvantage of using plant fibers is the durability of these fibers in the cement matrix and the compatibility between the two phases. Alkaline media tend to mineralize the fibers, which is associated with long-term loss of toughness of the composite material attributed to interfacial damage caused by the continuous volume change of the porous plant fibers in the cement matrix. In addition, the fibers in contact with the reaction water release water-soluble substances that interfere with the hydration of the cementitious material, decreasing its potential toughness [18].

The present research study proposes to evaluate sustainable plant-based cementitious composites: (1) a composite based on a cementitious matrix with natural lechuguilla fibers (LFC) and flax fibers (FFC) as reinforcement; and (2) a cementitious matrix composite with wood shavings (WSC) and hemp shavings (HSC) as bio-aggregates, and whose hygrothermal, mechanical, and durability properties were evaluated. The results are expected to have an important benefit for the environment, for a comfortable habitat (improvements in the hygrothermal performance of the home), and as sustainable materials for dignified housing.

2. Materials and Mixtures

2.1. Cementitious Materials

Two types of Portland cement were used, namely, type III from México, which complies with ASTM C150-21 [19], and type II from France, which complies with UNE-EN 197-1 [20], with a relative density of 3.1 and 3.0, respectively. The fly ash used was class F according to ASTM C-618-19 [21], which was obtained from a thermoelectric power plant in the city of Nava, Coahuila, and its relative density was 2.2. In addition, a metakaolin was used, since this type of pozzolan is very reactive in hydraulic cementitious matrices, and it had a relative density of 2.5. The fly ash came from Mexico and the metakaolin came from France, and both pozzolans were used additionally. The chemical compositions of the cementitious substances are shown in Table 1.

Table 1. Chemical composition of cementitious materials.

Material	Chemical Composition (%)					
	SiO_2	Al_2O_3	Fe_2O_3	CaO	MgO	Na_2O
Cement Type III	14.99	4.26	3.15	68.11	1.23	0.37
Cement Type II	15.20	3.82	2.12	59.90	-	-
Fly ash	59.95	25.98	4.92	2.66	0.12	-
Metakaolin	65.90	25.10	4.26	1.63	0.10	-

2.2. Aggregates

Two types of fine aggregates were used, namely, crushed limestone sand (Mexico) and river sand (France). The physical properties of the aggregates are shown in Table 2, and their grain size is presented in Figure 1. The limestone aggregate complied with ASTM C33-18 [22], and the river aggregate complied with NF P18-545 [23]. The grain size of the aggregates improves the workability of the mixture when plant components are added. Previous studies have shown that particle sizes with a maximum size of less than 10 mm are adequate when organic materials are used in a cementitious matrix, since they allow for a uniform distribution and reduce the agglomeration effect of the aggregates [24].

Table 2. Physical properties of aggregates.

	Bulk Density kg/m^3	Dry Weight kg/m^3	Surface Dry Saturated Weight kg/m^3	Water Absorption %	Specific Weight kg/m^3	Water Content %	Fineness Modulus
Limestone	1651	2600	2617	1.60	2600	0.10	2.71
River	1660	2628	2660	1.54	2660	0.15	2.70

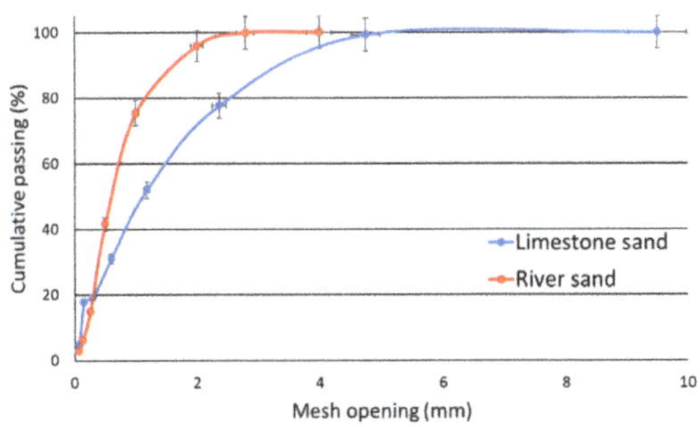

Figure 1. Particle grain size classification of aggregates.

2.3. Organic Materials

Two plant fibers were used as reinforcements; the first was the lechuguilla fiber (*Agave lechuguilla*) coming from the arid zones of Mexico, and the second was the flax fiber (*Linum*) from the south of France. Table 3 shows the physical and mechanical properties of the fibers. Pine wood shavings and commercial hemp shavings were used as bio-aggregates, and their properties are also shown in Table 3.

Table 3. Properties of organic materials [24–26].

Material	Physical and Mechanical Properties						
	Diameter/Width mm	Average Length mm	Absorption %	Specific Weight	Tensile Strength MPa	Elongation %	Porosity %
Lechuguilla	0.16–0.26	45	92.3	1.38	275–627	2.0–2.5	21–25
Flax	0.009–0.049	45	120.0	1.54	1015	1.8–2.0	3–21
Wood	1.0	4.1	180.0	1.05	710	2.0–3.0	56
Hemp	1.8	7.6	105.0	1.50	900	1.6–1.8	51

2.4. Composition of the Mixtures

To choose the composition of the mix, one was looked for that met adequate workability and compaction parameters, which would allow for a better distribution and prevent agglomeration of the fibers and bio-aggregates in the mixture. Control mixes were produced to optimize the concrete proportions; a w/c ratio = 0.5 was used for the composites with fibers, and a w/c ratio = 1.0 for the composites with bio-aggregates. The mixtures with different formulations of both plant fibers and bio-aggregates are shown in Table 4. Table 5 shows the proportions of the five definitive control mixtures; their consistency was measured with the ASTM C143-20 slump test [27].

Table 4. Formulations of the mixes for the plant-based cementitious composites, % in volume.

Mixture	Composite	LF	FF	WS	HS	MK	FA
M1	LFC	0.4				15	
M2		0.7				15	
M3		1.0				15	
M4	FFC		0.4			15	
M5			0.7			15	
M6			1.0			15	
M7	WSC			2.0			
M8				4.0			
M9				2.0			20
M10				4.0			20
M11				10.0			20
M12	HSC				4.0		
M13					4.0	20	
M14					10.0	20	

Table 5. Proportion of the control composite mixture.

Mixture	Mix Proportion (LFC–FFC) (kg/m^3)				W/C Ratio	Slump (mm)
	Type II Cement	Water	River Sand	Metakaolin		
C-1	383	169	1350	68	0.50	50
Mixture	**Mix Proportion (WSC) (kg/m^3)**				**W/C Ratio**	**Slump (mm)**
	Type III Cement	**Water**	**Limestone Sand**	**Fly Ash**		
C-2	150	150	1928	0	1.00	72
C-3	150	150	1893	30 (20%)	1.00	75
Mixture	**Mix Proportion (HSC) (kg/m^3)**				**W/C Ratio**	**Slump (mm)**
	Type II Cement	**Water**	**River Sand**	**Metakaolin**		
C-4	150	150	1973	0	1.00	70
C-5	150	150	1941	30 (20%)	1.00	73

3. Experimental Procedure

3.1. Testing of LFC and FFC Specimens

A deterioration procedure was performed to study the durability of specimens fabricated with the M1–M6 mixtures corresponding to the LFC and FFC, with Vf values of 0.4%, 0.7%, and 1.0%. This accelerated deterioration procedure consisted of 8 wetting and drying (AD) cycles; in each cycle, the composites were exposed to one day of wetting and 3 days of drying, and the specimens without fibers that were exposed to this procedure were the control specimens. The fibers were treated (TF) with a wax-based coating (Emulwax 3060) for protection in the alkaline cementitious environment. Uncoated fibers were also used as controls (UF), and for all mixtures the fiber length was 45 mm. The experimentation of the LFC and FFC is described below.

To evaluate the flexural strength, three 40 × 40 × 160 mm bars were fabricated for each test mixture. The specimens were demolded 24 h after fabrication, and during this period, water evaporation was prevented by protecting them with a plastic film, and they were kept at a room temperature of 20 °C. After specimens were extracted, they were kept for one week in a curing room at 20 °C with 100% relative humidity. Both fabrication and curing were performed using the procedure indicated in NF EN 196-1 [28]. The flexural strength of all specimens was evaluated at 40 days of age at the end of 8 exposure cycles according to NF EN 196-1 [28]. The compressive strength was obtained from the three specimens previously tested in flexure according to NF EN 196-1 [28]. Before testing, the specimens were dried in an oven at a temperature of 50 °C for 7 days to ensure that the specimens were completely dry.

Porosity and density are physical properties that are correlated and that have an impact on the durability of the material. The three samples from the specimens previously tested in compression were obtained, and their values were determined as described in NF P18-459 [29]. Finally, the interaction mechanisms of the fiber/matrix interface were determined by scanning electron microscopy. The observations were performed with a scanning electron microscope (JEOL JSM-6380LV). In order to limit the deterioration of the plant material, the observations were performed without coating the sample and at a vacuum of 60 Pa (LV) with an accelerating voltage of 15 kV.

3.2. Testing of WSC and HSC Specimens

Consumption values of 2%, 4%, and 10% substitution by volume of the aggregate were proposed, according to the method of absolute volumes. The following tests were carried out to evaluate the physical and mechanical behavior of these composites based on wood and hemp shavings.

To determine the compressive strength, 6 cylindrical specimens of 100 × 200 mm each were fabricated for each test formulation based on the specifications of ASTM C192-19 [30]. After 24 h of initial curing, all specimens were placed in the curing and maturing room at a temperature of 23° ± 2 °C at a relative humidity above 95% based on the recommendations of ASTM C511-21 [31]. Compression tests were performed at the corresponding ages according to ASTM C39-18 [32].

The bulk density was obtained in the 6 cylindrical and 3 prismatic specimens from the compressive strength and thermal conductivity tests, respectively. This was done by placing the specimens inside an oven at a temperature of 105° ± 5 °C in order to obtain a constant weight, which was related to the volume of each specimen, and thus the bulk density was obtained.

To measure the thermal conductivity, three prismatic specimens of 50 × 150 × 150 mm each were manufactured for each test mixture according to the specifications of the Lambda Meter-500 equipment, intended for the hot plate method test according to RILEM recommendations in the RILEM-13 report [33]. The conductivity tests were performed according to the specifications of the measuring equipment and standards specified in the manual [34].

The MBV test consists of subjecting a specimen to cycles of high and low relative humidity to determine the humidity content that the composite material absorbs and releases under these conditions. Three 50 × 150 × 150 mm prismatic specimens were prepared for each mixture. This test was performed in a climatic chamber at a preset temperature and relative humidity for a defined time. All specimens completed more than 20 cycles, which in total was more than 480 h of testing. The test was based on the research work "Moisture Buffering of Building Materials in Nordisk Innovations Center" [35].

For the measurement of water vapor permeability, two cylindrical specimens of 100 × 200 mm were produced, from which three slices of 100 × 25 mm were cut to be placed in metallic cups for the wet cup method. Additionally, two specimens of 110 × 220 mm were manufactured, from which three slices of 110 × 25 mm were cut to be placed in the metallic cups for the dry cup method. These tests were performed according to the specifications of NF EN ISO 12572:2001-10 [36].

The pore network of the WSC and HSC was analyzed by scanning electron microscopy, which allowed the complementing of the experimental results obtained for compressive strength, bulk density, heat transfer, moisture regulation in absorption and desorption cycles, and water vapor permeability. A scanning electron microscope (JEOL JSM-6380LV) was used.

4. Analysis and Discussion of the Results

4.1. Results of LFC and FFC Specimens

4.1.1. Flexural Strength

Figure 2 shows the results obtained for flexural strength at 40 days of age. The treatment of the fibers adversely affected the flexural strength for the LFC and FFC. In all cases, the untreated fibers showed better mechanical behavior; this was due to the better adhesion with the cementitious matrix without the presence of a wax-based coating [37]. Additionally, it is evident that this coating on the fibers did not provide adequate protection when the composites were exposed to accelerated deterioration. The LFC/TF resulted in lower flexural strength values than the 0% fiber volume, while the LFC/UF had an increase in flexural strength with respect to the 0% fiber volumes of 38%, 28%, and 16% for Vf = 0.4%, 0.7%, and 1.0%, respectively. The FFC composites with and without coatings were more affected by accelerated deterioration for all fiber volumes exhibiting lower flexural strength than the control; this may have been due to flax fiber being more susceptible to degradation in alkaline environments. In all cases, a reduction in strength was observed as Vf increased [37].

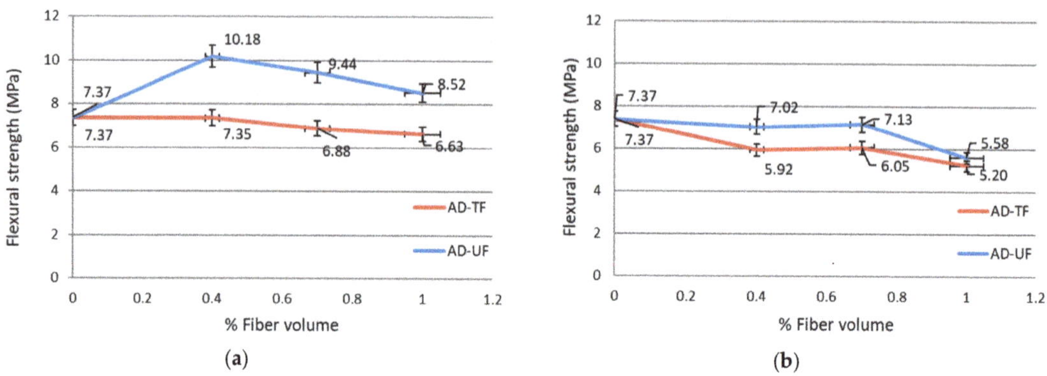

Figure 2. Flexural strength for different Vf values with exposure to accelerated deterioration. (**a**) LFC with and without fiber treatment. (**b**) FFC with and without fiber treatment.

4.1.2. Compressive Strength

Figure 3 shows the results of the compressive strength at 40 days of age. The behavior was similar to that of the flexural strength. The LFC composites with Vf = 0.4% without wax coating had higher compressive strength than the control in 14%, and for all the Vf, the strengths were higher than those corresponding to the FFC, and the accelerated deterioration seemed to affect more the latter composites. The tendency for compressive strength to decrease with increasing Vf was maintained as in the bending behavior. The chemical composition of both LF and FF had an important role in their mechanical behavior, since by their nature they are mainly composed of cellulose, hemicellulose, and lignin, which provide strength, stability, and stiffness in the fiber [18] and can contribute to the mechanical properties of the developed composites. Their cellulose and lignin composition has an impact on the hydration process and finally on the development of flexural and compressive strength [38].

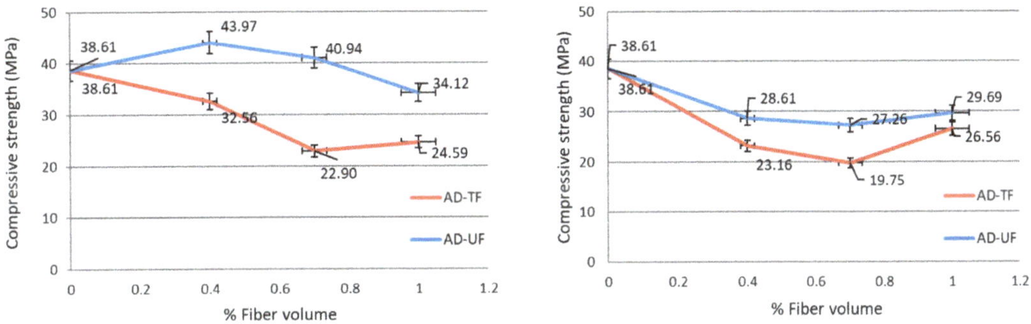

(a) (b)

Figure 3. Compressive strength with exposure to accelerated deterioration. (**a**) LFC with and without fiber treatment. (**b**) FFC with and without fiber treatment.

4.1.3. Porosity and Density

The open porosity of a material is the ratio between the volume of pores communicating to the exterior and its total volume expressed in percentage. Porosity and dry density are properties that are related to each other. Figure 4 shows that porosity increased while density decreased with the increase of fibers in the matrix [39]. For LFC (Figure 4a), the results showed a slight increase in porosity of 1%, 7%, and 9% for fiber volumes of 0.4, 0.7, and 1.0%, respectively; these results agree with Siddique [40]. Figure 4b for FFC shows higher porosity than LFC, which may be due to the physical properties of flax fiber that have a higher absorption capacity than agave fibers, and this could cause a higher porosity in the composite.

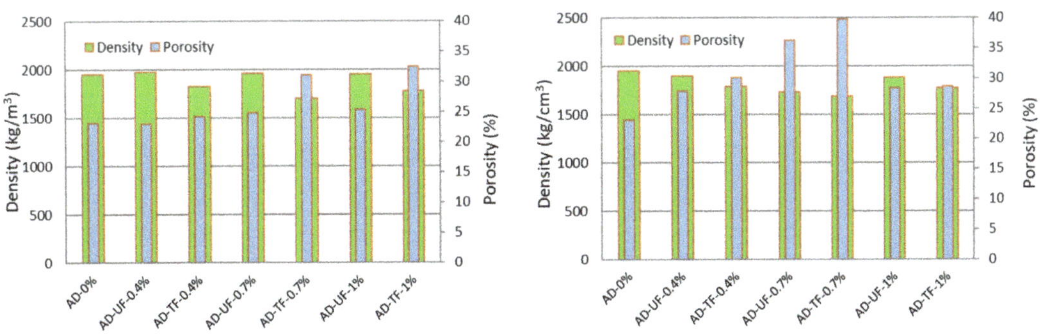

(a) (b)

Figure 4. Relationship between porosity and density for specimens with exposure to accelerated deterioration. (**a**) LFC with and without fiber treatment. (**b**) FFC with and without fiber treatment.

Figure 4 shows that the treatment affects in a more significant manner the porosity/density than do the presence and the percentage of fibers in both composites. On the other hand, porosity and density are important properties of concrete, as they affect mechanical behavior. Figure 5 shows how compressive strength is affected by density and porosity, showing that the compressive strength increases with higher density and decreases with higher porosity [39]. The FFC showed lower density and higher porosity than LFC, which achieved higher compressive strength values.

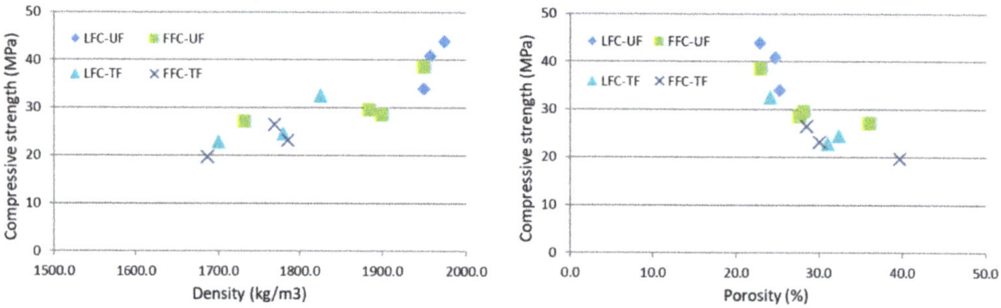

Figure 5. Effect of porosity and density on compressive strength at 40 days of age, for LFC–FFC specimens, with and without fiber treatment.

4.1.4. Microstructure of LFC and FFC with Deterioration

Figure 6 shows the micrographs of the LFC with accelerated deterioration. Deposits of mineral products on the fiber walls can be observed in Figure 6a,b, which were in the form of small plaques (white deposits), but these were not present in the depths of the fiber. This may be because matrix products could adhere to the fiber, but these did not embrittle the fiber, even in fibers without wax coatings. Figure 6c,d shows the failure mode of the fibers. In Figure 6c, the rupture of the untreated fiber caused by the adhesion to the cementitious matrix can be observed. While the treated fibers showed a pull-out type of failure due to adhesion failure, as can be seen in Figure 6d, it is clear that the adhesion between the LFs and the matrix is an important factor in the mechanical strength of the LFCs. The SEM observations show the interface with the cementitious matrix, where volumetric variations produced by water absorption could be observed because an amount of water absorbed by the fiber during mixing did not diffuse into the paste, and upon drying, it produced a reduction in volume. These dimensional variations could result in adhesion defects such as those observed in the micrographs (Figure 6d), which can lead to a loss of strength. Similar results of pull-out failures in FF have been reported in other studies [41].

Similarly, micrographs of the FFC with accelerated deterioration are shown in Figure 7. Figure 7b,d show that the flax fiber bundles did not disperse into individual fibers; this could have occurred when the fibers were incorporated into the mixture. The FF had a smaller diameter than the LF, which gave it more bonding surface area for the same volume; however, this was not evident in the mechanical results, as relatively good bonding between the FF and the matrix was lost due to the agglomeration of the FF. Additionally, in Figure 7a,b, some mineral deposits can be observed that are visible on the surface of the FF; after the accelerated deterioration procedure, the specimens were at least 60 days old at the time of the observations conducted, and with this elapsed time there were no signs of deep mineralization for the FF with and without treatment. This was evident by the fact that no embrittlement was observed in the fibers; on the contrary, they had a certain degree of ductility. SEM observations showed greater chemical interaction between the LFs with the mineral deposits compared to the FF, as they tended to have greater embrittlement, and their color changed. Other studies have used a metakaolin matrix as the formulation of this research, and they agree that this pozzolan can help to avoid the mineralization of plant fibers [42]. On the other hand, minimal volumetric variations of the untreated fiber are observed in Figure 7c. The agglomeration shown in Figure 7d makes it evident that the flax fiber was not adequately distributed, which may have influenced the low flexural and compressive strength that the FFC showed. This also explains, however, the higher porosity of these composites.

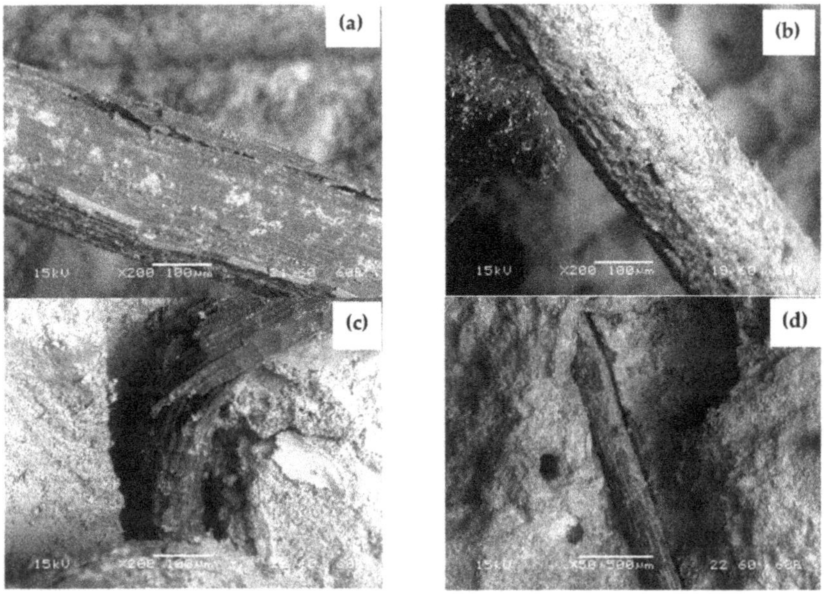

Figure 6. SEM observations: (**a**) (×200) untreated LFC; (**b**) (×200) treated LFC; (**c**) (×200) untreated LFC, failure mode rupture type; (**d**) (×50) treated LFC, failure mode pull-out type.

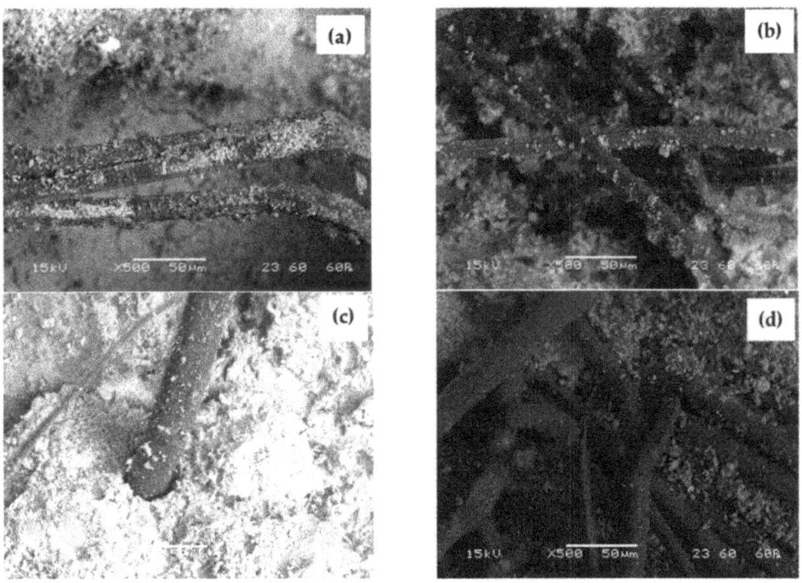

Figure 7. SEM observations (×500): (**a**) untreated FFC; (**b**) treated FFC; (**c**) untreated FFC, matrix–fiber interface; (**d**) treated FFC, matrix–fiber interface.

4.2. Results of WSC and HSC Specimens

4.2.1. Compressive Strength

The addition of porous plant aggregates led to a decrease in compressive strength, which was the expected behavior. The results of the compressive strength at 84 days of age of the composites are shown in Figure 8, where it is shown that the WSCs had better mechanical behavior than the HSCs. For WSC, the formulations M7 and M8 with 0% FA were similar to C2 (0% FA), but M9, M10, and M11 (20% FA) formulations did not exceed C3 (20% FA). Comparing C2 and C3, it was observed that the addition of FA produced a 59% increase in resistance; this contribution of FA had a positive impact on formulations M9 and M10. For HSC, the difference in resistance with respect to the control formulations C4 (0% MK) and C5 (20% MK) was not very significant; M12 (0% MK) was lesser than C4, and M13 and M14 with 20% MK were lesser than C5. In the case of HSC, it was mainly due to the presence of bio-aggregates (HS) with high water absorption capacity in the cementitious matrix, as well as the impact of hydrosolubility coming from HS on the hydration and hardening of the mineral matrix, which promoted a slower evolution in strength growth [43]. It was observed that both FA and MK behaved as fillers and also densified the matrix of WSC–HSC specimens.

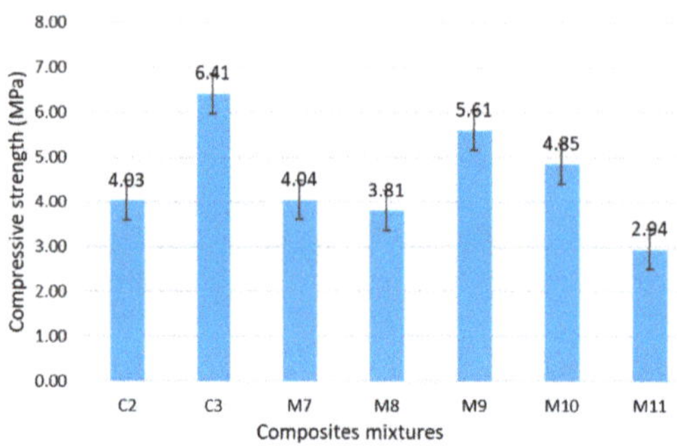

(a)

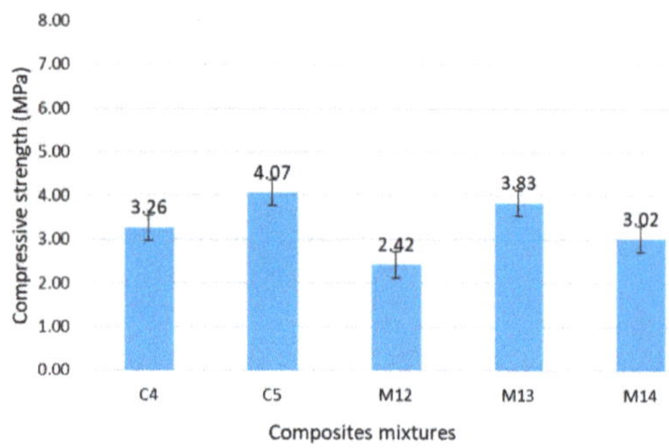

(b)

Figure 8. Variation of the compressive strength for the different formulations: (**a**) WSC, (**b**) HSC.

Figure 9 shows the effect of shavings replacement within the cementitious matrix; as the percentage of bio-aggregate replacement increased, there was a decrease in compressive strength. Figure 9a shows that for WSCs, for 10% WS replacement there was a 54% reduction with respect to the C3 control formulation with an FA-based matrix. HSC specimens with the MK matrix in contrast had a lower strength reduction of 26% for the maximum 10% replacement of HS compared to C5. It is evident that the presence of bio-aggregates affected the matrices with FA more than the matrices with MK and even the matrices that did not have cementitious agents, as can be seen in Figure 9b. However, the matrices with FA without bio-aggregates had a higher resistance to compression. A previous study found different mechanical behaviors, reporting that FA mortars had the lowest compressive strength values up to 180 days compared to MK mortars [44]. It is well known that if a sufficient amount of calcium hydroxide is not available in the early stages of cement hydration, the initial strength of the pozzolanic compounds forming additional calcium silicate may be lower than the strength of the control, which may explain the mechanical behavior observed in Figures 8 and 9.

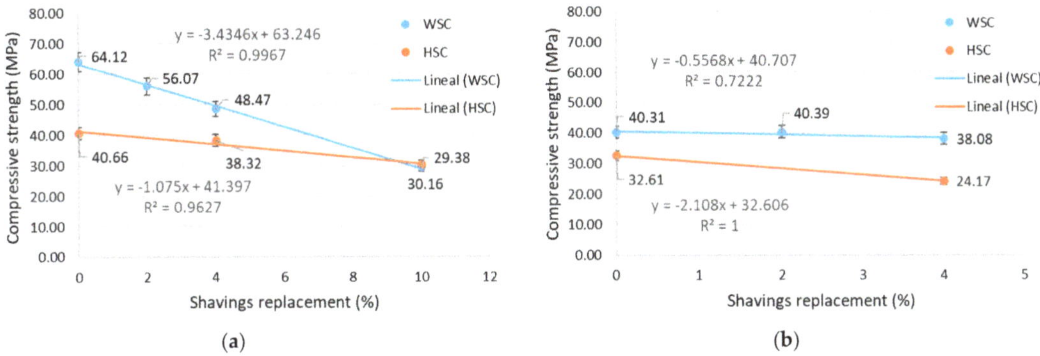

Figure 9. Effect of the shaving volume on the compressive strength for WSC–HSC specimens: (**a**) cementitious matrix with FA and MK; (**b**) cementitious matrix without FA and MK.

4.2.2. Density

The results of the bulk density are shown in Table 6. It was observed that all WSC formulations reduced their density with respect to the controls (0% WS) C2 and C3, with formulation M11 (10% WS) being the one that decreased up to 19% and 16%, respectively. The HSC had similar behavior; all formulations had lower densities than the C4 control (0% HS), and the formulation with higher shaving content of M14 (10% HS) decreased up to 13% with respect to the C4 control. With respect to the C5 control with 20% MK, the reduction was not very significant (2%), and even the formulations M12 and M13 had a slightly higher density, namely, 5% and 3%, respectively, with 4% HS. Higher percentages of WS and HS would be necessary to obtain even lower bulk density values; however, they must be related to the compressive strength in order not to affect its mechanical behavior.

Table 6. Variation of the bulk density for the different formulations (kg/m^3).

Composite						Mixtures						
	C-2	C-3	C-4	C-5	M7	M8	M9	M10	M11	M12	M13	M14
WSC	2121	2052			1786	1777	1775	1792	1728			
HSC			1900	1699						1788	1742	1658

Bulk density is affected by the accommodation of different shavings sizes within the cementitious matrix; thus, the pore content will have an impact on the reduction of bulk density in the specimens. The unit weight of the WSC and HSC decreased compared to the control specimens, as observed in Table 6. This reduction was also related to the fact that WS and HS were lighter than the fine aggregate, and this lightened the cementitious matrix [45]. Another aspect to consider is that during compaction, water is expelled from the shavings, leaving an unoccupied void within it, and subsequently these voids in the shavings are occupied by air [46]. The use of wood shavings has been found to substantially reduce the bulk density in cementitious composites and has comparable values with wood agglomerates [47].

4.2.3. Thermal Conductivity

The thermal behavior of the WSC and HSC are shown in Figure 10. It was observed that as the density decreased, the thermal conductivity also decreased. The highest density and the thermal conductivity of 0.7 W/(mK) were given by the control values, and all formulations produced lower results. As seen in Figure 10a, for the WSCs, the values showed that the use of WS decreased from 0.7 to 0.4 W/(mK), leading to a thermal conductivity decrease of approximately 43% with respect to the control, which is comparable to the materials classified as insulators [48]. As seen in Figure 10b, HSC specimens showed a reduction in thermal conductivity from 0.7 to 0.4 W/(mK), which was lower than the control and which represents 31%, and its density was lower than WSCs. It is evident that for the same % of replacement, HSC is just as efficient as WSC in reducing the thermal conductivity, and the difference in the evolution of the thermal conductivity with respect to the shaving replacement, as shown in Figure 11, can also be attributed to the origin of the bio-aggregates, since the different morphological nature of the hemp for HSC and the wood used for the WSC affects the thermal conductivity.

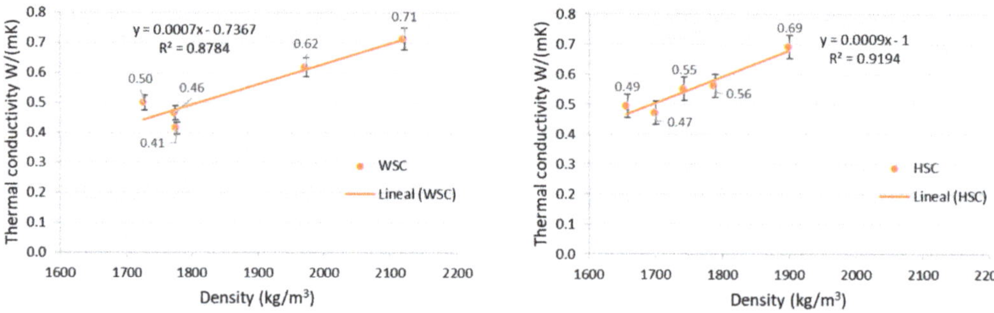

(a) (b)

Figure 10. Effect of the bulk density with the thermal conductivity: (**a**) WSC, (**b**) HSC.

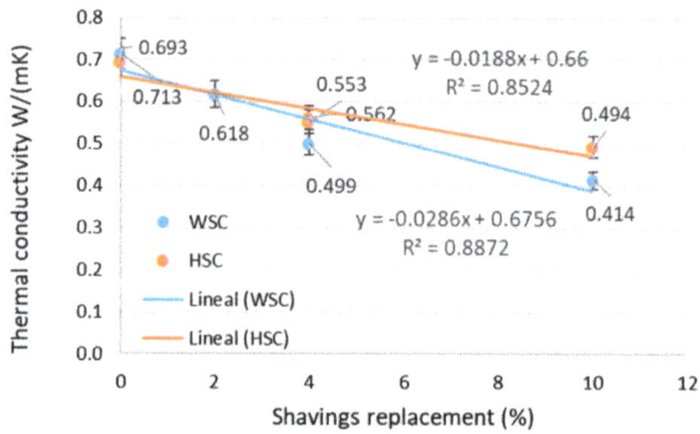

Figure 11. Effect of the shaving volume on the thermal conductivity for the WSC–HSC specimens.

As seen in Figure 11, increasing the content of shavings decreased the thermal conductivity of the composites and thus increased their insulating capacity. With the proportions of shavings used, the HSC showed slightly higher thermal conductivities than the WSC [49], and the tendency was to increase further with the increase of the replacement percentage. The low thermal conductivity of WSC was also related to the large number of voids, as reported in other studies [50]. The obtained values of thermal conductivity for WSC and HSC were relatively low and within the limits of lightweight and low thermal conductivity materials [50].

4.2.4. Moisture Buffering Value (MBV)

Figure 12 shows that the bulk density modified the MBV; as the density decreased, the MBV increased for WSC and HSC. Figure 12a shows that for WSC, the MBV values varied from 0.4 to 0.6 g/(m^2%RH), being higher for HSC, which varied from 0.8 to 1.0 g/(m^2%RH) (see Figure 12b). The control formulations were those with the highest density and lowest MBV, and it was observed that the presence of WS and HS led to an improvement in the hygroscopic behavior by increasing the MBV, and the effect of HS was more evident. For the same density, HSC mixtures exhibited higher MBV than WSC; we can thus conclude that the nature of the cementitious matrix and the nature of the plant aggregate significantly affected the buffering behavior of the composite. The decrease in density was produced mainly by the porous structure of the WSC and HSC, which was primarily due to the nature of the bio-aggregates themselves and to the effect of the inclusion of these materials in a cementitious matrix. It is evident that porosity influenced the decrease in density and also the improvement of the hygroscopic behavior and water vapor permeability, as will be seen later [51].

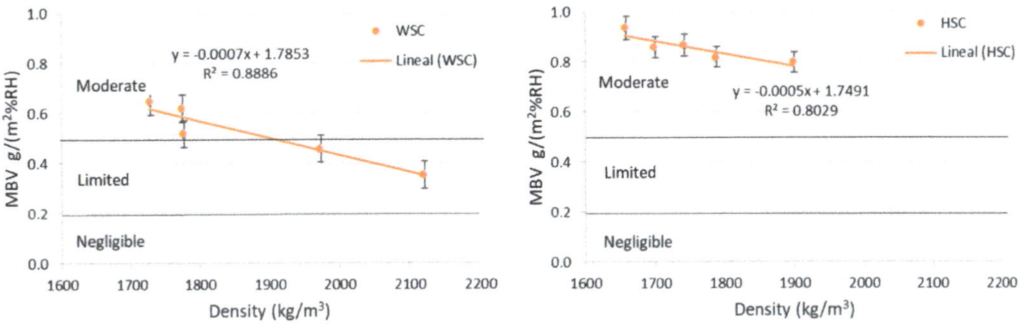

(**a**) (**b**)

Figure 12. Effect of the bulk density with MBV: (**a**) WSC, (**b**) HSC.

Table 7 shows ranges of values for the MBV classification; according to these value, the WSC and HSC could be classified in the moderate category (0.5–1.0) for moisture regulation, which for a material with a cementitious matrix is adequate. The aforementioned can be seen in Figure 13, where the average MBV values are shown for different materials. For example, concrete and brick had a value of approximately 0.4 g/(m^2%RH), similar to the control formulation of WSCs. For HSCs with a higher percentage of HS (10%), the MBV values were approximate to the order of cellular concrete [35].

Table 7. Ranges for practical moisture buffer value classes [35].

MBV Practical Class	Minimum MBV Level	Maximum MBV Level
	g/(m^2%RH)	
Negligible	0	0.2
Limited	0.2	0.5
Moderate	0.5	1.0
Good	1.0	2.0
Excellent	2.0	-

It is important to know the moisture-regulating properties of the materials to classify them in accordance with Table 7 and Figure 13; in this manner, one can have an idea of the behavior they will have in their specific application. Thus, a material with excellent conditions to regulate moisture will be able to absorb moisture and release it quickly. This will allow one to know if the WSC and HSC are able to regulate the humidity of the environment (absorption and desorption) to which they are exposed.

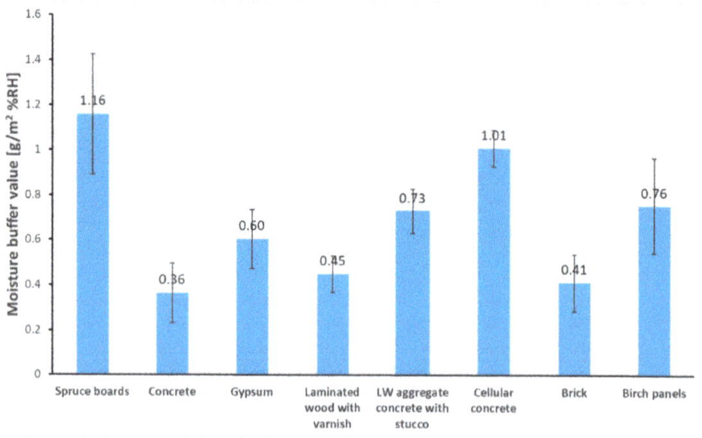

Figure 13. Average MBV values for different materials [35].

4.2.5. Water Vapor Permeability

Figure 14 shows the behavior of WSC and HSC with respect to the water vapor resistance factor. Figure 14a for WSC shows that the WS effect caused a reduction in the factor with respect to the C2 and C3 controls, allowing the transfer of moisture in greater quantities. Formulation M8 was the one that had a significant reduction with respect to the other formulations and the controls. Figure 14b shows that all the HSC formulations, including the controls, had higher water vapor permeability than those obtained for the WSCs. The HSCs were able to release more moisture to the environment due to the porous microstructure of HS, which caused an increase in the flow of water vapor passing through the capillary tubes, leading to higher moisture transfer. Other studies on hygroscopic materials have found that the presence of palm fibers in these composites decreases the resistance to water vapor diffusion and improves moisture transfer, and also in concretes with hemp there is a high permeability to water vapor, regardless of age [52]. These results can be positive depending on the place of application, since it can serve as a buffer to sudden changes in humidity; however, in cold places, it can affect the thermal performance of the envelope, and therefore it is a factor that should be considered at the time of application.

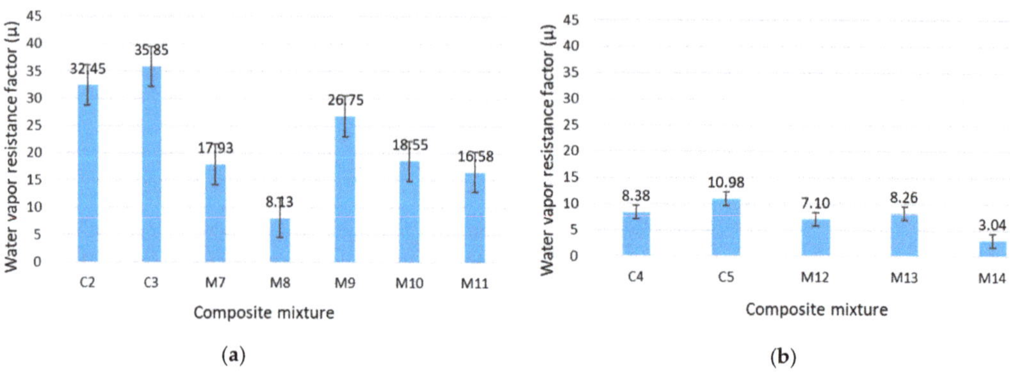

Figure 14. Variation of the water vapor permeability for different formulations: (**a**) WSC, (**b**) HSC.

4.2.6. Microstructures of WS and HS

Figure 15a,b show micrographs of WS, where high porosity can be observed from the micrographs. WS was found to be composed of 60% cellulose, and cellulose has a fibrous structure originating from compact microfibrils that constitute the plant cell wall [53]. WS had a three-dimensional cellular microstructure, with an orientation of stiff cellulose microfibrils in a matrix of hemicellulose and lignin (see) Figure 15a,b. Hemicelluloses are closely associated with cellulose microfibrils, and these are embedded with lignin, which gives it its mechanical and physical properties [54]. HS, as can be observed in Figure 15c,d, had a distribution of cells or micropores with dimensions of 20–25 μm wide by 40–50 μm long in a tubular shape. Figure 15c shows a separation between the large parenchyma cells seen in the upper section and the lower part with smaller longitudinal parenchyma cells. In Figure 15d, the micropore walls that are not pitted can be seen. In other studies, it has been observed that some walls show cell-to-cell pitting [55]. HS is a highly porous material, as opposed to WS, as can be seen in the micrographs. HS has an approximate porosity 25 times larger than WS, due to the species type and stem size. The higher porosity of HS observed in its microstructure explains the lower compressive strength and bulk density, along with a better performance in thermal conductivity, moisture regulation in absorption and desorption cycles (MBV), and water vapor permeability, as shown in the experimental results of HSC.

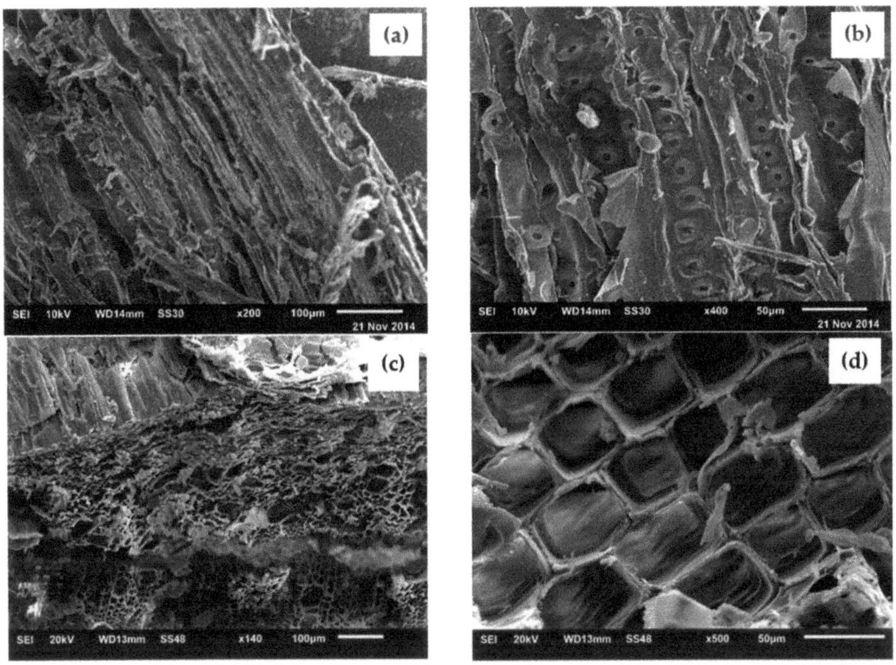

Figure 15. SEM observations of WS at (**a**) ×200 and (**b**) ×400, and HS at (**c**) ×140 and (**d**) ×500.

5. Conclusions

The present study aimed to determine the physical, mechanical, microstructural, and durability properties of sustainable plant-based concretes in order to reduce the environmental impact of conventional concretes, and the conclusions for each of the proposals are presented below:

(1) LFC and FFC with wax treatment on the fibers had an adverse effect on flexural strength; on the other hand, untreated fibers and those with accelerated deterioration showed better mechanical behavior, since they had an increase in flexural strength with respect to the control of 38% and 19%, respectively, for a Vf = 0.4%. Flexural and compressive strength decreased with increasing Vf. Although the fibers may not

improve the flexural strength still they contribute to reducing the brittleness of the material by allowing performance to be maintained after the first cracks appear.

(2) Porosity increases while density decreases when there is the presence of fibers in the matrix. The FFC presented higher porosity than LFC, which could be due to the higher porosity of the FF; in addition, the agglomeration shown in the micrographs is evidence that the FF does not have a uniform distribution within the matrix, and this could cause larger voids in the matrix, which influenced the low flexural and compressive strength shown by the FFC.

(3) Micrographs of LFC and FFC with accelerated deterioration showed no signs of fiber embrittlement in any case, even in the untreated fibers. The failure mode for the untreated LF was the rupture type and with treatment was the pull-out type, which explains the flexural and compressive behavior of the composites.

(4) The microstructure of WS and HS had a significant effect on all physical and mechanical properties of the composites. The high porosity influenced the experimental results obtained since it decreased the compressive strength and bulk density; however, for the thermal conductivity, hygroscopicity, and vapor resistance, it showed a behavior in most cases better than the controls that did not have bio-aggregates.

(5) The incorporation of WS and HS was important in improving the hygrothermal properties of the composites. On the other hand, WSC and HSC showed reductions in thermal conductivity values of 43% and 31%, respectively, with respect to the controls without bio-aggregates. These results are comparable with those established in the current regulations for low conductivity or insulating materials. The WSC and HSC are classified in the moderate category according to the MBV classification, and in comparison with values established for other materials with respect to the efficiency of moisture regulation, the HSC with the highest volume of bio-aggregate (10%); the MBV values are of the order of cellular concrete.

(6) Although the results demonstrate the positive effects of using vegetable raw materials as reinforcement or aggregate in cementitious matrices, future research should focus on the variability of the raw materials by incorporating local agricultural residues and on the long-term durability of the developed composites considering the incompatibility that may exist on the use of natural fibers in alkaline environments.

Author Contributions: Conceptualization, C.A.J.-A.; Formal analysis, C.M.; Investigation, C.A.J.-A.; Methodology, C.M. and G.E.; Validation, C.A.J.-A., C.M. and G.E.; Writing—original draft, C.A.J.-A., B.T.T.-T. and F.R.-D.; Writing—review and editing, C.M., G.E. and P.L.V.-T. All authors have read and agreed to the published version of the manuscript.

Funding: This research was funded by the Consejo Nacional de Ciencia y Tecnología (CONACYT), grant number 320051, supported within the framework of the Convocatoria de Ciencia Básica y/o Ciencia de Frontera. Modality: Paradigms and Controversies of Science 2022.

Institutional Review Board Statement: Not applicable.

Informed Consent Statement: Not applicable.

Data Availability Statement: Not applicable.

Acknowledgments: The present study could not have been possible without the support of the Institute of Civil Engineering of the Facultad de Ingeniería Civil at UANL, and the Laboratoire Matériaux et Durabilité des Constructions (LMDC) at Université de Toulouse, INSAT, UPS. In addition, we would like to thank the Consejo Nacional de Ciencia y Tecnología (CONACYT), the governing body of scientific research in Mexico, for the economic resources granted with project number 320051, supported within the framework of the Convocatoria de Ciencia Básica y/o Ciencia de Frontera—Modalidad: Paradigmas y Controversias de la Ciencia 2022.

Conflicts of Interest: The authors declare no conflict of interest.

References

1. United Nations Department of Economic and Social Affairs, Population Division. *World Population Prospects 2022: Summary of Results*; United Nations Publication: New York, NY, USA, 2022.
2. Silva, F.D.A.; Mobasher, B.; Filho, R.T. Cracking mechanisms in durable sisal fiber reinforced cement composites. *Cem. Concr. Compos.* **2009**, *31*, 721–730. [CrossRef]
3. García, G.A. Energía en Edificaciones. *Rev. Mex. Física* **2013**, *59*, 44–51.
4. Agudelo, H.A.; Hernández, A.V.; Cardona, D.A.R. Sostenibilidad: Actualidad y necesidad en el sector de la construcción en Colombia. *Gestión Y Ambiente* **2012**, *15*, 105–118.
5. Masson-Delmotte, V.; Zhai, P.; Pirani, A.; Connors, S.L.; Péan, C.; Berger, S.; Caud, N.; Chen, Y.; Goldfarb, L.; Gomis, M.I.; et al. IPCC: Summary for Policymakers. In *Climate Change 2021: The Physical Science Basis. Contribution of Working Group I to the Sixth Assessment Report of the Intergovernmental Panel on Climate Change*; Cambridge University Press: Cambridge, UK, 2021; Volume 32.
6. Becchio, C.; Corgnati, S.P.; Kindinis, A.; Pagliolico, S. Improving environmental sustainability of concrete products: Investigation on MWC thermal and mechanical properties. *Energy Build* **2009**, *41*, 1127–1134. [CrossRef]
7. Rosas-Díaz, F.; García-Hernández, D.G.; Mendoza-Rangel, J.M.; Terán-Torres, B.T.; Galindo-Rodríguez, S.A.; Juárez-Alvarado, C.A. Development of a Portland Cement-Based Material with *Agave salmiana* Leaves Bioaggregate. *Materials* **2022**, *15*, 6000. [CrossRef] [PubMed]
8. Mora, E.P. Life cycle, sustainability and the transcendent quality of building materials. *Build. Environ.* **2007**, *42*, 1329–1334. [CrossRef]
9. Mohanty, A.K.; Misra, M.; Drzal, L.T.; Selke, S.E.; Harte, B.R.; Hinrichsen, G. *Natural Fibers, Biopolymers, and Biocomposites*; Taylor & Francis Group: Boca Raton, FL, USA, 2016. [CrossRef]
10. Krishna, N.K.; Prasanth, M.; Gowtham, R.; Karthic, S.; Mini, K.M. Enhancement of properties of concrete using natural fibers. *Mater. Today Proc.* **2018**, *5*, 23816–23823. [CrossRef]
11. Amziane, S.; Sonebi, M. Overview on Biobased Building Material made with plant aggregate. *RILEM Tech. Lett.* **2016**, *1*, 31. [CrossRef]
12. Mohammadkazemi, F.; Doosthoseini, K.; Ganjian, E.; Azin, M. Manufacturing of bacterial nano-cellulose reinforced fiber-cement composites. *Constr. Build. Mater.* **2015**, *101*, 958–964. [CrossRef]
13. Mythili, R.; Venkatachalam, P.; Subramanian, P.; Uma, D. Characterization of bioresidues for biooil production through pyrolysis. *Bioresour. Technol.* **2013**, *138*, 71–78. [CrossRef]
14. Cury R, K.; Aguas M, Y.; Martinez M, A.; Olivero V, R.; Chams Ch, L. Residuos agroindustriales su impacto, manejo y aprovechamiento. *Rev. Colomb. De Cienc. Anim.—RECIA* **2017**, *9*, 122. [CrossRef]
15. Mohanty, A.K.; Vivekanandhan, S.; Pin, J.-M.; Misra, M. Composites from renewable and sustainable resources: Challenges and innovations. *Science* **2018**, *362*, 536–542. [CrossRef]
16. Restrepo, S.M.V.; Arroyave, G.J.P.; Vásquez, D.H.G. Uso de Fibras Vegetales en Materiales Compuestos de Matriz Polimérica: Una Revisión Con Miras a SU Aplicación en El Diseño de Nuevos Productos Use of Vegetable Fibers in Polymer Matrix Composites: A Review. *Informador Técnico* **2016**, *80*, 77–86. [CrossRef]
17. Chabannes, M.; Nozahic, V.; Amziane, S. Design and multi-physical properties of a new insulating concrete using sunflower stem aggregates and eco-friendly binders. *Mater. Struct. Constr.* **2015**, *48*, 1815–1829. [CrossRef]
18. Sood, M.; Dwivedi, G. Effect of fiber treatment on flexural properties of natural fiber reinforced composites: A review. *Egypt. J. Pet.* **2018**, *27*, 775–783. [CrossRef]
19. *ASTM C150/C150M-2021*; Standard Specification for Portland Cement. American Society for Testing and Materials International: West Conshohocken, PA, USA, 2021; Volume I, pp. 1–10.
20. *UNE-EN 197-1*; Cemento. Parte 1. Composición, Especificaciones y Criterios de Conformidad de los Cementos Comunes. Asociación Española de Normalización y Certificación: Madrid, Spain, 2011; pp. 1–38.
21. *ASTM C-618-19*; Standard Specification for Coal Fly Ash and Raw or Calcined Natural Pozzolan for Use in Concrete. American Society for Testing and Materials International: West Conshohocken, PA, USA, 2019; pp. 1–4.
22. *ASTM C-33-18*; Standard Specification for Concrete Aggregates. American Society for Testing and Materials International: West Conshohocken, PA, USA, 2018; pp. 1–8.
23. *NF P18-545*; Granulats—Elément de Définition, Conformité et Codification, Normes Nationales et Documents Normatifs Nationaux. Association Française de Normalisation Editions: Paris, France, 2011; pp. 1–11.
24. *ACI Committe 544.1R-96*; State-of-the-Art Report on Fiber Reinforced Concrete. ACI Manual of Concrete Practice. American Concrete Institute: Farmington Hills, MI, USA, 1996; pp. 544.1R-1–544.1R-66.
25. Bederina, M.; Marmoret, L.; Mezreb, K.; Khenfer, M.; Bali, A.; Quéneudec, M. Effect of the addition of wood shavings on thermal conductivity of sand concretes: Experimental study and modelling. *Constr. Build. Mater.* **2007**, *21*, 662–668. [CrossRef]
26. Jiang, Y.; Ansell, M.P.; Jia, X.; Hussain, A.; Lawrence, M. Physical characterisation of hemp shiv: Cell wall structure and porosity. *Acad. J. Civ. Eng.* **2017**, *35*, 22–28.
27. *ASTM C143/C143M-20*; Standard Test Method for Slump of Hydraulic-Cement Concrete. American Society for Testing and Materials International: West Conshohocken, PA, USA, 2020; pp. 1–4.
28. *NF-EN 196-1*; Méthodes D'essais des Ciments—Partie 1: Détermination des Résistances Mécaniques. Normes Nationales et Documents Normatifs Nationaux. Association Française de Normalisation Editions: Paris, France, 2011; pp. 1–12.

29. *NF P 18-459*; Béton—Essai Pour Béton Durci—Essai de Porosité et de Masse Volumique. Normes Nationales et Documents Normatifs Nationaux. Association Française de Normalisation Editions: Paris, France, 2011; pp. 1–8.
30. *ASTM C192/C192M-19*; Standard Practice for Making and Curing Concrete Test Specimens in the Laboratory. American Society for Testing and Materials International: West Conshohocken, PA, USA, 2019; pp. 1–8.
31. *ASTM C511-21*; Standard Specification for Mixing Rooms, Moist Cabinets, Moist Rooms, and Water Storage Tanks Used in the Testing of Hydraulic Cements and Concretes. American Society for Testing and Materials International: West Conshohocken, PA, USA, 2021; pp. 1–3.
32. *ASTM C39/C39M-18*; Standard Test Method for Compressive Strength of Cylindrical Concrete Specimens. American Society for Testing and Materials International: West Conshohocken, PA, USA, 2018; pp. 1–7.
33. *RILEM-13*; Round Robin Test for Hemp Shiv Characterisation: Committee Report of tc—236 Bio-Based Building Materials. Springer: Berlin, Germany, 2013.
34. Dresden, L.-M.G. *Système de Mesure de Conductivité Thermique*; Université Paul Sabatier—Toulouse III: Toulouse, France, 2008.
35. Peuhkuri, R.H.; Mortensen, L.H.; Hansen, K.K.; Time, B.; Gustavsen, A.; Ojanen, T.; Ahonen, J.; Svennberg, K.; Arfvidsson, J.; Harderup, L.-E. *Moisture Buffering of Building Materials*; Rode, C., Ed.; BYG Report No. R-127; Technical University of Denmark, Department of Civil Engineering: Lyngby, Denmark, 2005; pp. 1–79.
36. *NF EN ISO 12572:2001-10*; Détermination des Propriétés de Transmission de la Vapeur D'eau. Norme Fracaise—European Norme: Association Française de Normalisation Editions: Paris, France, 2010; pp. 1–33.
37. Juarez, C.; Duran, A.; Valdez, P.; Fajardo, G. Performance of "*Agave lecheguilla*" natural fiber in portland cement composites exposed to severe environment conditions. *Build. Environ.* **2007**, *42*, 1151–1157. [CrossRef]
38. Kesikidou, F.; Stefanidou, M. Natural fiber-reinforced mortars. *J. Build. Eng.* **2019**, *25*, 100786. [CrossRef]
39. Lertwattanaruk, P.; Suntijitto, A. Properties of natural fiber cement materials containing coconut coir and oil palm fibers for residential building applications. *Constr. Build. Mater.* **2015**, *94*, 664–669. [CrossRef]
40. Siddique, R.; Klaus, J. Influence of metakaolín on the properties of mortar and concrete: A review. *Appl. Clay Sci.* **2009**, *43*, 342–400. [CrossRef]
41. Sabathier, V.; Magniont, C.; Escadeillas, G.; Juarez, C.A. Flax and hemp fibre reinforced pozzolanic matrix: Evaluation of impact of time and natural weathering. *Eur. J. Environ. Civ. Eng.* **2016**, *21*, 1403–1417. [CrossRef]
42. Melo Filho, J.D.A.; Silva, F.D.A.; Filho, R.T. Degradation kinetics and aging mechanisms on sisal fiber cement composite systems. *Cem. Concr. Compos.* **2013**, *40*, 30–39. [CrossRef]
43. Arnaud, L.; Gourlay, E. Experimental study of parameters influencing mechanical properties of hemp concretes. *Constr. Build. Mater.* **2012**, *28*, 50–56. [CrossRef]
44. Mardani-Aghabaglou, A.; Sezer, G.İ.; Ramyar, K. Comparison of fly ash, silica fume and metakaolin from mechanical properties and durability performance of mortar mixtures view point. *Constr. Build. Mater.* **2014**, *70*, 17–25. [CrossRef]
45. Li, Z.; Wang, X.; Wang, L. Properties of hemp fibre reinforced concrete composites. *Compos. Part A Appl. Sci. Manuf.* **2006**, *37*, 497–505. [CrossRef]
46. Mohammed, B.S.; Abdullahi, M.; Hoong, C.K. Statistical models for concrete containing wood chipping as partial replacement to fine aggregate. *Constr. Build. Mater.* **2014**, *55*, 13–19. [CrossRef]
47. Coatanlem, P.; Jauberthie, R.; Rendell, F. Lightweight wood chipping concrete durability. *Constr. Build. Mater.* **2006**, *20*, 776–781. [CrossRef]
48. *NMX-AA-164-SCFI-2013*; Edificación Sustentable: Criterios Y Requerimientos Ambientales Mínimos. Normas Mexicanas, Secretaría de Economía: Cd. de México, Mexico, 2013; pp. 1–158.
49. Uysal, H.; Demirbog, R. The effects of different cement dosages, slumps, and pumice aggregate ratios on the thermal conductivity and density of concrete. *Cem. Concr. Res.* **2004**, *34*, 845–848. [CrossRef]
50. Lagouin, M.; Magniont, C.; Sénéchal, P.; Moonen, P.; Aubert, J.-E.; Laborel-préneron, A. Influence of types of binder and plant aggregates on hygrothermal and mechanical properties of vegetal concretes. *Constr. Build. Mater.* **2019**, *222*, 852–871. [CrossRef]
51. Haba, B.; Agoudjil, B.; Boudenne, A.; Benzarti, K. Hygric properties and thermal conductivity of a new insulation material for building based on date palm concrete. *Constr. Build. Mater.* **2017**, *154*, 963–971. [CrossRef]
52. Bennai, F.; Issaadi, N.; Abahri, K.; Belarbi, R.; Tahakourt, A. Experimental characterization of thermal and hygric properties of hemp concrete with consideration of the material age evolution. *Heat Mass Transf.* **2017**, *54*, 1189–1197. [CrossRef]
53. Thoemen, H.; Irle, M.; Sernek, M. *Wood-Based Panels an Introduction for Specialists*; Brunel University Press: London, UK, 2010; ISBN 978-1-902316-82-6.
54. Ansell, M.P. (Ed.) 1—Wood microstructure—A cellular composite. In *Wood Composites*; Woodhead Publishing Ltd.: Cambridge, UK, 2015; pp. 3–26. [CrossRef]
55. Jiang, Y.; Lawrence, M.; Ansell, M.P.; Hussain, A. Cell wall microstructure, pore size distribution and absolute density of hemp shiv. *R. Soc. Open Sci.* **2018**, *5*, 171945. [CrossRef] [PubMed]

sustainability

MDPI

Article

Potential Use of Water Treatment Sludge as Partial Replacement for Clay in Eco-Friendly Fired Clay Bricks

Masoud Ahmadi [1,*], Babak Hakimi [2], Ahmadreza Mazaheri [1] and Mahdi Kioumarsi [3]

1 Department of Civil Engineering, Ayatollah Boroujerdi University, Boroujerd 6919969737, Iran; a.mazaheri@abru.ac.ir
2 Department of Civil and Environmental Engineering, Tarbiat Modares University, Tehran 14115397, Iran; babak_8090@yahoo.com
3 Department of Built Environment, OsloMet-Oslo Metropolitan University, Pilestredet 35, 0166 Oslo, Norway; mahdi.kioumarsi@oslomet.no
* Correspondence: masoud.ahmadi@abru.ac.ir

Abstract: The traditional production process of clay bricks involves the extraction of significant amounts of raw materials and consumes considerable energy, leading to anthropogenic greenhouse gas emissions and environmental degradation. Using environmentally friendly materials in the construction industry has become an attractive alternative for mitigating sustainability issues. One such alternative is incorporating waste materials, such as water treatment sludge (WTS), into clay brick production. This research aims to assess the viability of using WTS as a replacement for conventional clay in fired clay brick production, thereby mitigating environmental pollution. Five distinct mixtures were created, with WTS replacing clay at 0, 20, 40, 60, and 80% ratios. The mechanical properties and durability of the produced bricks were analyzed through various tests, such as Atterberg limits, optimum water content, unconfined compression, apparent porosity, compressive strength, flexural strength, density, water absorption, and efflorescence. The results demonstrated that as WTS content increased, Atterberg limits and apparent porosity increased. The bulk density, compressive strength, and bending capacity of the specimens were found to decrease as the WTS replacement ratio increased. Additionally, moderate efflorescence was observed in samples with higher sludge ratios.

Keywords: eco-friendly; brick; water treatment sludge; environmental pollution

Citation: Ahmadi, M.; Hakimi, B.; Mazaheri, A.; Kioumarsi, M. Potential Use of Water Treatment Sludge as Partial Replacement for Clay in Eco-Friendly Fired Clay Bricks. *Sustainability* **2023**, *15*, 9389. https://doi.org/10.3390/su15129389

Academic Editor: Agostina Chiavola

Received: 28 April 2023
Revised: 7 June 2023
Accepted: 9 June 2023
Published: 11 June 2023

1. Introduction

Due to their cost effectiveness, accessibility, and versatility, clay bricks have been a prevalent construction material for centuries [1]. These products possess high density, compressive strength, freeze–thaw cycle resistance, and low water absorption values. Clay bricks are environmentally friendly because of their durability and long lifespan, allowing them to be recycled and resold. Bricks from demolition can be cleaned, reused, or repurposed, making them more sustainable than using new ones. However, traditional clay brick production involves extracting significant amounts of raw materials and consuming large amounts of energy, resulting in human-induced greenhouse gas emissions and environmental degradation [2,3]. There has been growing interest in discovering sustainable and eco-friendly alternatives to conventional clay brick production to mitigate greenhouse gas emissions and reduce the carbon footprint of the construction industry [4,5]. Employing recycled materials in brick manufacturing can reduce landfill waste and lower the construction industry's carbon footprint. Furthermore, producing bricks from recycled materials can help preserve natural habitats and minimize the environmental impact of resource extraction [6].

Numerous investigations have explored the use of waste materials in producing clay bricks. For instance, Monteiro and Vieira's study [7] provides a crucial update on the manufacture of fired clay bricks using waste materials. The research discusses the benefits and limitations of employing various waste types, including fly ash, sludge, glass, rubber, plastics, sawdust, and more, as raw materials for brickmaking. It also compares the properties of bricks made from waste materials to those of traditional clay bricks and other alternative bricks. The study concludes that incorporating waste materials into brickmaking can enhance the quality, durability, and sustainability of bricks while conserving energy and reducing emissions. Similarly, Velasco et al.'s investigation [8] reviews research on fired clay bricks produced by adding waste materials as a sustainable construction material. The paper summarizes the effects of different wastes, such as marble powder, rice husk ash, textile sludge, ceramic waste, paper sludge, and others, on the physical, mechanical, and thermal properties of bricks. The research also emphasizes the environmental advantages of using waste materials in brickmaking, including reduced consumption of natural resources, decreased energy demand, and minimized waste disposal challenges.

Globally, sewage sludge production is estimated to be approximately 75–100 million tons each year. By 2030, this figure is expected to increase to 127.5 million tons [9]. The annual sewage sludge production in regions such as the European Union, the United States, China, and Iran fluctuates between 18 and 50 million tons when measured in dry weight [10,11]. Incorporating waste materials such as water treatment sludge (WTS) into clay brick production offers an alternative. WTS, a water treatment process by-product, is typically disposed of in landfills, contributing to environmental pollution. Using WTS in clay brick production reduces raw material consumption and alleviates pressure on landfill sites.

Several studies have investigated the use of WTS in fired or sintered clay brick production, revealing key findings. The utilization of WTS in brick production has proven to be an effective way to manage waste, reducing the volume of sludge that would otherwise be disposed of in landfills or incinerated, thereby mitigating environmental pollution [12,13]. The incorporation of WTS in clay bricks yields satisfactory physical and mechanical properties, including compressive strength, water absorption, and density [14–18]. Most of the time, the bricks met or surpassed relevant building standard requirements. However, the ideal sludge content in the brick mixture varies based on sludge and clay properties, with typical ranges between 5% and 30%. Bricks made with WTS have been observed to have different colors compared to conventional bricks, often with a darker hue. The color change is primarily due to the presence of iron and other transition metals in the sludge, which can influence the brick's final color after firing [19]. Studies have indicated that the leaching of heavy metals and other potentially toxic elements from bricks containing WTS is generally within acceptable limits [20,21]. This suggests that using sludge in bricks does not pose significant environmental risks. However, it is crucial to perform leaching tests on a case-by-case basis, as the specific compositions of sludge and clay can affect the leaching behavior. The long-term durability of bricks containing WTS is an area that requires further research [14,17,22]. However, preliminary studies have shown promising results, with some bricks demonstrating resistance to freezing–thawing cycles and other forms of physical and chemical weathering.

Despite these encouraging outcomes, we still need to examine in greater detail the optimal rates of WTS replacement and how it affects the various mechanical properties and durability of the bricks produced. It is equally vital to supervise the firing conditions meticulously to yield the right kind of ash for the specific application. Our study aims to investigate further the best WTS substitution levels and understand their impact on the mechanical strength and durability of the bricks created. As a result, this method reduces the need for clay extraction, thereby conserving natural resources.

2. Research Significance

The utilization of water treatment sludge in the industry of construction products is promising and economically reasonable, and the products produced are not contaminated with hazardous impurities. By analyzing the effects of different WTS replacement ratios on the bricks' mechanical properties and durability, this research offers valuable insights into the optimal WTS incorporation levels in clay brick production. Additionally, this research has practical implications for the brick manufacturing and waste management industries. The study's findings can guide brick manufacturers in adopting more sustainable production practices by incorporating WTS into their processes. Moreover, using WTS in brick production presents a viable disposal solution, reducing the environmental impact of landfill WTS disposal.

3. Materials and Methods

The subject of this section pertains to the materials used to make eco-friendly bricks and the testing methodologies that have been conducted on them.

3.1. Materials

Sludge for this study was provided by two wastewater treatment plants (WTPs) in Tehran, Iran, supplied by the Jajrood River. The WTPs use physical, chemical, and biological treatments to reduce raw water's turbidity and microbial load, along with the modification and reduction of certain chemical compounds and total dissolved solids. The sludge produced by both plants, undergoing the same coagulation and flocculation processes, exhibits similar characteristics. The sludge density, determined using a pycnometer, was found to be 17.1 kN/m^3. Laboratory analysis, including influent water quality data (biochemical oxygen demand (BOD), chemical oxygen demand (COD), and total suspended solids (TSS)) along with metal content, is presented in Table 1. The results show that the percentage of heavy metals, including chromium (Cr), cobalt (Co), cadmium (Cd), copper (Cu), lead (Pb), nickel (Ni), zinc (Zn), and magnesium (Mg), is negligible and the amount of ammonia, nitrite, and nitrate is below 5 mg/L. The pH was specified using a sample-to-distilled-water ratio of 1:10. The sample was agitated for 30 min. After agitation, the pH was determined. The method used is based on the process recommended in [23]. In addition, similar trends have been used in other research [24,25].

Table 1. The characteristics and metal content of sludge.

Ingredient	Weight (mg/L)
BOD	8
COD	1470
TSS	5680
Fe	10.97
Mn	5.91
Cr	1.05
Cu	0.35
Zn	0.23
Cd	0.15

The local clay raw material was sourced from a reputed brick industry in Tehran, Iran. The chemical composition of the clay, with a significant quantity of silicon dioxide and aluminum oxide, is displayed in Table 2. Historically, clay with a silicon dioxide content of 40 to 60% and aluminum oxide content of 10 to 20% has been preferred for brickmaking [8,26]. In this investigation, these values are 54.12% and 14.73%, respectively.

Table 2. The chemical composition of clay.

Ingredient	Weight Percent
SiO_2	56.12
Al_2O_3	14.73
Fe_2O_3	7.35
CaO	5.21

Table 2. *Cont.*

Ingredient	Weight Percent
MgO	1.32
Na_2O	0.51
K_2O	1.35
TiO_2	0.47
P_2O_5	0.09

3.2. Specimen Preparation

Table 3 depicts the details of the mixture proportions used to investigate the influence of WTP sludge replacement on the behavior of clay–sludge bricks. The control mixture did not contain any sludge. To remove all moisture from the sludge, it was heated at 110 degrees Celsius in an oven for 24 h [27,28]. The WTS was crushed to a sufficient particle size while it cooled. Because the dried sludge contained coarse particles and plant leaves, the crushed sludge was passed through a No. 16 mesh (1.18 mm) to guarantee correct mixing with the clay. The brick preparation steps are shown in Figure 1.

Table 3. The mixture compositions of samples.

Specimen	Clay (%)	Water Treatment Sludge (%)
C100	100	0
C80S20	80	20
C60S40	60	40
C40S60	40	60
C20S80	20	80

Five sets of mixtures were prepared, designated as C100, C80S20, C60S40, C40S60, and C20S80. In the C80S20, C60S40, C40S60, and C20S80 samples, 20, 40, 60, and 80% clay, respectively, was used to partially replace the required sludge. The optimal water-to-binder ratio was determined through numerous trials with various amounts of mixing water. The mixing process was carried out as follows: Initially, clay and WTS were combined in a blender and dry mixed for 60 s. Half of the mixing water was then added, and blending of the mixture was continued for approximately 120 s. Upon the addition of the remaining mixing water, the mixture was blended for an additional 150 s. The fresh mixtures were then placed into molds sized 210 mm × 100 mm × 55 mm. To prevent deformation and cracking post baking, a hydraulic piston was utilized to compact the fresh mixture. These samples were kept at room temperature (20–25 °C) for a week to allow moisture to evaporate. Ultimately, they were fired in an electric furnace with temperatures ranging from 800 to 1000 °C at an average rate of 5 °C/min.

3.3. Testing Program

This study comprised a two-part experimental work. The first part focused on the impact of WTP sludge addition on the mixture's properties, including Atterberg limits, optimum water content, and unconfined compression tests, based on ASTM D4318 [29], ASTM D698 [30], and ASTM D2166 [31], respectively. The second part assessed the durability and mechanical properties of the eco-friendly bricks, involving tests for apparent porosity [32], compressive capacity [33], bending strength [33], density [32], water absorption [32], and efflorescence [33].

Figure 1. The brick preparation steps.

4. Results and Discussion

4.1. Characteristics of Sludge–Clay Mixtures

4.1.1. Atterberg Limits

Previous research [15,27] has emphasized the usefulness of Atterberg limits as a reliable indicator of soil plasticity characteristics. These limits determine the moisture content at which soil transitions from a liquid state to a plastic state, then to a semi-solid state, and finally to a solid state. Table 4 presents the results from examining the Atterberg limits for clay and WTS mixtures. The values shown are averages derived from at least four measurements. As the proportion of sludge increases, both the liquid limit (LL) and the plastic limit (PL) also increase. A PL value of 19 for pure clay indicates that it can be classified as having low plasticity. The results suggest that mixtures containing 20% and

40% sludge closely resemble the reference sample in terms of their Atterberg limits. Higher percentages of sludge in the mixtures result in a softer consistency and improved plasticity, enabling greater water absorption without losing the plastic behavior. The PL value of a mixture helps determine the amount of water that needs to be added, although in practice the actual amount of water used is typically lower than the calculated value.

Table 4. The Atterberg limits of clay–WTP sludge combinations.

Specimen		C100	C80S20	C60S40	C40S60	C20S80
LL	Average	41	43	46	48	49
	SD *	2.76	2.78	4.09	6.27	8.38
	CoV ** (%)	6.73	6.46	8.89	13.06	17.1
PL	Average	22	23	24	25	25
	SD *	2.05	1.74	2.33	3.91	3.51
	CoV ** (%)	9.32	7.57	9.71	15.64	14.04
PI	-	19	20	22	23	24

* SD = standard deviation, ** CoV = coefficient of variation.

4.1.2. Optimum Water Content

Figure 2 depicts the influence of water content, ranging from 14 to 21%, on the relative dry density. Each test was conducted three times, and the average data are reported. The results indicate that the optimal water content for the various mixtures ranged between 16.2 and 19.5%. The control sample reached a maximum dry density of 22.38 kN/m^3 at a moisture content of 16.2%, while a sample with 40% sludge had a maximum dry density of 21.7 kN/m^3 at a moisture content of 17.6%. These results suggest that the finer particle size of WTS compared to clay leads to a reduction in the maximum dry density and an increase in the optimal water content as sludge content increases. When moisture content surpasses the optimal level, water occupies the additional space, leading to reduced dry density.

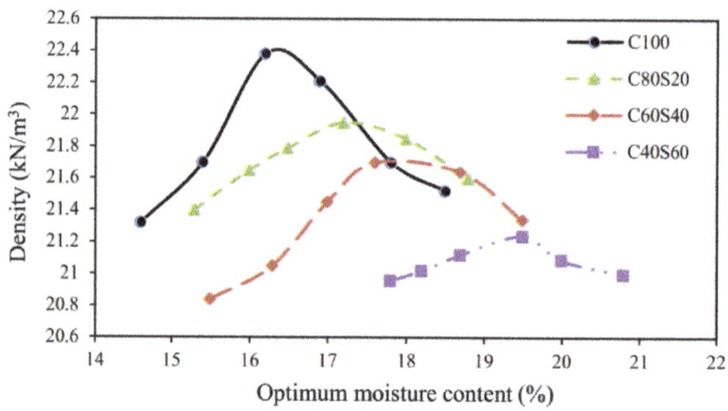

Figure 2. Moisture content vs. dry density results.

4.1.3. Unconfined Compression Strength

The outcomes of the unconfined compression test are displayed in Figure 3. Data were averaged from three experiments, and the coefficient of variation (CoV) was less than 10%. An observed trend was that an increase in WTS content led to a decrease in the unconfined compression strength of the mixtures. The control mixture exhibited an unconfined compression strength of 234 kPa, while mixtures with varying sludge content showed unconfined compression strengths between 184 and 223 kPa. If WTS was substituted into clay soil in the proportions of 20, 40, 60, and 80%, the unconfined compression values were found to be 223, 209, 196, and 184 kPa, respectively. It should be mentioned that adding up to 40% sludge has caused a decrease of about 11% in the unconfined compression strength. Furthermore, the strain corresponding to the unconfined compression strength of the mixes increased with higher sludge content compared to the control mixture. For instance, incorporation of 20, 40, 60, and 80% sludge resulted in an increase of 20%, 37%, 50%, and 71% in failure strain, respectively, when compared to the control mixture.

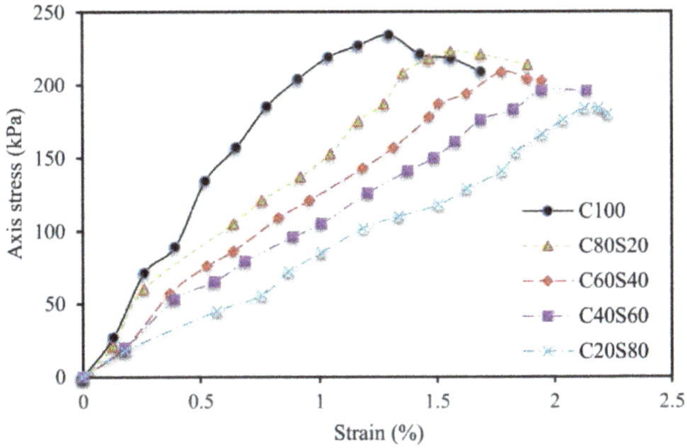

Figure 3. Unconfined compressive strength against axial strain.

4.2. Mechanical and Durability Properties of Bricks

4.2.1. Apparent Porosity

Figure 4 illustrates the apparent porosity test results for conventional clay bricks and eco-friendly bricks with various sludge levels. The results, representing the mean of five brick specimens (CoV below 7%), show an inverse relationship between apparent porosity and the quantity of waste treatment sludge (WTS) in the mixture. The lower porosity was observed for bricks fired at 1000 °C compared to those fired at 800 °C. Relative to the control sample fired at 1000 °C, the porosity of C80S20, C60S40, C40S60, and C20S80 was higher by 6.79%, 29.43%, 61.77%, and 70.64%, respectively. Similarly, the porosity of the samples compared to the control sample at 800 °C increased by 11.82%, 41.21%, 56.8%, and 82.36%, respectively. These results underscore the significant influence of firing temperature and WTS replacement ratio on pore formation.

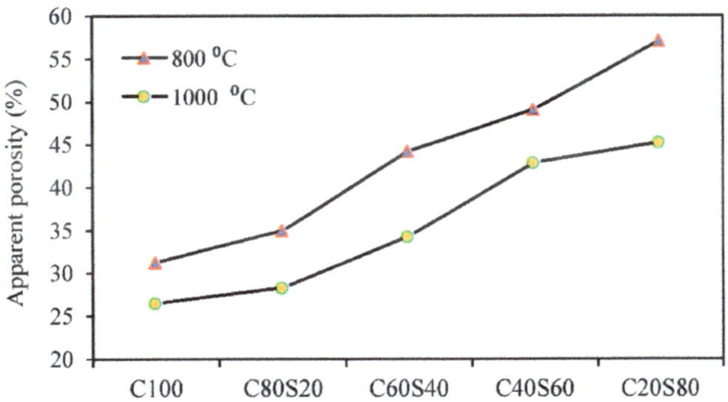

Figure 4. Apparent porosity test results.

The WTP sludge utilized in this study contained more calcium oxide than clay, and under high temperatures it led to the formation of new phases, such as calcium alumino-silicates and clay minerals. Thus, replacing clay with WTP sludge in eco-friendly samples resulted in increased porosity due to gas formation at high firing temperatures. These observations align with previous research findings [16,28].

4.2.2. Water Absorption

A brick's ability to absorb water is an important factor in its durability. It is believed that bricks that are less susceptible to water infiltration will be more durable and more resilient to aggressive environments [25,34]. Thus, the brick's interior structure should be designed to minimize water absorption. Due to the smaller size of the sludge particles compared to the clay particles, a greater specific surface area is created, which may explain why a greater amount of sludge replacement increases water absorption. Based on Figure 5, water absorptions for control samples fired at temperatures of 800 °C and 1000 °C were 16.38% and 14.51%, respectively. Each water absorption value is the average of values obtained on five similar brick specimens with a CoV of less than 6%. It also demonstrates that while WTS addition increased water absorption, raising the sintering temperature mitigated this effect. The value of water absorption is directly proportional to the quantity of sludge added. The previously reported plastic limit values have revealed that sludge addition augments the plasticity of the mixture and diminishes its bonding ability, increasing the pore size and water absorption. When the mixture contains a rather higher amount of sludge, the adhesiveness of the mixture decreases, but the internal pore size of the brick increases.

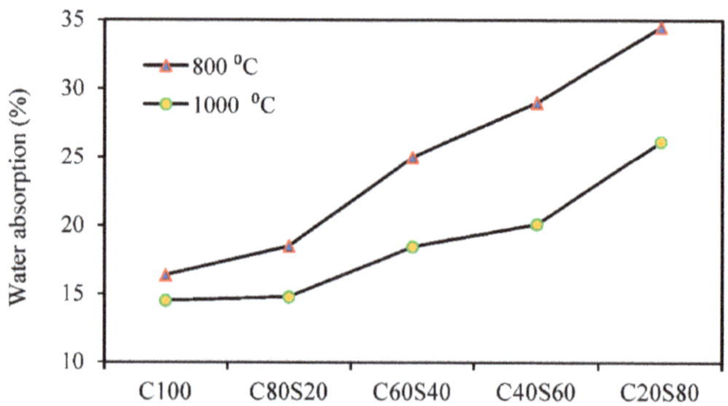

Figure 5. Water absorption test results.

By increasing the sintering temperature from 800 to 1000 °C, the water absorption fell from 25% to 18.46% in the case of 40% sludge replacement. Bricks that absorb more water have a larger pore size than bricks that do not absorb much water. In accordance with ASTM C62 [35], brick specimens with water absorption values of less than 17% and 22% are considered suitable for use in severe and moderate exposure conditions, respectively. Only samples sintered at 1000 °C with a 20% replacement ratio achieved absorption rates below 17%. Bricks with up to 20% WTP sludge had absorption rates below 22%, indicating suitability for moderate exposure.

4.2.3. Density

According to Figure 6, the WTS replacement ratio and fire temperature affect the bulk density of samples. For each combination, results are reported as the average of three brick specimens with CoV less than 9%. Clay brick density is influenced by the manufacturing process, sintering temperature, and specific gravity of raw materials [36]. Researchers found that burned clay bricks had a bulk density of approximately 17 to 21 kN/m^3 [15,37,38]. Based on the results from eco-friendly samples, bulk density was found to be inversely correlated with the amount of WTP sludge added to the mixture. As the sludge replacement ratio increased, the bulk density of eco-friendly bricks decreased. When the fire temperature was raised, the density of the samples also increased. For samples sintered at 1000 °C with increasing WTS replacement from 0% to 80%, bulk densities decreased from 20.4 to 12.8 kN/m^3. Similarly, by comparison to the control sample at 800 °C, the bulk density for C80S20, C60S40, C40S60, and C20S80 was reduced by 7.78%, 30.57%, 41.96%, and 49.22%, respectively. According to the findings of the study, bulk density is directly proportional to compressive strength. It should also be noted that when the sintering temperature was raised, the color of the sintered samples changed as well.

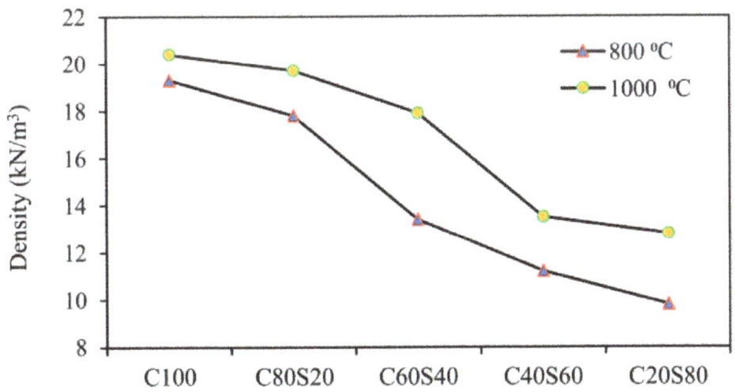

Figure 6. Density test results.

4.2.4. Efflorescence

Efflorescence results in the appearance of a thin, salty, white layer on the brick's surface [39]. A slight efflorescence is defined as a condition where less than 10% of the exposed brick surface is covered with this salt layer. The efflorescence is categorized as moderate if the coverage extends to 50% without any surface flaking. Severe efflorescence is designated as when significant salt deposits cover 50% or more of the exposed area [25]. Calcium oxide has been identified as the primary cause of efflorescence, although the presence of Fe_2O_3 in raw materials may sometimes influence it [40,41].

Figure 7 presents the results of the percentage of brick specimens affected by efflorescence, averaged over three replicates for each mixture. The area affected by efflorescence is only estimated. Samples sintered at 800 °C and 1000 °C, the C40S60 and C20S80 samples (containing 60% and 80% WTS, respectively), demonstrated moderate efflorescence. Compared to the control brick, which exhibited efflorescence on merely 4% of the surface, bricks with 20% WTS showed 7% efflorescence. It is inferred that the substitution of clay with WTS increased the CaO content, leading to an enlarged area of efflorescence. It is important to note that firing temperature did not significantly effect the efflorescence.

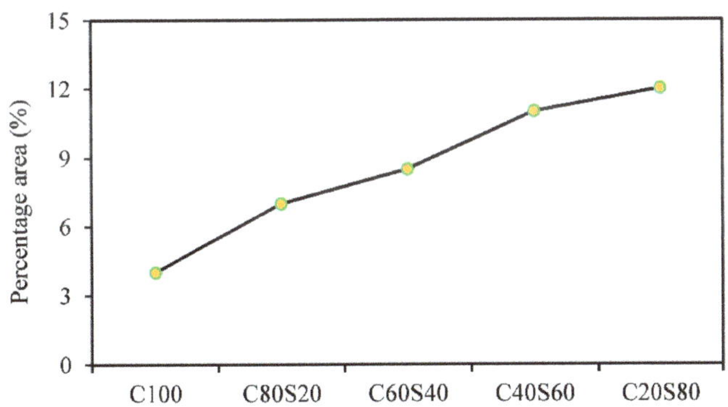

Figure 7. Efflorescence test results.

4.2.5. Compressive Strength

The compressive strength test is crucial for assessing the mechanical characteristics of eco-friendly bricks. Figure 8 displays the results of compressive strength tests conducted on bricks made from WTP sludge–clay. These results were calculated based on the examination of five brick samples per mixture. The CoV was less than 8% across all samples. The results demonstrated that the sludge content and firing temperature significantly affect brick strength. While an increase in the sludge content decreases the brick's compressive strength, higher firing temperatures increase it.

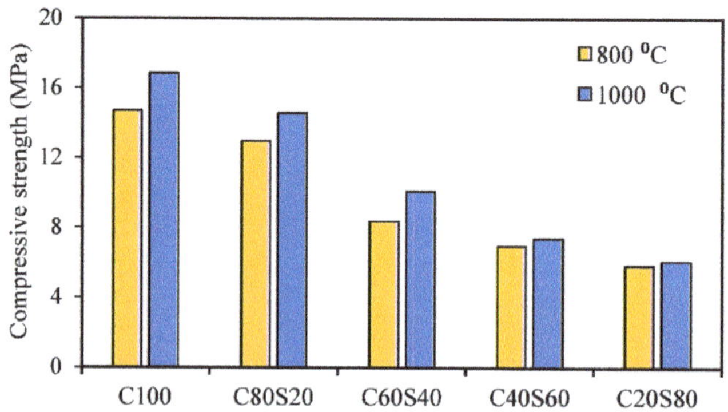

Figure 8. Compressive strength test results.

For bricks incorporating WTS, the compressive strength was found to be lower than that of the control samples. The principal factors governing the compressive strength of the brick samples were density, porosity, and pore size distributions. At a firing temperature of 1000 °C, the compressive strength of specimens C80S20, C60S40, C40S60, and C20S80 was reduced by 13.46%, 40.02%, 56.11%, and 63.86%, respectively, compared to the control. Furthermore, when the specimen was fired at 800 °C, the compressive strength was reduced by 11.84%, 42.95%, 52.55%, and 60.18% compared to the control specimen.

This trend aligns with previous studies [15,42]. Completing the crystallization process and closing open pores at high firing temperatures may explain this trend. The impact of the WTS ratio is likely due to the reduced silica content and increased number of open pores as the sludge ratio increases, leading to a decrease in compressive strength. According to the building codes of various countries, bricks should maintain a certain compressive strength: 8 MPa for Iran [43] and Pakistan [44] and 5 MPa for Kenya [45]. Hence, eco-friendly bricks with up to 40% clay substitution by WTS satisfied these requirements.

4.2.6. Flexural Strength

Figure 9 displays the modulus of rupture of eco-friendly clay bricks containing WTP sludge. These data were averaged from five bricks with a less than 10% CoV. Control specimens displayed average flexural strengths of 1.72 MPa and 1.95 MPa when fired at 800 °C and 1000 °C, respectively. The specimens with WTS addition had flexural strengths between 0.54 and 1.46 MPa and 0.56 and 1.66 MPa, respectively, for firing temperatures of 800 °C and 1000 °C.

Firing at 1000 °C resulted in a decrease in flexural strength for bricks containing 20%, 40%, 60%, and 80% WTS replacement ratios of 12.72%, 29.24%, 57.47%, and 65.86%, respectively, compared to WTS-free bricks. It is notable that the rate of decrease in flexural strength accelerated with higher sludge replacement percentages. The number of voids in burnt clay bricks plays a vital role in flexural strength; fewer voids typically correlate with higher flexural strength. However, adding sludge increased the clay bricks' porosity, thereby reducing their flexural strength. The lower concentration of SiO2 in sludge-containing samples might contribute to this decrease in flexural strength. Prior studies also support the observation that brick specimens' flexural strength is typically inversely proportional to their porosity level [46,47].

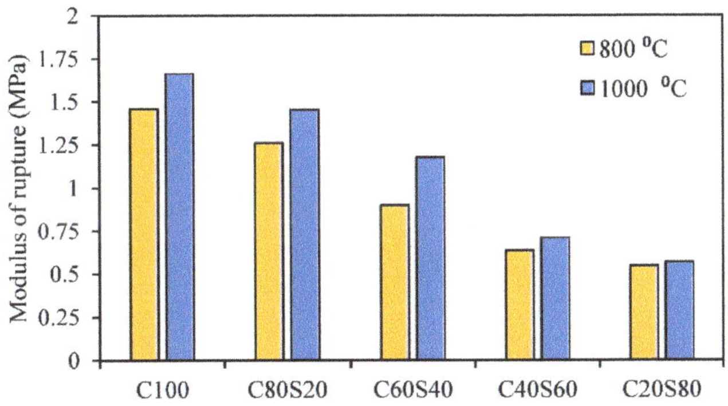

Figure 9. Modulus of rupture test results.

The ASTM standard [33] stipulates a minimum flexural strength of 0.65 MPa. Therefore, it can be concluded that bricks containing up to 40% WTP sludge meet this requirement, providing a sustainable and environmentally friendly solution for masonry constructions.

5. Conclusions

In this investigation into the impact of WTS on the properties of environmentally friendly clay bricks, several critical observations have been made. An increase in WTS content has been shown to enhance the plasticity and water absorption of the mixtures, but has concurrently compromised their dry density and strength. In particular, the unconfined compression strength decreased with an increase in WTS content, and the strain associated with this strength escalated with higher sludge inclusion. Simultaneously, we observed an increase in porosity and a decrease in bulk density as the WTS content rose. Despite these challenges, bricks containing up to 20% WTS and fired at 1000 °C have demonstrated improved properties, making them suitable for both severe and moderate exposure conditions. This has underscored the potential benefits of sludge incorporation. Interestingly, efflorescence has been found to be predominantly influenced by the WTS content, with bricks containing higher sludge ratios exhibiting moderate efflorescence. Contrary to expectations, the firing temperature did not significantly affect this property. Despite the observed reduction in compressive and bending strength with increasing WTS content, it is noteworthy that bricks with up to 40% WTS adhered to standard requirements when subjected to higher firing temperatures. The evidence collected in this study has suggested that achieving an optimal balance between WTS content and firing temperature could be instrumental in the production of eco-friendly bricks.

Author Contributions: Conceptualization, M.A. and B.H.; Methodology, M.A., B.H. and A.M.; Validation, A.M. and M.K.; Formal analysis, M.A., B.H. and A.M.; Investigation, B.H.; Resources, B.H. and M.K.; Data curation, B.H. and A.M.; Writing—original draft, A.M.; Writing—review & editing, M.A. and M.K.; Visualization, B.H. and A.M.; Supervision, M.A. and A.M.; Project administration, M.K. All authors have read and agreed to the published version of the manuscript.

Funding: This research received no external funding.

Institutional Review Board Statement: Not applicable.

Informed Consent Statement: Not applicable.

Data Availability Statement: The data are available from the corresponding author upon reasonable request.

Conflicts of Interest: The authors declare no conflict of interest.

References

1. Ebadi Jamkhaneh, M.; Ahmadi, M.; Shokri Amiri, M. Sustainable Reuse of Inorganic Materials in Eco-Friendly Clay Bricks: Special Focus on Mechanical and Durability Assessment. *J. Mater. Civ. Eng.* **2021**, *33*, 04021111. [CrossRef]
2. Raut, S.P.; Ralegaonkar, R.V.; Mandavgane, S.A. Development of Sustainable Construction Material Using Industrial and Agricultural Solid Waste: A Review of Waste-Create Bricks. *Constr. Build. Mater.* **2011**, *25*, 4037–4042. [CrossRef]
3. Azarhomayun, F.; Haji, M.; Shekarchi, M.; Kioumarsi, M. Investigating the Effectiveness of the Stable Measurement Tests of Self-Compacting Concrete. *Constr. Build. Mater.* **2023**, *383*, 131262. [CrossRef]
4. De Silva, G.S.; Hansamali, E. Eco-Friendly Fired Clay Bricks Incorporated with Porcelain Ceramic Sludge. *Constr. Build. Mater.* **2019**, *228*, 116754. [CrossRef]
5. Farahzadi, L.; Kioumarsi, M. Application of Machine Learning Initiatives and Intelligent Perspectives for CO2 Emissions Reduction in Construction. *J. Clean. Prod.* **2022**, *384*, 135504. [CrossRef]
6. Ebadi-Jamkhaneh, M.; Ahmadi, M.; Kontoni, D.P.N. Experimental Study of the Mechanical Properties of Burnt Clay Bricks Incorporated with Plastic and Steel Waste Materials. In Proceedings of the IOP Conference Series: Earth and Environmental Science, Athens, Greece, 23–24 October 2021; IOP Publishing: Bristol, UK, 2021; Volume 899, p. 012042.
7. Monteiro, S.N.; Vieira, C.M.F. On the Production of Fired Clay Bricks from Waste Materials: A Critical Update. *Constr. Build. Mater.* **2014**, *68*, 599–610. [CrossRef]
8. Velasco, P.M.; Ortíz, M.P.M.; Giró, M.A.M.; Velasco, L.M. Fired Clay Bricks Manufactured by Adding Wastes as Sustainable Construction Material–A Review. *Constr. Build. Mater.* **2014**, *63*, 97–107. [CrossRef]
9. AL-Huqail, A.A.; Kumar, P.; Abou Fayssal, S.; Adelodun, B.; Širić, I.; Goala, M.; Choi, K.S.; Taher, M.A.; El-Kholy, A.S.; Eid, E.M. Sustainable Use of Sewage Sludge for Marigold (*Tagetes Erecta* L.) Cultivation: Experimental and Predictive Modeling Studies on Heavy Metal Accumulation. *Horticulturae* **2023**, *9*, 447. [CrossRef]
10. Akbarzadeh, A.; Valipour, A.; Meshkati, S.M.H.; Hamnabard, N. Municipal Wastewater Treatment in Iran: Current Situation, Barriers and Future Policies. *J. Adv. Environ. Health Res.* **2023**, *11*, 60–71.
11. Ferrentino, R.; Langone, M.; Fiori, L.; Andreottola, G. Full-Scale Sewage Sludge Reduction Technologies: A Review with a Focus on Energy Consumption. *Water* **2023**, *15*, 615. [CrossRef]
12. Gencel, O.; Kazmi, S.M.S.; Munir, M.J.; Sutcu, M.; Erdogmus, E.; Yaras, A. Feasibility of Using Clay-Free Bricks Manufactured from Water Treatment Sludge, Glass, and Marble Wastes: An Exploratory Study. *Constr. Build. Mater.* **2021**, *298*, 123843. [CrossRef]
13. Harja, M.; Gencel, O.; Sarı, A.; Sutcu, M.; Erdogmus, E.; Hekimoğlu, G. Production and Characterization of Natural Clay-Free Green Building Brick Materials Using Water Treatment Sludge and Oak Wood Ash. *Arch. Civ. Mech. Eng.* **2022**, *22*, 79. [CrossRef]
14. Gencel, O.; Kizinievic, O.; Erdogmus, E.; Kizinievic, V.; Sutcu, M.; Muñoz, P. Manufacturing of Fired Bricks Derived from Wastes: Utilization of Water Treatment Sludge and Concrete Demolition Waste. *Arch. Civ. Mech. Eng.* **2022**, *22*, 78. [CrossRef]
15. Benlalla, A.; Elmoussaouiti, M.; Dahhou, M.; Assafi, M. Utilization of Water Treatment Plant Sludge in Structural Ceramics Bricks. *Appl. Clay Sci.* **2015**, *118*, 171–177. [CrossRef]
16. Sutcu, M.; Gencel, O.; Erdogmus, E.; Kizinievic, O.; Kizinievic, V.; Karimipour, A.; Velasco, P.M. Low Cost and Eco-Friendly Building Materials Derived from Wastes: Combined Effects of Bottom Ash and Water Treatment Sludge. *Constr. Build. Mater.* **2022**, *324*, 126669. [CrossRef]
17. Erdogmus, E.; Harja, M.; Gencel, O.; Sutcu, M.; Yaras, A. New Construction Materials Synthesized from Water Treatment Sludge and Fired Clay Brick Wastes. *J. Build. Eng.* **2021**, *42*, 102471. [CrossRef]
18. Amin, F.; Abbas, S.; Abbass, W.; Salmi, A.; Ahmed, A.; Saeed, D.; Sufian, M.; Sayed, M.M. Potential Use of Wastewater Treatment Plant Sludge in Fabrication of Burnt Clay Bricks. *Sustainability* **2022**, *14*, 6711. [CrossRef]
19. Ling, Y.P.; Tham, R.-H.; Lim, S.-M.; Fahim, M.; Ooi, C.-H.; Krishnan, P.; Matsumoto, A.; Yeoh, F.-Y. Evaluation and Reutilization of Water Sludge from Fresh Water Processing Plant as a Green Clay Substituent. *Appl. Clay Sci.* **2017**, *143*, 300–306. [CrossRef]
20. Hassan, K.M.; Fukushi, K.; Turikuzzaman, K.; Moniruzzaman, S.M. Effects of Using Arsenic–Iron Sludge Wastes in Brick Making. *Waste Manag.* **2014**, *34*, 1072–1078. [CrossRef]
21. Gomes, S.D.C.; Zhou, J.L.; Li, W.; Long, G. Progress in Manufacture and Properties of Construction Materials Incorporating Water Treatment Sludge: A Review. *Resour. Conserv. Recycl.* **2019**, *145*, 148–159. [CrossRef]
22. Tantawy, M.A.; Mohamed, R.S.A. Middle Eocene Clay from Goset Abu Khashier: Geological Assessment and Utilization with Drinking Water Treatment Sludge in Brick Manufacture. *Appl. Clay Sci.* **2017**, *138*, 114–124. [CrossRef]
23. Vice Presidency for Strategic Planning and Supervision. *Water Guidelines for Determination and Selection of Chemicals, Glasswares and Equipment Needed for a Wastewater Treatment Laboratory (Publication No. 285)*; Management and Planning Organization of Iran: Kerman, Iran, 2010.
24. Chiang, K.-Y.; Chou, P.-H.; Hua, C.-R.; Chien, K.-L.; Cheeseman, C. Lightweight Bricks Manufactured from Water Treatment Sludge and Rice Husks. *J. Hazard. Mater.* **2009**, *171*, 76–82. [CrossRef]
25. Shathika Sulthana Begum, B.; Gandhimathi, R.; Ramesh, S.T.; Nidheesh, P. V Utilization of Textile Effluent Wastewater Treatment Plant Sludge as Brick Material. *J. Mater. Cycles Waste Manag.* **2013**, *15*, 564–570. [CrossRef]
26. Kazmi, S.M.S.; Munir, M.J.; Wu, Y.-F.; Hanif, A.; Patnaikuni, I. Thermal Performance Evaluation of Eco-Friendly Bricks Incorporating Waste Glass Sludge. *J. Clean. Prod.* **2018**, *172*, 1867–1880. [CrossRef]
27. Wolff, E.; Schwabe, W.K.; Conceição, S.V. Utilization of Water Treatment Plant Sludge in Structural Ceramics. *J. Clean. Prod.* **2015**, *96*, 282–289. [CrossRef]

28. Heniegal, A.M.; Ramadan, M.A.; Naguib, A.; Agwa, I.S. Study on Properties of Clay Brick Incorporating Sludge of Water Treatment Plant and Agriculture Waste. *Case Stud. Constr. Mater.* **2020**, *13*, e00397. [CrossRef]
29. *ASTM D4318*; Standard Test Methods for Liquid Limit, Plastic Limit, and Plasticity Index of Soils. American Society for Testing and Materials: West Conshohocken, PA, USA, 2010.
30. *ASTM D698*; Standard Test Methods for Laboratory Compaction Characteristics of Soil Using Modified Effort (56,000 ft-lbf/ft^3 (2,700 kN-m/m^3)). American Society for Testing and Materials: West Conshohocken, PA, USA, 2009.
31. *ASTM D2166*; Standard Test Method for Unconfined Compressive Strength of Cohesive Soil. American Society for Testing and Materials: West Conshohocken, PA, USA, 2006.
32. *ASTM C20*; Standard Test Methods for Apparent Porosity, Water Absorption, Apparent Specific Gravity, and Bulk Density of Burned Refractory Brick and Shapes by Boiling Water. American Society for Testing and Materials: West Conshohocken, PA, USA, 2015.
33. *ASTM C67*; Standard Test Methods for Sampling and Testing Brick and Structural Clay Tile. American Society for Testing and Materials: West Conshohocken, PA, USA, 2007.
34. Hegazy, B.E.-D.E.; Fouad, H.A.; Hassanain, A.M. Reuse of Water Treatment Sludge and Silica Fume in Brick Manufacturing. *J. Am. Sci.* **2011**, *7*, 569–576.
35. *ASTM C62*; Standard Specification for Building Brick (Solid Masonry Units Made from Clay or Shale). American Society for Testing and Materials: West Conshohocken, PA, USA, 2017.
36. Karaman, S.; Gunal, H.; Ersahin, S. Quantitative Analysis of Pumice Effect on Some Physical and Mechanical Properties of Clay Bricks. *J. Appl. Sci.* **2008**, *8*, 1340–1345.
37. Phonphuak, N.; Kanyakam, S.; Chindaprasirt, P. Utilization of Waste Glass to Enhance Physical–Mechanical Properties of Fired Clay Brick. *J. Clean. Prod.* **2016**, *112*, 3057–3062. [CrossRef]
38. Abi, C.B.E. Effect of Borogypsum on Brick Properties. *Constr. Build. Mater.* **2014**, *59*, 195–203.
39. Kazmi, S.M.S.; Abbas, S.; Nehdi, M.L.; Saleem, M.A.; Munir, M.J. Feasibility of Using Waste Glass Sludge in Production of Ecofriendly Clay Bricks. *J. Mater. Civ. Eng.* **2017**, *29*, 4017056. [CrossRef]
40. Ngayakamo, B.H.; Bello, A.; Onwualu, A.P. Development of Eco-Friendly Fired Clay Bricks Incorporated with Granite and Eggshell Wastes. *Environ. Chall.* **2020**, *1*, 100006. [CrossRef]
41. Vračević, M.; Ranogajec, J.; Vučetić, S.; Netinger, I. Evaluation of Brick Resistance to Freeze/Thaw Cycles According to Indirect Procedures. *J. Croat. Assoc. Civ. Eng.* **2014**, *66*, 197–209.
42. Weng, C.-H.; Lin, D.-F.; Chiang, P.-C. Utilization of Sludge as Brick Materials. *Adv. Environ. Res.* **2003**, *7*, 679–685. [CrossRef]
43. Formulation of National Building Codes Department. *Iranian National Building Code (Part 8): Design and Construction of Masonry Buildings*; Ministry of Road and Urban Development: Tehran, Iran, 2014.
44. Building Code of Pakistan. *Seismic Hazard Evaluation Studies: National Engineering Services of Pakistan*; Ministry of Housing and Works, Government of Pakistan: Islamabad, Pakistan, 2007.
45. *Kenya Standard (DKS 2802-1)*; Specification for Masonry Units–Part 1: Clay Masonry Units. iTeh Standards: Etobicoke, ON, Canada, 2019.
46. Kazmi, S.M.S.; Abbas, S.; Saleem, M.A.; Munir, M.J.; Khitab, A. Manufacturing of Sustainable Clay Bricks: Utilization of Waste Sugarcane Bagasse and Rice Husk Ashes. *Constr. Build. Mater.* **2016**, *120*, 29–41. [CrossRef]
47. Bilgin, N.; Yeprem, H.A.; Arslan, S.; Bilgin, A.; Günay, E.; Marşoglu, M. Use of Waste Marble Powder in Brick Industry. *Constr. Build. Mater.* **2012**, *29*, 449–457. [CrossRef]

sustainability

Review

Advancements in Exploiting *Sporosarcina pasteurii* as Sustainable Construction Material: A Review

Shiva Khoshtinat

Department of Materials, Chemistry and Chemical Engineering "Giulio Natta", Politecnico di Milano, 20133 Milan, Italy; shiva.khoshtinat@polimi.it

Abstract: With the development of bioinspired green solutions for sustainable construction over the past two decades, bio-cementation, which exploits the naturally occurring phenomenon of calcium carbonate precipitation in different environments, has drawn a lot of attention in both building construction and soil stabilization. Various types of microorganisms, along with specific enzymes derived from these microorganisms, have been utilized to harness the benefits of bio-cementation. Different application methods for incorporating this mechanism into the production process of the construction material, as well as a variety of experimental techniques for characterizing the outcomes of bio-cementation, have been developed and tested. Despite the fact that the success of bio-cementation as a sustainable method for construction has been demonstrated in a significant body of scientific literature at the laboratory scale, the expansion of this strategy to construction sites and field application remains a pending subject. The issue may be attributed to two primary challenges. Firstly, the complexity of the bio-cementation phenomenon is influenced by a variety of factors. Secondly, the extensive body of scientific literature examines various types of microorganisms under different conditions, leading to a wide range of outcomes. Hence, this study aims to examine the recent advancements in utilizing the most commonly employed microorganism, *Sporosarcina pasteurii*, to emphasize the significance of influential factors identified in the literature, discuss the findings that have been brought to light, and outline future research directions toward scaling up the process.

Keywords: bio-cementation; sustainable construction materials; building construction; soil stabilization; *Sporosarcina pasteurii*

check for updates

Citation: Khoshtinat, S. Advancements in Exploiting *Sporosarcina pasteurii* as Sustainable Construction Material: A Review. *Sustainability* **2023**, *15*, 13869. https://doi.org/10.3390/su151813869

Academic Editors: Mahdi Kioumarsi and Vagelis Plevris

Received: 14 August 2023
Revised: 29 August 2023
Accepted: 12 September 2023
Published: 18 September 2023

1. Introduction

Calcium carbonate ($CaCO_3$) precipitation is a naturally occurring phenomenon in different environments, including marine water, freshwater, and soils. Microbially induced calcium carbonate precipitation Induced Calcium Carbonate Precipitation (MICP), which exploits the microbial metabolic processes for bio-cementation to enhance the durability of construction materials, has drawn attention not only in soil stabilization [1–11] and building construction [12–22] but also in wind-induced desertification [23–26], stone artwork conservation [27], and even subsurface-related applications [28,29]. Microorganisms engaged in the nitrogen cycle, sulphate-reducing bacteria, and photosynthetic microorganisms have all been reported to induce calcium carbonate [30,31]. Although the bio-cementation process produces ammonia gas that is undesirable [32], in comparison to the traditional methods, which make use of Portland cement, it offers a variety of benefits, including the following:

- Bio-cementation can be produced at room temperature; thus, the amount of energy needed for its manufacturing is significantly less than that required for conventional cement, resulting in a 43–95% decrease in embodied energy [33];
- The carbon footprint from the bio-cementation process is about 18–49.6% lower than that from traditional cement [33];

- Due to the relatively low viscosity of the cementation solution and bacterial suspension, the bacterium can flow like water during the bio-cementation process and move through the pores of the concrete [34];
- Bacterial sizes are less than 10 μm, which is considerably smaller than the sizes of cement (<40 μm); therefore, the pore openings can be as small as 6 mm [35];
- The cost-effectiveness of the MICP treatment compared to conventional treatments demonstrated a lifecycle cost reduction of about 98% [36,37].

1.1. Bio-Cementing Agents

Over the last two decades, researchers have investigated various techniques to exploit bio-cementation. These strategies are founded on three primary approaches, as follows:

(i) Using allochthonous or autochthonous alive cells to exploit their metabolism [7,11,12,21,38,39];
(ii) The cell-free approach, which uses bacterial fraction components in the absence of viable cells [40];
(iii) Isolating specific enzymes from microorganisms or plants for Enzymatically Induced Calcium Carbonate Precipitation (EICP) [7,9,41–46].

A wide range of bacteria [38], including *Sporosarcina pasteurii* [2,46–55], *Bacillus Subtilis* [14,16–18,40,56–59], *Bacillus Cereus* [20,60,61], and *Bacillus Megaterium* [23,24,62,63], as well as enzymes isolated from plants, such as Jack bean [42,45,64] and soybean [36,41,44], have been investigated in the literature. Navigating through the literature, however, it can be observed that *Sporosarcina pasteurii* (*S. pasteurii*) is the most widely employed bacteria, as it demonstrates a higher rate and quality of calcium carbonate precipitation due to its species characteristics [65,66]. Parallel comparison investigations comparing the results of native bacteria present in soil with *S. pasteurii* cementation proved that this bacterium is highly efficient for bio-cementation and outperforms indigenous bacteria [47,67,68].

1.2. Application Methods of Bio-Cementation

Various application strategies, including mixing, injection, spraying, immersion [20,69–71], and even 3D printing [72–74], have been explored in the literature for introducing the bio-cementing agent into cementitious materials for MICP or EICP. However, the majority of the literature primarily uses three methods: injection or grouting, surface percolation or spraying, and mixing [11].

Injection is the most popular method to introduce bio-cementing agents into the system, in particular for soil stabilization applications [11]. Using a peristaltic pump, bacterial suspension and cementation solution are injected into the sand in either a vertical [10,75] or horizontal [76] trajectory. The advantage of this approach is that testing conditions, such as injection flow, pressure, and hydraulic gradient, may be easily modified for optimal treatment [10,77]. The main disadvantages of this method are unequal bacterial dispersion and inhomogeneity of $CaCO_3$ concentration, which result in non-uniform treatment along the injection path. This problem derives from the fact that when bacteria are injected via the pore space, they are likely to be filtered by sand, resulting in a linear drop in bacterial concentration along the injection direction [78].

Surface percolation and spraying are superficial treatments in which the bacterial suspension and cementation solution are alternately dripped or sprayed over the soil surface, followed by solution penetration into the soil by gravity [37,79–82]. Due to the similarity of these methods to the injection technique, they share the drawbacks of uneven precipitation and surface clogging, which are mostly influenced by the sand's particle size. A comparative study on the effect of sand dimension in 2 m columns of one-dimensional trials conducted by Cheng and Cord-Ruwisch [83] revealed that, after repeated treatments, the column of fine sand (size 0.3 mm) showed blockage at the injection end, resulting in a limited cementation depth of less than 1 m, but this issue was not observed in the column of coarse sand (size > 0.5 mm).

In the mixing method, which is relatively new, the bacteria are first cultured in a nutrient broth medium, and the broth is mechanically mixed with cementitious material until the desired homogeneity is attained. This method provides the opportunity to control the growth of bacteria and optimize the bacteria culture prior to the mixing procedure in order to achieve the highest possible concentration of bacteria. However, this approach has the potential downside of causing disturbance to the soil, which is significant since it may cause false stress to form in the soil sample due to the vigorous mixing of the soil and the cementing agent [11]. Although this method can open the doors to the development of "smart-living concrete", both the protection or immobilization of bacteria and the source of nutrients for establishing the long-term and repetitive self-healing effect are still open questions [17].

1.3. *Parameters That Influence Bio-Cementation*

Bio-cementation treatment is a complex phenomenon influenced by several factors, including bacteria type (e.g., high, medium, or low urease activity), bacteria condition (e.g., live, bacterial fraction, or enzyme extracted from bacteria or plant) [27], application method (e.g., injection, mixture, spraying) [3,19], as well as environmental factors, such as pH [84], temperature [20], nutrition media [9,85,86], and grain size and nucleation sites [46,49,87], all of which affect the precipitation quantity and quality of calcium carbonate crystals and, consequently, alter the permeability and mechanical properties of the treated construction material. Figure 1 depicts a summary of factors that affect the bio-cementation process. For instance, a study on the morphology and evolution of crystals by two *Escherichia Coli* (*E. Coli*) strains altered to have either high or low urease activity over the course of seven days revealed that the strain with low urease activity produced nanocrystalline sheets during the first day, which after seven days of precipitation resulted in well-crystallized and extraordinarily large crystals that were 1924% larger than those with high urease activity due to the difference in kinetics of mineralization between the two strains [88].

Like any other chemical process, temperature significantly impacts the kinetics of the enzymatic activity of the bacteria in bio-cementation, and it can have a direct impact on crystal dissolution behavior following the Ostwald law [89]. A modest adjustment of 1 or 2 °C in the reaction temperature might cause changes in the outcomes of 10% to 20%, and most enzymes' activity increases by 50% to 100% when the temperature rises by 10 °C [90]. However, this rise only lasts as long as the elevated temperature does not denature the enzyme and change its structure. Similar to temperature, the pH of the environment (construction material in this case) substantially impacts bacterial enzymatic activity. At the microlevel, pH variation affects microbial cells by two different mechanisms: it alters the enzymatic reaction rate and the transport of nutrients into the cell [91]. Changing the pH level outside of its optimum range may slow down or even stop the bacteria's enzymatic activity and $CaCO_3$ precipitation, as the bacteria will not survive [43].

One additional factor that impacts the bio-cementation process and contributes to its complexity is the nutrition medium on which the bacteria rely for food [20,85]. Calcium carbonate crystal types include calcite, vaterite, and aragonite, with calcite being the most common and stable [92,93]. The composition of the nutritional medium has a significant effect on the $CaCO_3$ crystal morphology [93]. Even though yeast extract slows down the rate of hydration by a large amount, it is often used as a carbon source in bio-cementation applications as a nutritional medium. As a result, ongoing research is being conducted to explore alternative solutions [64,85,86].

Grain size has a significant impact on the outcomes of bio-cementation, as it can lead to non-homogeneity in large-scale applications. The particle's granularity directly affects the spacing between particles, which in turn acts as nucleation sites for the precipitation of $CaCO_3$ by the bio-cementing agent. In the case of a mixture containing predominantly fine grains, when the particles are excessively small, the bio-cementing agent will fill the narrow gaps between the particles and become restricted within these confined spaces. In the case of a mixture containing predominantly coarse grains with excessive distance between the particles, on the other hand, bacteria will have unrestricted movement between the pores. However, over time, gravity will cause the bacteria to migrate towards the lower level of the specimen, resulting again in non-homogeneity [46]. The urease activity of bacteria or enzymes and the kinetics of $CaCO_3$ precipitation also play an additional role in this matter. In the case of a low urease activity bacteria that exhibits a slower precipitation rate of $CaCO_3$, the reduced kinetic activity may allow for the potential migration of the bacteria to other pores, resulting in a more uniform outcome.

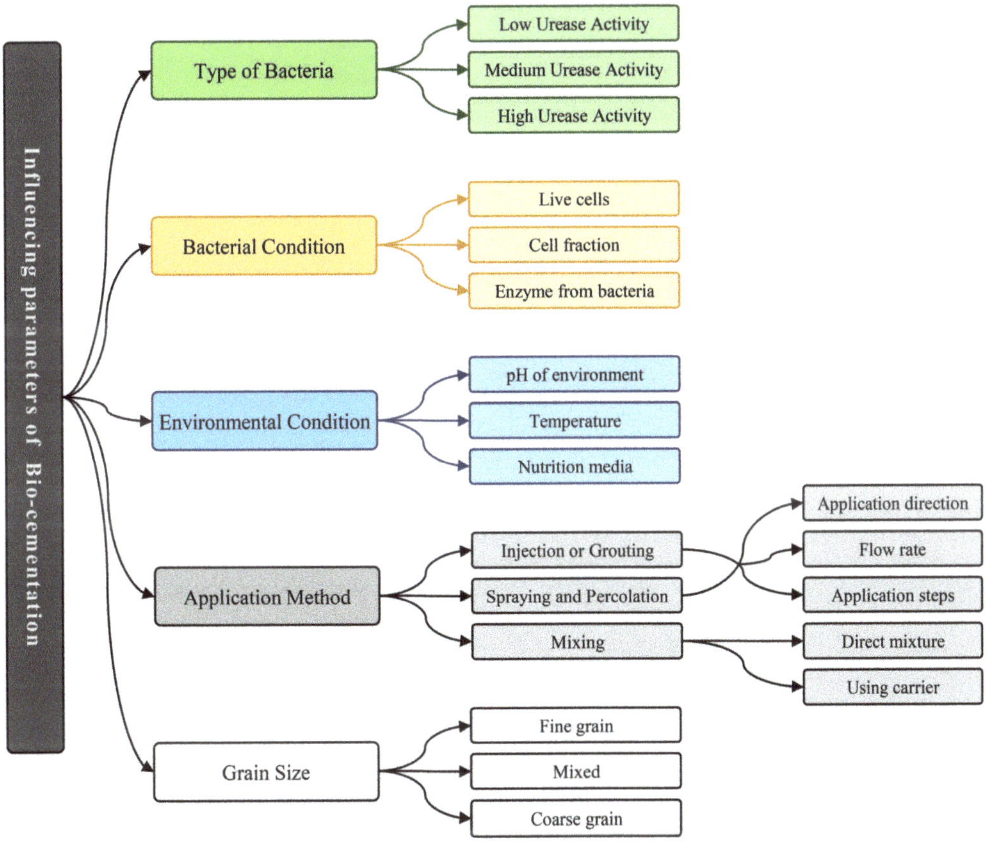

Figure 1. Schematic representation of influencing parameters on bio-cementation.

1.4. Characterization Techniques

Various experimental methodologies for qualitative and quantitative characterization of treated material and bio-cementation efficiency were employed. Scanning electron microscopy (SEM) has been employed for morphological analysis, which offers a wide range of information, including the efficiency of bio-cementation in filling nucleation sites [46], as well as the effect of the bacterial strain and its urease activity rate [21,39,47,68,88], bacterial condition [40,67], nutritional solution [37,64], and the difference between MICP and EICP [46,84] on the shape and quantity of precipitated $CaCO_3$ crystals. For instance, Phua and Røyne [64] used SEM to highlight the influence of nutritional solution on the $CaCO_3$ crystal shape precipitated by Jack bean (*Canavalia Ensiformis*) urease, revealing that calcium chloride solution resulted in rhombohedral $CaCO_3$ crystals with dimensions ranging from 20 to 80 μm, while calcium lactate and dissolved chalk solution precipitated spherical $CaCO_3$ crystals with a diameter between 100 and 250 μm (Figure 2).

(a) (b) (c)

Figure 2. SEM images of $CaCO_3$ crystals formed in (**a**) calcium chloride, (**b**) calcium lactate, and (**c**) dissolved chalk solution, adapted with permission from Ref. [64] (2023, Elsevier).

Several studies took advantage of the elemental mapping capability of energy-dispersive X-ray spectroscopy (EDS or EDX) to study the surface composition and polymorphs of calcium carbonate (calcite, vaterite, and aragonite) in treated samples [46,55,64,68]. X-ray diffraction analysis (XRD) has been used to determine the crystallographic structure of the newly formed precipitated $CaCO_3$ as well as the intensity of each polymorph compared to the other [84,94].

One of the most important outcomes of bio-cementation is the $CaCO_3$ concentration, which has a significant impact on the mechanical properties of the treated construction material. The distribution of $CaCO_3$ content after the completion of the curing time or treatment is used to determine whether sand columns are adequately solidified. In a comparative study conducted by Choi et al. [95], six different methods, namely titration, inductively coupled plasma (ICP), X-ray diffraction (XRD) TOPAS, thermogravimetric analysis (TGA), ASTM, and washing methods, were adopted to measure the calcium carbonate content, which highlighted that titration and ICP techniques gave the lowest value and the washing method the highest value for $CaCO_3$ content.

Regarding the mechanical properties characterization, experimental techniques, such as unconfined compressive strength tests (UCSs), California bearing ratio (CBR), and cone penetration testing (CPT), based on ASTM D2166/D2166M-16 [96], ASTM D1883-21 [97], and ASTM D3441-16 [98], respectively, as well as direct shear tests (DSTs), have been employed. Permeability is another important property that indicates the efficiency of bio-cementation treatment. The significance of this property may vary across different application fields; however, it is evident that achieving low permeability is the objective across all application fields. In the context of building construction applications, high permeability leads to the rapid penetration of water containing harmful substances, thereby compromising durability and causing gradual deterioration. In soil stabilization, on the other hand, the elevated permeability characteristic of certain soils can facilitate the infiltration of liquids, such as water, thereby potentially causing the erosion and removal of particles. Depending on the application field and sand properties, the permeability properties of treated samples have been characterized using a variety of methods, including the rapid chloride permeability test (RCPT) [20,61], water absorption, constant head test method [8,10], falling head test [94], and standard protocols, such as ASTM D2434-22 [99], ASTM D2435/ D2435-11 (2020) [100], ASTM C 1585-20 [101], and ASTM 5856-15 [102].

1.5. Focus of This Review and Bibliometric Analysis

As a result of technological advancement and actions towards multidisciplinarity with the involvement of microbiology and material engineering, the window of opportunity to exploit this phenomenon has widened and provided an infinite number of possibilities, which in turn makes it more difficult to reach the optimized solution. Although a large body of scientific literature has been devoted to this topic, the viability of scaling up this technology to the construction site- and field-scale remains to be extensively explored.

Therefore, an overview of the most recent advances in this context will be required to emphasize the advantages, drawbacks, and knowledge gaps, and to shed light on the road towards scaling up this process. Although the mechanism of bio-cementation has been described in many studies devoted to exploiting bio-cementation, it is important to elaborate on the chemical reactions occurring during this phenomenon. In light of this, the first topic that will be covered in this review is the process by which bio-cementing agents precipitate calcium carbonate. Herein, the biochemical reactions for $CaCO_3$ precipitation by microorganisms involved in the nitrogen cycle are described as an example. For the sake of brevity and in order to have a more concise view on the influencing factors and different application methods, this review concentrates on the bacteria that has been referenced most frequently across the literature, namely *Sporosarcina pasteurii*. Thus, a section of this review article is devoted to the properties of *S. pasteurii* from a microbiological standpoint, as well as how various parameters influence the outcomes of bio-cementation processes that are assisted by this bacterium in particular. Following that, recent advances in employing *Sporosarcina pasteurii* as a sustainable construction material, specifically in soil stabilization and the building construction sector, will be presented. This section will examine studies that presented systematic comparative analysis in these application sectors, and the most significant results will be highlighted. Last but not least, concluding remarks, knowledge gaps, and a future perspective will be presented.

The process of selecting the cited literature in this review article has been conducted according to the following steps:

1. The four keywords, namely "bio-cementation", "*Sporosarcina pasteurii*", "Microbially Induced Calcium Carbonate Precipitation (MICP)", and "Enzymatic Induced Calcium Carbonate Precipitation (EICP)", were searched on Google Scholar and Scopus.
2. The pre-screening process was conducted to determine the relevance of the search results. A total of 194 articles were identified as primarily relevant within the scope of this review article.
3. The articles that were chosen were classified into two categories: "review articles" and "original research".
4. The original research articles were classified according to two criteria: "the type of bacteria or enzyme studied", and the specific "application field" in which the research was conducted.
5. For the selected original articles, the "application method", the primary "experimental methodology" used for characterization, and the "key outcomes" were highlighted. Following this secondary screening, 140 publications were chosen to be included in this manuscript.
6. The review articles and original research articles that were not specifically related to *Sporosarcina pasteurii* but presented results that highlighted specific outcomes for bio-cementation, such as the effect of influencing parameters or application methods or specific outcomes for experimental methods used in a creative or critical manner, were used in the introduction section to depict a clear background of bio-cementation and governing parameters for the reader.
7. Articles that focus on the specific characteristics of *S. pasteurii* and its behavior in various environments, as well as the results obtained from its application in soil stabilization and building construction, are used in Sections 3 and 4.

2. Bio-Cementation Mechanism

Bio-cementation is a complex multi-step process that starts with the production of urease, a multi-subunit, nickel-containing enzyme that is synthesized by ureolytic bacteria. Figure 3 depicts a schematic representation of the chemical reaction for microbially induced calcium carbonate precipitation by a microorganism involved in the nitrogen cycle as an example. The stepwise chemical reactions of MICP are as follows:

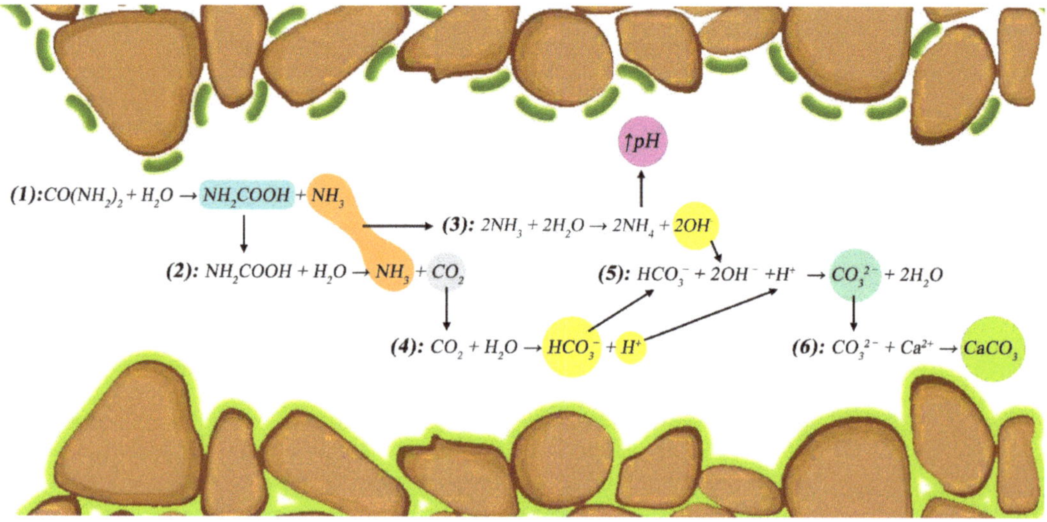

Figure 3. Chemical reactions of microbially induced calcium carbonate precipitation by nitrogen cycle bacteria.

1. The hydrolysis of urea ($CO(NH_2)_2$) under the catalysis of the bacteria's urease produces ammonia (NH_3) and carbamic acid (NH_2COOH);
2. Carbamic acid hydrolysis leads to ammonia and carbon dioxide;
3. Ammonia interacts with water and generates hydroxide (OH^-), and ammonium (NH_4) leading to a pH increase of about 1–2 pH;
4. Meanwhile, carbon dioxide in the system reacts with water, resulting in bicarbonate (HCO_3^-) and hydrogen ions (H^+);
5. The system reaches equilibrium and generates carbonate ions (CO_3^{2-}) and water;
6. Eventually, the reaction of carbonate with calcium ions (Ca^{2+}) in the environment results in the precipitation of $CaCO_3$ crystals.

It is important to keep in mind that Ca^{2+} and other cations in this system are bonded with water dipoles in an aqueous environment (nutrient media). As a result, these ions need to be dehydrated before they can attract CO_3^{2-} and precipitate $CaCO_3$ [103].

Although the EICP method employs urease obtained directly from plants or bacterial cells rather than live urease-producing bacteria (MICP) for the purpose of urea hydrolysis, it relies on identical biochemical reactions. The alterations seen in EICP treatment pertain primarily to the reaction rate and, as a result, the characteristics of the precipitated calcium carbonate [46]. In addition, it should be noted that the cost of EICP is higher compared to MICP. This is primarily attributed to the significant expenses involved in acquiring commercially purified urease enzymes derived from plants or bacteria, even for small-scale laboratory experiments. According to previous research findings, this expense can account for up to 60% of the overall operating costs, which is almost twice the cost of the growth medium ingredients themselves [10,104].

3. *Sporosarcina pasteurii* (*S. pasteurii*)

Sporosarcina pasteurii, previously referred to as *Bacillus Pasteurii* in older classifications, is an aerobic, mesophilic, rod-shaped (0.5–1.2 μm in width and 1.3–4.0 μm in length), gram-positive bacterium that is the most dominant microorganism used in MICP and EICP. *S. pasteurii* is able to form endospores, which are non-reproductive structures generated by bacteria, or the dormant form to which the bacterium can reduce itself [65]. Endospore development is frequently driven by a lack of nutrition. Endospores allow the bacterium to remain dormant for extended periods of time in the absence of nutrition, surviving ultraviolet radiation, desiccation, high temperatures, freezing temperatures, and chemical disinfectants until the environment improves and the endospore can revive itself [66]. This feature improves *S. pasteurii*'s ability to survive in hostile environments and gives it the opportunity to have its metabolic system enabled or disabled depending on the requirements of the engineers. Various strains of *S. pasteurii* have been utilized in the literature, including ATCC 11859 [2,46–55], ATCC 1376 [105,106], PTCC 1645 [107,108], DSM 33 [8,76,81,109], and NB28 (SUTS) [1,4]. Being an alkaliphile, this bacterium grows best in an alkaline environment with a pH between 9 and 10 [110]; however, it can survive in moderately harsh circumstances up to a pH of 11.2 [111], making it also a suitable admixture component for building construction applications.

S. pasteurii, like all other *Sporosarcina* species, is heterotrophic, which means it cannot make its own food and must rely on other sources for nutrition. *S. pasteurii* needs, in particular, urea and ammonium as nutrients for growth [112]. Depending on the type of nutrient provided for bacteria, its metabolism and, as a consequence, its urease activity and the rate at which calcium carbonate precipitates will differ, which emphasizes the significance of nutrition media in bio-cementation [38]. An investigation into the optimum concentrations of urea, calcium, and nickel to enhance calcium carbonate precipitation and *S. pasteurii* growth rate revealed that high urea and nickel concentrations and low calcium concentrations are required [113].

As mentioned before (Section 1.3), yeast extract is a commonly employed nutritional medium in bio-cementation applications, despite its significant impact on hydration rate reduction. However, the study on the effect of nutrient medium for optimizing *S. pasteurii* growth and cement hydration using six different nutrition media, namely yeast extract, lactose mother liquor, corn steep liquor, meat extract, glucose, and sodium acetate, revealed that the combination of meat extract and sodium acetate is a suitable replacement for yeast extract, as it reduced retardation by 75% (as compared to yeast extract) without compromising bacterial growth, urea hydrolysis, cell zeta potential, or the ability to promote calcium carbonate formation [86]. However, additional research is required to investigate the impact of these different growth media on other characteristics of concrete, such as its cost, strength, shrinkage, resistance to corrosion, and workability. Another investigation on the viability of *S. pasteurii* under extreme conditions (30, 45, and 55 °C) and pH (12.5–13.6) conditions demonstrated that exposing the bacteria to extreme temperatures (specifically 55 °C for 4 h) and extreme pH (specifically 13.6 for 4 h) led to the most significant decrease in both the concentration of viable cells and the initial rate of ammonia production through urea hydrolysis [114]. However, to the best of the author's knowledge, a comprehensive parametric study in the literature providing insights on the effect of temperature within a larger range (−5–50 °C) and extreme pH (1–14) on variation of $CaCO_3$ precipitation or denaturation of bacteria or enzymes, which could help the simulation of this process for optimization, does not exist.

As mentioned before, for the same microorganism, the agent condition (live, enzyme extracted, or cell fraction) affects the result of precipitated $CaCO_3$ and, consequently, the properties of the treated material. Hoang et al. [46] compared the crystallography, mechanical, and permeability properties of bio-cementation via MICP and EICP by *S. pasteurii* after 12 cycles of treatment. The results of this comprehensive investigation concluded the following:

- The MICP strategy creates large crystal clusters with a thickness ranging from 50 to 100 μm, whereas the size of crystals via EICP are significantly smaller, ranging from 5 to 20 μm.
- The same number of treatments resulted in a higher $CaCO_3$ content for MICP, and at a constant percentage of $CaCO_3$, the mechanical properties of EICP-treated samples were superior to those treated by MICP.
- While $CaCO_3$ precipitation levels differed (2.5–16%) and (1.5–8%), the peak stress for UCS tests of treated samples in both cases ranged from 200–2400 kPa.
- Permeability levels for EICP-treated samples were slightly lower in the range of 1.5–4% $CaCO_3$ precipitation compared to MICP-treated, while the MICP-treated permeability decreased by 3 to 4 orders of magnitude between 13–16% $CaCO_3$ content.

The comparison of MICP and EICP treatment by *S. pasteurii* in relation to pH fluctuation by Lai et al. [84] highlighted three key aspects. First, in acid environments (pH = 4–7), MICP treatment appears to be faster than EICP treatment since the delay time for visible flocculation—an indication of precipitation kinetics—is shorter in MICP treatment, and MICP appears to be less influenced by pH variation as it begins to precipitate $CaCO_3$, even at a pH of about 4.5 when the EICP process does not even start (Figure 4a). Second, EICP treatment appears to outperform MICP treatment in terms of average calcium carbonate content (CCC) at a pH above 5.5; by decreasing the pH below 5.5, CCC decreases and declines dramatically at pH = 4.5, while MICP treatment appears to have an almost consistent precipitation up to pH = 4.5 and a decrease at pH = 4 (Figure 4b). The third outcome of this investigation concerns the morphology of calcium carbonate crystals. For the same initial pH, EICP treatment seems to predominantly precipitate calcite, which is the most stable polymorph of $CaCO_3$, while MICP treatment produces both calcite and vaterite. This aspect results in significantly higher mechanical properties for EICP treatment (Figure 4c,d). Despite the numerous advantages of EICP over MICP, the MICP treatment continues to receive more attention due to its significantly lower cost in comparison to EICP.

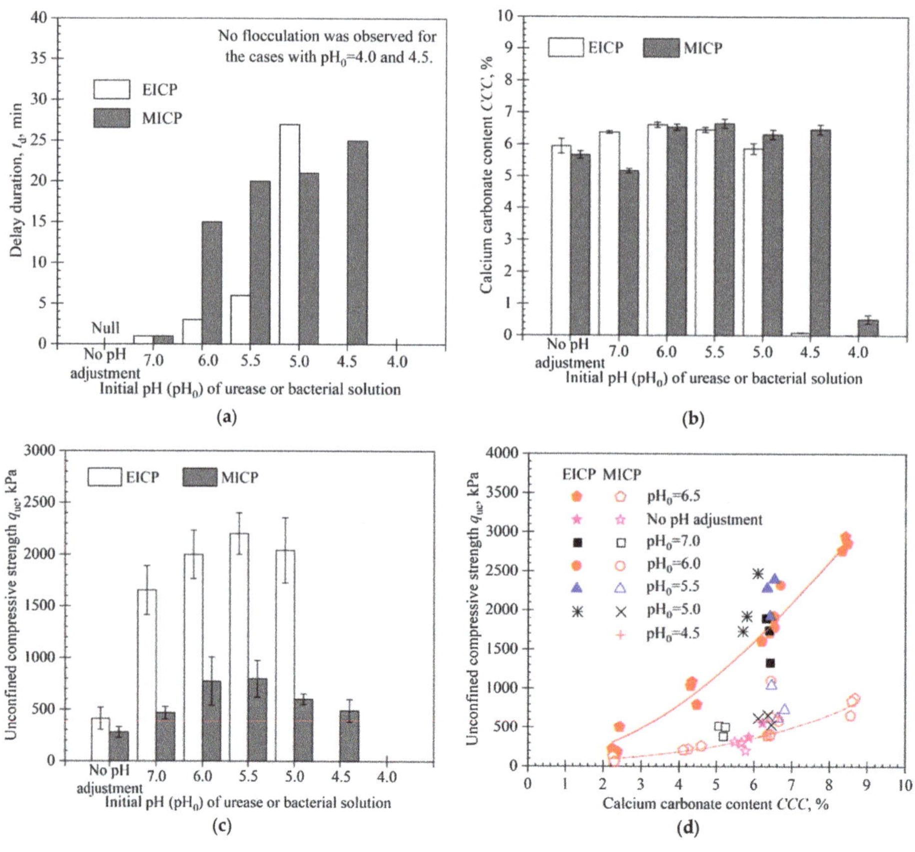

Figure 4. Effect of pH variation on EICP- and MICP-treated soil by *S. pasteurii*: (**a**) delay in flocculation at various initial pH conditions; (**b**) average calcium carbonate content in the whole sample; (**c**) unconfined compressive strength; (**d**) relation between strength and calcium carbonate content. Adapted with permission from Ref. [84] (2023, Springer Nature).

4. Recent Advancements in Exploiting *S. pasteurii*

As mentioned before, microbially induced calcium carbonate precipitation is already a complex phenomenon that can be influenced by a variety of parameters. On the other hand, a large body of literature has been devoted to the study of *S. pasteurii* as a sustainable construction material, investigating one or two parameters at a time. Hence, this section focuses only on the literature that presents parallel comparative methodologies and systematic characterizations and assessments. The preceding section (Section 3) has covered the latest developments concerning the factors that impact the distinct characteristics of *S. pasteurii*, as well as its behavior and viability in different environments. This section is dedicated to discussing the in situ application of bio-cementation. It will cover relevant literature that explores the effects of different application methods and the challenges encountered in various application sectors. Among the application domains of bio-cementation, the soil stabilization and building construction sector has investigated this strategy the most.

Although researchers were able to employ bio-cementation to develop commercially available building construction materials, such as bio-based bricks that received a patent [115] and reached mass production level (e.g., Biolith® tiles by Biomason, Inc. (Durham, NC, USA) [116]), the viability of scaling up this technology to the construction site and field scale remains to be extensively explored [1]. Since application methods of bio-cementation (Section 1.2), in particular injection or grouting, are already widely used in soil stabilization, exploiting MICP in field-scale applications in this sector is much more advanced than building construction. Hence, a closer look at the latest developments in soil stabilization can shed light on the most recent findings and areas of knowledge that require further exploration for scaling up this technology in this field as well as in building construction applications.

4.1. Soil Stabilization

Traditional soil improvement strategies, such as adding natural and synthetic materials (e.g., recycled glass fibers, tires, fruit brunches, polypropylene, and polyester), injection of chemical grouting or deep mixing using cement and/or lime, and application of sand or stone columns, rely on synthetic substances that require significant energy for production and application and raise the pH of groundwater, causing serious environmental problems and contributing to ecosystem disturbance [11]. With the widespread implementation of bio-cementation in soil stabilization applications, researchers have investigated the effects of different parameters, such as application method, nutrition media, and sand granularity, on the properties of *S. pasteurii*-treated soil at both laboratory and in situ field scales. Regarding the application method, injection is the most common method for MICP-based soil stabilization.

Table 1 summarizes some of the findings from systematic investigations in the literature that used *S. pasteurii* for soil stabilization applications by injection. A table of the literature that used other application methods to introduce *S. pasteurii* for soil stabilization is presented in the Supplementary Materials, Table S1.

Although the injection method has proven to be quite effective in small specimens at the laboratory scale, significant problems have been noted at larger scales. Whiffin et al. [10] investigated the impact of vertical injection of *S. pasteurii* along a 5-meter-long sand column. Their findings revealed that while calcium carbonate precipitation occurred throughout the entire treatment length, the concentration profile exhibited non-uniformity. In particular, higher quantities of calcium carbonate were observed at the injection points, with a subsequent decline in concentration along the length of the column. The impact of injection direction was assessed by Paassen et al. [76] in a large-scale experiment involving the horizontal injection of *S. pasteurii* in a soil bed with a volume of 100 m^3. The study revealed a considerable variation in the peak strength of the unconfined compressive strength values, again confirming non-uniformity along the path of injection within the treated sand volume. In general, as the distance from the injection points increases, there is a decrease in the calcium carbonate content, which leads to an increase in porosity and a subsequent decline in the mechanical properties. This limitation of MICP application on a large scale arises from the occurrence of system clogging near the injection sites. This phenomenon is caused by the rapid precipitation of calcium carbonate by the bio-cementing agent, which leads to the closure of pores and hinders the migration of the bio-cementing agent throughout the entire treatment area, resulting in non-homogeneity. Hence, it is anticipated that the utilization of bio-cementing agents with low urease activity or the regulation of bio-cementation kinetics to achieve a slower rate would result in a reduced level of non-homogeneity in treatment, as it would facilitate the effective distribution of the bio-cementing agent along the entire length of the treatment. This disadvantage can be mitigated by employing a two-phase vertical injection procedure (alternating the injection of bacterial suspension and cementation or nutritional solution) and fine-tuning the injection parameters, such as decreasing the injection rate of bacterial suspension and increasing the flow rate of cementation solution [10,77].

Among the several studies that evaluated the impact of particle size on the effectiveness of bio-cementation treatment aided by *S. pasteurii*, three stand out due to their systematic approaches [46,49,117]. Lin et al. [117] studied the effect of grain size on MICP-treated Ottawa sand by *S. pasteurii*, considering fine grains ($D_{10\%}$ = 0.26 mm and $D_{50\%}$ = 0.33 mm) and coarse grains ($D_{10\%}$ = 0.58 mm and $D_{50\%}$ = 0.71 mm). The findings of this investigation revealed that the coarse-grain soil had a lower $CaCO_3$ content in comparison to the fine-grain soil. However, it was observed that the coarse-grain soil demonstrated greater increases in S-wave velocity, peak shear strength, and cohesiveness when compared to the fine-grain soil. Moreover, the results obtained from triaxial testing indicated that the peak deviator stress of the coarse-grained soil containing 1.6% $CaCO_3$ and the fine-grained soil containing 1% $CaCO_3$ exhibited an average increase of 93% and 171%, respectively, as compared to their respective untreated specimens.

Hoang et al. [46], on the other hand, investigated this effect on EICP-treated silica (SiO_2)/quartz sand, considering fine grains ($D_{10\%}$ = 0.26 mm and $D_{50\%}$ = 0.36 mm) and coarse grains ($D_{10\%}$ = 0.61 mm and $D_{50\%}$ = 0.72 mm). The outcomes of this study highlighted that, similar to MICP-treated sand, the mechanical properties of EICP-treated sand, such as Young's modulus and UCS test, are higher in coarse-grained sand than in fine-grained sand.

As in Table 1, the trend of permeability reduction influenced by grain size indicated that permeability reduction in fine-grained bio-cemented sands steadily declined with the increase in $CaCO_3$ content, while for the same increase in $CaCO_3$, the trend of reduction in permeability declination was comparatively less pronounced in coarse-grained sand [46]. This phenomenon is often attributed to the smaller distances between particles, known as nucleation sites, in fine-grained sand. The precipitation of calcium carbonate fills the small pores in fine grain rapidly, leading to the obstruction of pathways through which bacteria can move to continue the cementation in other pores. In coarse-grained sand, the interparticle spacing is greater. Through the precipitation of $CaCO_3$, the contact points between the particles are initially joined, followed by the deposition of additional layers of calcium carbonate on top. This process facilitates the movement of bacteria or enzymes through the pores, enabling them to reach more profound or distant locations.

Mahawish et al. [49] studied the effect of fine (0.075–9.5 mm) and coarse (2.36–16 mm) grain percentages in mixed Pakenham Blue Metal columns on the mechanical properties of treated sand. Five columns with different percentages of fine/coarse grains (A = 0/100, B = 25/75, C = 50/50, D = 75/25, E = 100/0%) were subjected to MICP treatment by *S. pasteurii*. The average calcium carbonate precipitation for columns A and E was determined to be about 7% and 6.5%, respectively, and about 6% for other mixed columns (B, C, and D). However, Column B exhibited the most uniform distribution of $CaCO_3$ along its length. The unconfined compressive strength test results for all three mixed columns (B, C, and D) were determined to be approximately 0.6 MPa, while for Column A, despite having the highest $CaCO_3$ content, this value was lowered by half to 0.3 MPa, and for Column E, it was dropped to nearly 0.5 MPa. These findings underscore a significant aspect regarding the impact of grain size. While a higher level of porosity (100% coarse grain in Column A) does facilitate the migration of the bio-cementing agent throughout the treated column and may result in the highest $CaCO_3$ content, it does not necessarily yield the best mechanical properties. This is due to the fact that the bacterial solution tends to migrate to the bottom of the column under the influence of gravity, resulting in non-homogeneity throughout the column. Consequently, the top of the column exhibits the lowest $CaCO_3$ content, while the bottom exhibits the highest, contributing to the overall inferior mechanical properties of the column due to non-homogeneity.

As previously stated, soil stabilization is more advanced than other application fields of bio-cementation when it comes to scaling up the process for field application. Few studies have been conducted on the scaling up of bio-cementation to field scale [1,118–121]; therefore, it is important to highlight the most significant findings from these investigations. Omoregie et al. [1] presented an economic strategy to scale up the production and cultivation of *S. pasteurii* under nonsterile condition using a custom-built stainless-steel stirred tank reactor with a capacity of 3 m^3. The scalability of the bacterial cells was investigated by increasing the volume of the seed cultures from 214 L to 2400 L and monitoring their growth for a duration of 90 h. The results of the in situ soil bio-cementation experiment demonstrated that bacterial cells cultivated using this method retained their ability to induce $CaCO_3$ precipitation even after being immobilized within sand specimens. The presence of $CaCO_3$ crystal formation within the treated sand particles was confirmed through the analysis of $CaCO_3$ content and microstructural data, which confirmed that the method presented in this study is as successful as the sterile condition.

Table 1. Selective literature employing *Sporosarcina pasteurii* for soil stabilization treatment via injection method.

Characterization								
SEM	**EDS**	**XRD**	**$CaCO_3$ Content**	**Mechanical Test (MPa)**	**Permeability Reduction**	**Treat. Time (Days)**	**Comments and Results**	**Ref.**
X			60 kg/m^3	UCS = 3.52		20	• $CaCO_3$ content, and ultrasonic wave velocity showed a linear relation	[4]
			60–105 kg/m^3	0.2–0.57	slight reduction	5	• Mechanical: single stage confined drained triaxial test • Decrease in porosity ranged from 10 to 2% along the distance from injection point	[10]
X	X		2.5–16%	UCS = 0.2–2.3	X	16	• MICP coarse sand (Young's modulus = 20–250 MPa)	[46]
			1.5–8%	UCS = 0.45–1.5			• EICP coarse sand (Young's modulus = 75–125 MPa)	
			2–6%	UCS = 0.2–0.9			• EICP fine sand (Young's modulus = 25–75 MPa)	
X	X		1.6–2.5%	X		1	• Fine sand	[117]
			0.9–1.1%	X			• Coarse sand	
X		X		UCS = 0.2–2.5	5–35%	1	• Treatment with soluble calcium is more efficient than calcium chloride	[94]
X			5.3%	CPT = 32.1		14	• MICP results in fewer but larger calcite crystals • Shear wave velocities increased by 600% • CPT increased by 500%	[67]
			X	X		16	• Horizontal injection	[76]
X			11%		Increased by 800%	14	• Bio-cemented clean sand	[8]
			11%	UCS = 1.3			• Bio-cemented oil-contaminated sand	
				UCS = 0.3			• Portland cement-treated oil-contaminated sand	

Table 1. *Cont.*

Characterization								
SEM	**EDS**	**XRD**	**$CaCO_3$ Content**	**Mechanical Test (MPa)**	**Permeability Reduction**	**Treat. Time (Days)**	**Comments and Results**	**Ref.**
X						13	• Various bacteria and several strains of *S. pasteurii* • Production rate of *S. pasteurii* is 2 orders of magnitude higher than the others	[122]
X			0.9%	X			• The liquefaction resistance and pre-triggering behaviors of loose sands improved significantly • Shear strengths and stiffness increased	[123]
				CPT = 3.6–5.1		15	• Shear wave velocities = 961 and 967 m/s • CPT improved by 419–543%	[124]
			3.6%	Shear strength increased up to 72%			• Increase in properties by number of treatments • Shear strength and slope of failure line in triaxial consolidated drained test improved	[125]
X	X	X		UCS = 1.47			• The presented method reduces ammonia discharge by more than 8 times.	[32]
X			X	UCS = 2.5			• 8 treatments with monitoring effect of each treatment on $CaCO_3$ content and mechanical properties.	[126]
X	X		13%	UCS = 0.9–2		8	• 16 treatments by solutions contacting different percentages of bacteria	[53]
X		X	0–4%	UCS = 0.16		10	• With modest $CaCO_3$ precipitation (0–4%) in intergranular sand voids, the UCS increased by more than 100 kPa.	[51]
X	X		7.5%	Shear strength = 0.15	98%	30	• 15 cycle of treatment (1 per 48 h) • The peak strengths increase by 266%	[68]
X	X	X	2.5–7.3%	Needle = 1.67–5.3			• Various strains of *S. pasteurii* • 25–44% reduction in riverbank erosion	[5]
X		X	15.2%	UCS = 1.74			• Injection with fixation	[3]
			7.6%	Needle = 0.5			• Injection without fixation	
X	X	X		UCS = 0.3	X	1	• Using seawater supplemented with urea concentrated cementation solution increased the mechanical properties	[6]

X: Experiment has been performed, UCS: unconfined compressive strength test, CPT: Cone penetration test.

4.2. Building Construction

Bio-cementation has been utilized in the field of building construction, primarily for producing new building materials rather than repairing damaged structures. The most common application method used in building construction is to employ bio-cementing solution as an admixture component of concrete. Table 2 presents an overview of key findings derived from systematic investigations documented in the literature, which have employed *S. pasteurii* to develop self-healing concrete through mixing techniques. As can be observed, there is a notable disparity in the quantity of relevant publications utilizing *S. pasteurii* in comparison to soil stabilization (Table 1 and Table S1). Moreover, it is worth mentioning that in the context of building construction (Table 2), the primary emphasis is placed on the mechanical and permeability characterization of the treated material rather than the assessment of calcium carbonate content, compared to soil stabilization (Table 1).

The experimental approach commonly used in the majority of the literature involves the following steps:

1. Introducing bacteria as an admixture component in concrete.

2. Curing the cast concrete for a period of 28 days.
3. Inducing cracks in the concrete by applying either uniaxial or flexural force.
4. Treating the damaged concrete with a bacterial nutrient media or treatment solution.
5. Monitoring the healing progress of the crack over time.
6. Conducting mechanical and permeability tests to characterize the healed concrete.

S. pasteurii has shown remarkable effectiveness in repairing concrete cracks. A study by Sohail et al. [127] utilizing this bacterium for the purpose of repairing cracks in concrete has demonstrated that not only does this bacterium have the capability to effectively fill cracks up to a width of 4 mm, but it can also successfully adhere two separate pieces of a concrete specimen that have become detached during a flexural test.

Various nutritional media, such as lactose mother liquor [128,129], sodium carbonate [129], calcium nitrate–urea, and calcium chloride–urea [129,130], have been employed in the cultivation of *S. pasteurii* within concrete mixtures. However, due to the variability in environmental conditions, concrete components, and treatment methods, it is unfortunate that a definitive conclusion cannot be reached on this matter. Kim et al. [131] investigated the influence of concrete type on the process of *S. pasteurii* bio-cementation by measuring the relative mass gained as a result of treatment on normal and lightweight concrete, concluding that the relative weight gains observed in both normal and lightweight concrete were comparable. This suggests that the precipitation of calcium carbonate by *S. pasteurii* is not dependent upon the specific type of concrete used, as long as the environmental conditions necessary for bacterial growth (such as concrete pH and temperature) remain consistent.

Table 2. Selective literature employing *S. pasteurii* for new construction material using mixing method.

Characterization					Treat. Time (Days)	Comments and Results	Ref.
SEM	EDS	XRD	Mechanical Test (MPa)	Permeability Reduction			
X		X	X	X	28	• Bacterial nutrient: calcium nitrate–urea and calcium chloride–urea	[130]
X		X			91	• The aggregates were immersed in bio-cementing solution before mixing • Visual analysis of induced crack healing	[132]
X	X	X	CS = 32	7%	91	• Bacterial concentration 10^3 cells/mL	[133,134]
			CS = 36	8%		• Bacterial concentration 10^5 cells/mL	
			CS = 31	6%		• Bacterial concentration 10^7 cells/mL	
X	X	X	0.2–0.3		28	• Higher optical density resulted in higher compressive strength (increased 33%) and lower water absorption	[136]
X	X		CS = 29–42		28	• Compressive strength (CS) increased by 79.2% • Flexural strength increased by 50%	[137]
X	X		28% increase	10.6%	120	• Bacteria proportion: 0.25% of cement weight	[138]
			41% increase	17.9%		• Bacteria proportion: 0.5% of cement weight	
			X		28	• Compressive strength increased by 45% • Flexural strength increased by 18%	[135]
X	X	X	CS = 46		28	• Cracks up to a width of 4 mm were healed • Two separate pieces of a concrete adhered again	[127]
			CS = 29.2		28	• Used rubber particles to immobilize the bacteria	[139]

X: Experiment has been performed, CS: compressive strength.

Although the straightforward method of adding bacterial and nutrient solutions into concrete mixtures is commonly employed, researchers have explored alternative approaches to introduce bacteria into concrete mixtures. Chen et al. [132] employed lightweight aggregates as carriers for *S. pasteurii* in concrete. In this study, the aggregates underwent several initial immersions in nutrient media (specifically, calcium lactate/yeast extract), followed by multiple drying cycles. Subsequently, the aggregates were immersed in a bacterial solution. After the aggregates were completely dried, they were combined with other components to produce concrete. The comparison of the healing process of induced cracks was conducted on three groups: the controlled group (A) without treated aggregate, the air-cured treated group (B) with treated aggregate but no additional supplement, and the immersion-cured treated group (C) with treated aggregate and immersion in a urea/calcium carbonate solution. The results showed that the precipitation of $CaCO_3$ by *S. pasteurii* can be reactivated in the presence of appropriate nutrition; in group C, the cracks were completely cured after 91 days, while in group B, only partial curing was observed, and no curing was observed in group A.

When using bio-cementing agents as a mixture component in concrete, it is important to keep in mind that an excessive bacterial concentration in the concrete solution may not always lead to the best permeability or mechanical properties. For instance, an investigation on the effect of *S. pasteurii* concentration (10^3, 10^5, and 10^7 cells/mL) in a concrete mixture on the mechanical properties of the concrete demonstrated that the compressive strength of concrete with 10^5 cells/mL bacterial concentration was 11% higher than that of 10^3 cells/mL and 13.5% higher than that of 10^7 cells/mL [133,134]. The observed decrease in compressive strength of concrete at higher cell concentrations of bacteria (10^7 cells/mL) can be attributed to the disruption of matrix integrity caused by excessive bacterial activity.

The study conducted by Zaerkabeh et al. [135] is among the limited number of research efforts that have examined the impact of crack size on the effectiveness of self-healing concrete utilizing *S. pasteurii*. Cracks with widths of 0.5 and 1 mm and depths of 5, 10, and 15 mm were investigated in this study. The study's findings indicate that the flexural strength and energy absorption capability of the specimens dropped as the depth and width of the cracks increased. Notably, the specimen with a crack depth and width of 0.5 exhibited the highest healing efficiency after 7 and 28 days. While the authors acknowledge that impeding the flow of nutrients and oxygen by clogging the pores could potentially lead to a drop in the healing rate among samples tested after 28 days, it is important to note that this study did not take into account the volume of the crack, which could possibly contribute to the observed reduction in healing in cracks with larger dimensions. It is comprehensible that in a suitable environment, the quantity of bio-cementation, like any other chemical reaction, is a function of time. Since the rate of the chemical reaction involving the precipitation of $CaCO_3$ by *S. pasteurii* for all the studied specimens is constant, the quantity of precipitated crystals will increase over time. Hence, in the case of larger cracks, a more extended period of monitoring the healing process may have yielded more precise conclusions.

5. Conclusions

5.1. Concluding Remarks

The goal of this review is to provide a concise summary of key elements pertinent to the current literature on using *Sporosarcina pasteurii* as a sustainable construction material. The main novelties of this review are that it offers a general overview of the factors to be considered when using bio-cementation (especially for *S. pasteurii*), provides a thorough summary of comparative studies on the subject that presented outstanding results, and sheds light on areas that have been overlooked in the literature and have the potential for future investigation. After a thorough examination and analysis of the most recent prominent works of literature, the concluding remarks of the review article may be summarized as follows:

- The application of bio-cementation has demonstrated notable efficiency in different engineering sectors, exhibiting enhancements in mechanical properties and a reduction in the permeability of construction materials at a laboratory scale. Moreover, the reduction in embodied energy (43–95%), carbon footprint (18–49.6%), and lifecycle cost-effectiveness render bio-cementation a significant step toward sustainable construction.

- Bio-cementation, whether produced by microbial or enzymatic agents (MICP or EICP), is a complex phenomenon whose kinetics can be influenced by various parameters related to the characteristics of bio-cementing agents (rate of urease activity and bio-cementing agent condition), environmental conditions (temperature, pH, and availability of nutrition and nucleation sites), and even the method that has been used to employ bio-cementation. The increase in urease activity rate of the bio-cementing agent, which is influenced by factors, such as the type of bio-cementing agent, nutrition media, pH, and temperature, leads to an increase in the number of precipitated crystals and a simultaneous decrease in the crystals' dimensions. The influence of pH and temperature on bio-cementation depends upon the specific bio-cementing agent employed, as each bacteria possesses its own distinct ideal pH and temperature conditions for enzymatic activity.
- Among the different methods for introducing the bio-cementation agent into the system, the injection method is the primary technique used in soil stabilization. Despite the presence of several drawbacks and obstacles, which appear to be resolvable through the fine-tuning of the injection process, this method appears to be better suited for large-scale applications in this sector. Building construction, on the other hand, mainly uses bacteria or enzyme solutions and nutrition media as admixture components of construction materials.
- *Sporosarcina pasteurii* is widely recognized as the predominant microorganism employed for bio-cementation due to its superior performance in terms of both quantity and quality of the precipitated $CaCO_3$, as well as the enhanced permeability and mechanical properties of the treated material. *S. pasteurii* has a biological mechanism that enables it to remain in a dormant state for an extended duration until it is exposed to a favorable environment where it can resume its activity of precipitating calcium carbonate. Moreover, it exhibits the ability to endure severe alkaline environments with a pH level as high as 13.6, as well as temperatures of 55 °C for a duration of 4 h, while persisting in the process of precipitation, although at a reduced rate.
- The results of the comparison between samples treated with microbially induced calcium carbonate precipitation (MICP) and samples treated with enzymatically induced calcium carbonate precipitation (EICP) by *S. pasteurii* indicate that the EICP treatment yields superior mechanical and permeability properties. However, the MICP process is favored by a larger number of investigations due to its cost-effectiveness. One of the reasons for this outcome is attributed to the disparity in size between the bacterium (on the order of micrometers) and the enzyme (on the order of nanometers), which facilitates the migration of enzyme within nucleation sites, leading to improved homogeneity and enhanced favorable properties.
- The granularity of the treated particles is a crucial factor that greatly influences the homogeneity of bio-cementation and the treatment outcomes. An excessive amount of either fine or coarse grains will lead to an uneven treatment outcome, resulting in poor mechanical and impermeability properties. The homogeneity and outcome of bio-cementation are enhanced when a wide range and variation of particle sizes are used.

5.2. Future Reasearch Directions

Upon thorough examination of the research studies included in this review article, as well as those initially selected but ultimately excluded from the bibliometric analysis, it becomes evident that further exploration is required to enhance our comprehension of various elements regarding exploiting bio-cementation. Despite the extensive study of *S. pasteurii* as the predominant bacterium used for bio-cementation, a comprehensive understanding of the process outcomes based on all the influencing parameters for optimization remains elusive due to the multitude of influencing parameters. This is due to the fact that the existing literature primarily focuses on the examination of one or two parameters at a time, while several other parameters vary between different studies. The knowledge gaps and the direction of future investigations for a better exploitation of bio-cementation can be outlined as follows:

- While there is a large body of scientific literature dedicated to using bio-cementation for new construction materials, exploiting this phenomenon for remediation of existing damaged construction (existing concrete structures) to extend their service lives and avert demolition, thereby decreasing a portion of the construction waste, has not progressed as much. For example, there is a scarcity of research that has explored the potential application of bio-cementation for repairing cracks in concrete structures that do not already contain a bio-cementing agent as an admixture.
- The current literature has approached employing bio-cementation from a construction standpoint, while the complexity of the bio-cementation phenomenon emphasizes the necessity of a multidisciplinary approach to exploiting bio-cementation at the intersection of biology, material engineering, and building construction, which is essential for addressing numerous unresolved inquiries pertaining to the exploitation of bio-cementation. Which bacterial strains or enzymatic agents demonstrate compatibility and suitability for applications in soil stabilization or building construction? Does the soil or construction material offer an ideal environment for the proper activity of the bio-cementing agent? Does achieving the optimal condition for the bio-cementing agent necessarily result in the attainment of optimal material qualities that are suited for the purposes of soil stabilization and construction applications?
- A key step in the process of scaling up bio-cementation is conducting a comprehensive and systematic analysis of the biological composition of bacterial urease activity and the impact of various environmental factors on the bacterial $CaCO_3$ precipitation quantity and quality to gain a better understanding of the influential factor for one specific bacterium. Being able to describe the kinetics of $CaCO_3$ precipitation as a function of pH, temperature, and nutrition media provides a more comprehensive understanding of potential results and constraints. For example, conducting a comprehensive investigation on the impact of temperature on the activity of urease in *S. pasteurii*, spanning from −5 °C to 55 °C at 5 °C intervals, can provide valuable insights into the consequences of freezing, the kinetics of precipitation in relation to temperature, and the temperature thresholds for bacterial viability and urease enzyme denaturation. In this context, using a multidisciplinary strategy in addressing this subject offers a broader and more extensive opportunity for experimental characterization. For instance, optical density (OD), which is a widely utilized method in microbiology for assessing cell viability in the context of microbiologically induced calcium carbonate precipitation (MICP), can offer valuable insights into the performance of bacteria and the kinetics of $CaCO_3$ precipitation as a function of temperature, pH, and nutrition media, paving the way for significant input information for future numerical models.
- Another underdeveloped aspect in the literature pertains to the numerical simulation models of bio-cementation. The significant constraints on studies are primarily attributed to the complex nature of the process, which occurs across different time and dimensional scales and entails interconnected biological, chemical, hydraulic, and mechanical phenomena. The extremely limited research on this subject concentrates mostly on greater scales beginning with the pore in the treated material, without taking into account interrelated parameters [15,140]. Figure 5 presents a schematic illustration of the proposed flowchart by the author for the interrelated elements in bio-cementation that need to be taken into account as inputs for numerical modeling of this phenomenon, including pertinent disciplines and dimensional scales. The dimensional scale ranges from the nano- or microscale, depending on the use of EICP (nanometers) or MICP (micrometers), to the metric scale of treated construction material or soil. The multidisciplinary nature of bio-cementation involves the incorporation of diverse interrelated parameters from several disciplines at different stages as inputs. First, it is fundamental to establish the rate of urease activity from a biological perspective in relation to environmental conditions, including the initial pH, temperature, and

nutrition media, to predict the rate and amount of $CaCO_3$ precipitation. As stated in Section 2, the bio-cementation process produces ammonium as a byproduct, leading to a pH increase of 1 to 2 units (Figure 3). This pH alteration subsequently influences the rate of urease activity. Hence, the starting pH and pH variation as a function of time with calcium carbonate precipitation must be addressed when determining the urease activity rate. On the other hand, as $CaCO_3$ precipitates, the availability of nutrition media for bio-cementing agents reduces with time, affecting urease activity once more. In parallel, parameters regarding treated material, such as application method and grain properties, that are specific for each application field, must be considered. For instance, in the case of soil stabilization by injection, the injection flow velocity and granularity of the treated soil are critical for fluid dynamic studies of injected solution propagation and bio-cementing agent migration. After taking into account all of these distinct characteristics from various disciplines on different dimensional scales, an accurate prediction of the bio-cementation process will be achievable.

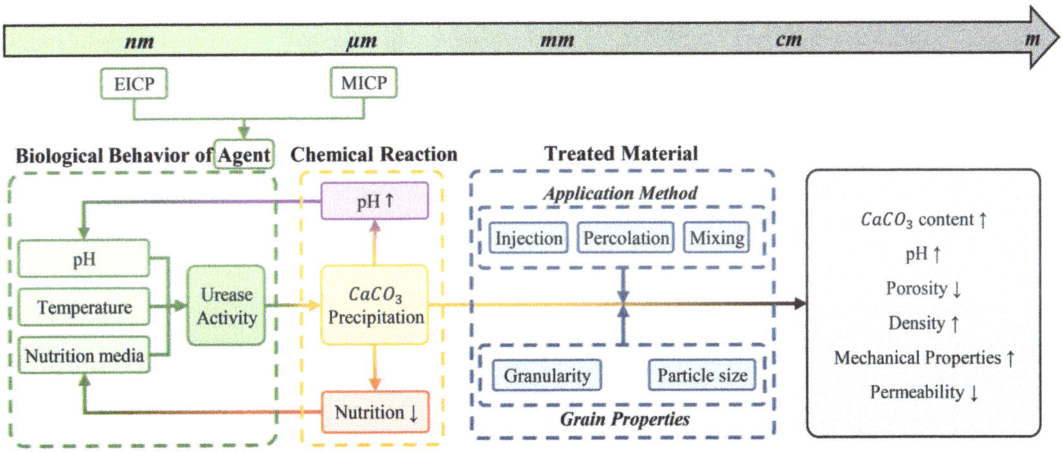

Figure 5. Schematic representation of interconnected parameters in bio-cementation for numerical modeling of this phenomenon at different dimensional scales.

Supplementary Materials: The following supporting information can be downloaded at: https://www.mdpi.com/article/10.3390/su151813869/s1, Table S1. Selective examples of studies employing *S. pasteurii* for soil stabilization via surface percolation and mixing techniques.

Funding: This research received no external funding.

Institutional Review Board Statement: Not applicable.

Informed Consent Statement: Not applicable.

Data Availability Statement: Data are contained within the article or supplementary material.

Conflicts of Interest: The author declares no conflict of interest.

References

1. Omoregie, A.I.; Palombo, E.A.; Ong, D.E.L.; Nissom, P.M. A Feasible Scale-up Production of Sporosarcina pasteurii Using Custom-Built Stirred Tank Reactor for in-Situ Soil Biocementation. *Biocatal. Agric. Biotechnol.* **2020**, *24*, 101544. [CrossRef]
2. Wani, K.M.N.S.; Mir, B.A. An Experimental Study on the Bio-Cementation and Bio-Clogging Effect of Bacteria in Improving Weak Dredged Soils. *Geotech. Geol. Eng.* **2021**, *39*, 317–334. [CrossRef]
3. Sharaky, A.M.; Mohamed, N.S.; Elmashad, M.E.; Shredah, N.M. Application of Microbial Biocementation to Improve the Physico-Mechanical Properties of Sandy Soil. *Constr. Build. Mater.* **2018**, *190*, 861–869. [CrossRef]
4. Xu, K.; Huang, M.; Xu, C.; Zhen, J.; Jin, G.; Gong, H. Assessment of the Bio-Cementation Effect on Shale Soil Using Ultrasound Measurement. *Soils Found.* **2023**, *63*, 101249. [CrossRef]

5. Dubey, A.A.; Ravi, K.; Mukherjee, A.; Sahoo, L.; Abiala, M.A.; Dhami, N.K. Biocementation Mediated by Native Microbes from Brahmaputra Riverbank for Mitigation of Soil Erodibility. *Sci. Rep.* **2021**, *11*, 15250. [CrossRef] [PubMed]
6. Cheng, L.; Shahin, M.A.; Cord-Ruwisch, R. Bio-Cementation of Sandy Soil Using Microbially Induced Carbonate Precipitation for Marine Environments. *Géotechnique* **2014**, *64*, 1010–1013. [CrossRef]
7. Abdel-Aleem, H.; Dishisha, T.; Saafan, A.; AbouKhadra, A.A.; Gaber, Y. Biocementation of Soil by Calcite/Aragonite Precipitation Using *Pseudomonas Azotoformans* and *Citrobacter Freundii* Derived Enzymes. *RSC Adv.* **2019**, *9*, 17601–17611. [CrossRef]
8. Yin, J.; Wu, J.-X.; Zhang, K.; Shahin, M.A.; Cheng, L. Comparison between MICP-Based Bio-Cementation Versus Traditional Portland Cementation for Oil-Contaminated Soil Stabilisation. *Sustainability* **2022**, *15*, 434. [CrossRef]
9. Venda Oliveira, P.J.; Neves, J.P.G. Effect of Organic Matter Content on Enzymatic Biocementation Process Applied to Coarse-Grained Soils. *J. Mater. Civ. Eng.* **2019**, *31*, 04019121. [CrossRef]
10. Whiffin, V.S.; van Paassen, L.A.; Harkes, M.P. Microbial Carbonate Precipitation as a Soil Improvement Technique. *Geomicrobiol. J.* **2007**, *24*, 417–423. [CrossRef]
11. Mujah, D.; Shahin, M.A.; Cheng, L. State-of-the-Art Review of Biocementation by Microbially Induced Calcite Precipitation (MICP) for Soil Stabilization. *Geomicrobiol. J.* **2017**, *34*, 524–537. [CrossRef]
12. Omoregie, A.I.; Palombo, E.A.; Nissom, P.M. Bioprecipitation of Calcium Carbonate Mediated by Ureolysis: A Review. *Environ. Eng. Res.* **2020**, *26*, 200379. [CrossRef]
13. Iqbal, D.M.; Wong, L.S.; Kong, S.Y. Bio-Cementation in Construction Materials: A Review. *Materials* **2021**, *14*, 2175. [CrossRef]
14. Huynh, N.N.T.; Imamoto, K.; Kiyohara, C. A Study on Biomineralization Using *Bacillus Subtilis* Natto for Repeatability of Self-Healing Concrete and Strength Improvement. *ACT* **2019**, *17*, 700–714. [CrossRef]
15. Bagga, M.; Hamley-Bennett, C.; Alex, A.; Freeman, B.L.; Justo-Reinoso, I.; Mihai, I.C.; Gebhard, S.; Paine, K.; Jefferson, A.D.; Masoero, E.; et al. Advancements in Bacteria Based Self-Healing Concrete and the Promise of Modelling. *Constr. Build. Mater.* **2022**, *358*, 129412. [CrossRef]
16. Huynh, N.N.T.; Phuong, N.M.; Toan, N.P.A.; Son, N.K. Bacillus Subtilis HU58 Immobilized in Micropores of Diatomite for Using in Self-Healing Concrete. *Procedia Eng.* **2017**, *171*, 598–605. [CrossRef]
17. Huynh, N.N.T.; Imamoto, K.; Kiyohara, C. Biomineralization Analysis and Hydration Acceleration Effect in Self-Healing Concrete Using *Bacillus Subtilis Natto*. *ACT* **2022**, *20*, 609–623. [CrossRef]
18. Huynh, N.; Imamoto, K.; Kiyohara, C. Compressive Strength Improvement and Water Permeability of Self-Healing Concrete Using Bacillus Subtilis Natto. In Proceedings of the XV International Conference on Durability of Building Materials and Components, Barcelona, Spain, 20–23 October 2020.
19. Joshi, S.; Goyal, S.; Reddy, M.S. Corn Steep Liquor as a Nutritional Source for Biocementation and Its Impact on Concrete Structural Properties. *J. Ind. Microbiol. Biotechnol.* **2018**, *45*, 657–667. [CrossRef] [PubMed]
20. Wu, M.; Hu, X.; Zhang, Q.; Xue, D.; Zhao, Y. Growth Environment Optimization for Inducing Bacterial Mineralization and Its Application in Concrete Healing. *Constr. Build. Mater.* **2019**, *209*, 631–643. [CrossRef]
21. Krishnapriya, S.; Venkatesh Babu, D.L. Isolation and Identification of Bacteria to Improve the Strength of Concrete. *Microbiol. Res.* **2015**, *174*, 48–55. [CrossRef]
22. Castro-Alonso, M.J.; Montañez-Hernandez, L.E.; Sanchez-Muñoz, M.A.; Macias Franco, M.R.; Narayanasamy, R.; Balagurusamy, N. Microbially Induced Calcium Carbonate Precipitation (MICP) and Its Potential in Bioconcrete: Microbiological and Molecular Concepts. *Front. Mater.* **2019**, *6*, 126. [CrossRef]
23. Devrani, R.; Dubey, A.A.; Ravi, K.; Sahoo, L. Applications of Bio-Cementation and Bio-Polymerization for Aeolian Erosion Control. *J. Arid. Environ.* **2021**, *187*, 104433. [CrossRef]
24. Fattahi, S.M.; Soroush, A.; Huang, N. Biocementation Control of Sand against Wind Erosion. *J. Geotech. Geoenviron. Eng.* **2020**, *146*, 04020045. [CrossRef]
25. Dubey, A.A.; Devrani, R.; Ravi, K.; Dhami, N.K.; Mukherjee, A.; Sahoo, L. Experimental Investigation to Mitigate Aeolian Erosion via Biocementation Employed with a Novel Ureolytic Soil Isolate. *Aeolian Res.* **2021**, *52*, 100727. [CrossRef]
26. Zomorodian, S.M.A.; Ghaffari, H.; O'Kelly, B.C. Stabilisation of Crustal Sand Layer Using Biocementation Technique for Wind Erosion Control. *Aeolian Res.* **2019**, *40*, 34–41. [CrossRef]
27. Marvasi, M.; Mastromei, G.; Perito, B. Bacterial Calcium Carbonate Mineralization in Situ Strategies for Conservation of Stone Artworks: From Cell Components to Microbial Community. *Front. Microbiol.* **2020**, *11*, 1386. [CrossRef]
28. Phillips, A.J.; Eldring, J.; Hiebert, R.; Lauchnor, E.; Mitchell, A.C.; Cunningham, A.; Spangler, L.; Gerlach, R. Design of a Meso-Scale High Pressure Vessel for the Laboratory Examination of Biogeochemical Subsurface Processes. *J. Pet. Sci. Eng.* **2015**, *126*, 55–62. [CrossRef]
29. Cunningham, A.B.; Class, H.; Ebigbo, A.; Gerlach, R.; Phillips, A.J.; Hommel, J. Field-Scale Modeling of Microbially Induced Calcite Precipitation. *Comput. Geosci.* **2019**, *23*, 399–414. [CrossRef]
30. Ariyanti, D. Feasibility of Using Microalgae for Biocement Production through Biocementation. *J. Bioproces Biotech.* **2012**, *2*, 1000111. [CrossRef]
31. Jansson, C.; Northen, T. Calcifying Cyanobacteria—The Potential of Biomineralization for Carbon Capture and Storage. *Curr. Opin. Biotechnol.* **2010**, *21*, 365–371. [CrossRef]
32. Yu, X.; Chu, J.; Yang, Y.; Qian, C. Reduction of Ammonia Production in the Biocementation Process for Sand Using a New Biocement. *J. Clean. Prod.* **2021**, *286*, 124928. [CrossRef]

33. Porter, H.; Mukherjee, A.; Tuladhar, R.; Dhami, N.K. Life Cycle Assessment of Biocement: An Emerging Sustainable Solution? *Sustainability* **2021**, *13*, 13878. [CrossRef]
34. Wu, C.; Chu, J.; Wu, S.; Hong, Y. 3D Characterization of Microbially Induced Carbonate Precipitation in Rock Fracture and the Resulted Permeability Reduction. *Eng. Geol.* **2019**, *249*, 23–30. [CrossRef]
35. Wang, Y.; Soga, K.; Dejong, J.T.; Kabla, A.J. Microscale Visualization of Microbial-Induced Calcium Carbonate Precipitation Processes. *J. Geotech. Geoenvironmental Eng.* **2019**, *145*, 04019045. [CrossRef]
36. Oktafiani, P.G.; Putra, H.; Erizal; Yanto, D.H.Y. Application of Technical Grade Reagent in Soybean-Crude Urease Calcite Precipitation (SCU-CP) Method for Soil Improvement Technique. *Phys. Chem. Earth Parts A/B/C* **2022**, *128*, 103292. [CrossRef]
37. Omoregie, A.I.; Palombo, E.A.; Ong, D.E.L.; Nissom, P.M. Biocementation of Sand by Sporosarcina pasteurii Strain and Technical-Grade Cementation Reagents through Surface Percolation Treatment Method. *Constr. Build. Mater.* **2019**, *228*, 116828. [CrossRef]
38. Mutitu, K.D.; Munyao, M.O.; Wachira, M.J.; Mwirichia, R.; Thiong'o, K.J.; Marangu, M.J. Effects of Biocementation on Some Properties of Cement-Based Materials Incorporating *Bacillus Species* Bacteria—A Review. *J. Sustain. Cem.-Based Mater.* **2019**, *8*, 309–325. [CrossRef]
39. Montaño-Salazar, S.M.; Lizarazo-Marriaga, J.; Brandão, P.F.B. Isolation and Potential Biocementation of Calcite Precipitation Inducing Bacteria from Colombian Buildings. *Curr. Microbiol.* **2018**, *75*, 256–265. [CrossRef]
40. Perito, B.; Marvasi, M.; Barabesi, C.; Mastromei, G.; Bracci, S.; Vendrell, M.; Tiano, P. A Bacillus Subtilis Cell Fraction (BCF) Inducing Calcium Carbonate Precipitation: Biotechnological Perspectives for Monumental Stone Reinforcement. *J. Cult. Herit.* **2014**, *15*, 345–351. [CrossRef]
41. Shu, S.; Yan, B.; Ge, B.; Li, S.; Meng, H. Factors Affecting Soybean Crude Urease Extraction and Biocementation via Enzyme-Induced Carbonate Precipitation (EICP) for Soil Improvement. *Energies* **2022**, *15*, 5566. [CrossRef]
42. Rohy, H.; Arab, M.; Zeiada, W.; Omar, M.; Almajed, A.; Tahmaz, A. One Phase Soil Bio-Cementation with EICP-Soil Mixing. In Proceedings of the 4th World Congress on Civil, Structural, and Environmental Engineering (CSEE'19), Rome, Italy, 7–9 April 2019.
43. Zehner, J.; Røyne, A.; Sikorski, P. A Sample Cell for the Study of Enzyme-Induced Carbonate Precipitation at the Grain-Scale and Its Implications for Biocementation. *Sci. Rep.* **2021**, *11*, 13675. [CrossRef] [PubMed]
44. Guan, D.; Zhou, Y.; Shahin, M.A.; Khodadadi Tirkolaei, H.; Cheng, L. Assessment of Urease Enzyme Extraction for Superior and Economic Bio-Cementation of Granular Materials Using Enzyme-Induced Carbonate Precipitation. *Acta Geotech.* **2023**, *18*, 2263–2279. [CrossRef]
45. Miftah, A.; Khodadadi Tirkolaei, H.; Bilsel, H. Biocementation of Calcareous Beach Sand Using Enzymatic Calcium Carbonate Precipitation. *Crystals* **2020**, *10*, 888. [CrossRef]
46. Hoang, T.; Alleman, J.; Cetin, B.; Choi, S.-G. Engineering Properties of Biocementation Coarse- and Fine-Grained Sand Catalyzed By Bacterial Cells and Bacterial Enzyme. *J. Mater. Civ. Eng.* **2020**, *32*, 04020030. [CrossRef]
47. Murugan, R.; Suraishkumar, G.K.; Mukherjee, A.; Dhami, N.K. Influence of Native Ureolytic Microbial Community on Biocementation Potential of Sporosarcina pasteurii. *Sci. Rep.* **2021**, *11*, 20856. [CrossRef]
48. Cardoso, R.; Pedreira, R.; Duarte, S.O.D.; Monteiro, G.A. About Calcium Carbonate Precipitation on Sand Biocementation. *Eng. Geol.* **2020**, *271*, 105612. [CrossRef]
49. Mahawish, A.; Bouazza, A.; Gates, W.P. Effect of Particle Size Distribution on the Bio-Cementation of Coarse Aggregates. *Acta Geotech.* **2018**, *13*, 1019–1025. [CrossRef]
50. Cardoso, R.; Pires, I.; Duarte, S.O.D.; Monteiro, G.A. Effects of Clay's Chemical Interactions on Biocementation. *Appl. Clay Sci.* **2018**, *156*, 96–103. [CrossRef]
51. Ashraf, M.S.; Hassan Shah, M.U.; Bokhari, A.; Hasan, M. Less Is More: Optimising the Biocementation of Coastal Sands by Reducing Influent Urea through Response Surface Method. *J. Clean. Prod.* **2021**, *315*, 128208. [CrossRef]
52. Graddy, C.M.R.; Gomez, M.G.; DeJong, J.T.; Nelson, D.C. Native Bacterial Community Convergence in Augmented and Stimulated Ureolytic MICP Biocementation. *Environ. Sci. Technol.* **2021**, *55*, 10784–10793. [CrossRef]
53. Cui, M.-J.; Zheng, J.-J.; Zhang, R.-J.; Lai, H.-J. Soil Bio-Cementation Using an Improved 2-Step Injection Method. *Arab. J. Geosci.* **2020**, *13*, 1270. [CrossRef]
54. Sun, X.; Miao, L.; Wu, L.; Wang, C. Study of Magnesium Precipitation Based on Biocementation. *Mar. Georesources Geotechnol.* **2019**, *37*, 1257–1266. [CrossRef]
55. Mahawish, A.; Bouazza, A.; Gates, W.P. Unconfined Compressive Strength and Visualization of the Microstructure of Coarse Sand Subjected to Different Biocementation Levels. *J. Geotech. Geoenviron. Eng.* **2019**, *145*, 04019033. [CrossRef]
56. Wani, K.M.N.S.; Mir, B.A. A Laboratory-Scale Study on the Bio-Cementation Potential of Distinct River Sediments Infused with Microbes. *Transp. Infrastruct. Geotech.* **2021**, *8*, 162–185. [CrossRef]
57. Tri Huynh, N.N.; Quynh Nhu, N.; Khanh Son, N. Developing the Solution of Microbially Induced $CaCO_3$ Precipitation Coating for Cement Concrete. *IOP Conf. Ser. Mater. Sci. Eng.* **2018**, *431*, 062006. [CrossRef]
58. Rao, M.V.S.; Reddy, V.S.; Sasikala, C. Performance of Microbial Concrete Developed Using Bacillus Subtilus JC3. *J. Inst. Eng. India Ser. A* **2017**, *98*, 501–510. [CrossRef]
59. Nguyen, H.N.T.; Nguyen, S.K. Use of Bio-Active Bacillus Subtilis Bacteria to Form Self-Healing Concrete. *Sci. Tech. Dev. J.* **2014**, *17*, 76–86. [CrossRef]
60. Fang, C.; He, J.; Achal, V.; Plaza, G. Tofu Wastewater as Efficient Nutritional Source in Biocementation for Improved Mechanical Strength of Cement Mortars. *Geomicrobiol. J.* **2019**, *36*, 515–521. [CrossRef]

61. Bhutange, S.P.; Latkar, M.V.; Chakrabarti, T. Studies on Efficacy of Biocementation of Cement Mortar Using Soil Extract. *J. Clean. Prod.* **2020**, *274*, 122687. [CrossRef]
62. Sun, X.; Miao, L.; Wu, L.; Chen, R. Improvement of Bio-Cementation at Low Temperature Based on Bacillus Megaterium. *Appl. Microbiol. Biotechnol.* **2019**, *103*, 7191–7202. [CrossRef]
63. Kaur, G.; Dhami, N.K.; Goyal, S.; Mukherjee, A.; Reddy, M.S. Utilization of Carbon Dioxide as an Alternative to Urea in Biocementation. *Constr. Build. Mater.* **2016**, *123*, 527–533. [CrossRef]
64. Phua, Y.J.; Røyne, A. Bio-Cementation through Controlled Dissolution and Recrystallization of Calcium Carbonate. *Constr. Build. Mater.* **2018**, *167*, 657–668. [CrossRef]
65. Wormser, G.P.; Stratton, C. Manual of Clinical Microbiology, 9th Edition Edited by Patrick R. Murray, Ellen Jo Baron, James H. Jorgensen, Marie Louise Landry, and Michael A. Pfaller Washington, DC: ASM Press, 2007 2488 Pp., Illustrated. $209.95 (Hardcover). *Clin. Infect. Dis.* **2008**, *46*, 153. [CrossRef]
66. Bacterial Endospores | CALS. Available online: https://cals.cornell.edu/microbiology/research/active-research-labs/angert-lab/epulopiscium/bacterial-endospores (accessed on 4 August 2023).
67. Gomez, M.G.; Graddy, C.M.R.; DeJong, J.T.; Nelson, D.C. Biogeochemical Changes During Bio-Cementation Mediated by Stimulated and Augmented Ureolytic Microorganisms. *Sci. Rep.* **2019**, *9*, 11517. [CrossRef] [PubMed]
68. Behzadipour, H.; Sadrekarimi, A. Biochar-Assisted Bio-Cementation of a Sand Using Native Bacteria. *Bull. Eng. Geol. Env.* **2021**, *80*, 4967–4984. [CrossRef]
69. Zhao, Q.; Li, L.; Li, C.; Zhang, H.; Amini, F. A Full Contact Flexible Mold for Preparing Samples Based on Microbial-Induced Calcite Precipitation Technology. *Geotech. Test. J.* **2014**, *37*, 917–921. [CrossRef]
70. Wen, K.; Li, Y.; Liu, S.; Bu, C.; Li, L. Development of an Improved Immersing Method to Enhance Microbial Induced Calcite Precipitation Treated Sandy Soil through Multiple Treatments in Low Cementation Media Concentration. *Geotech. Geol. Eng.* **2019**, *37*, 1015–1027. [CrossRef]
71. Yu, X.; Wang, Z.; Wang, J. Mechanical Properties of Bio-Cementation Materials in Pre-Precipitation Mixing Process. *Env. Sci. Pollut. Res.* **2022**, *29*, 1314–1323. [CrossRef]
72. Nething, C.; Smirnova, M.; Gröning, J.A.D.; Haase, W.; Stolz, A.; Sobek, W. A Method for 3D Printing Bio-Cemented Spatial Structures Using Sand and Urease Active Calcium Carbonate Powder. *Mater. Des.* **2020**, *195*, 109032. [CrossRef]
73. Erdmann, N.; Kästner, F.; de Payrebrune, K.; Strieth, D. Sporosarcina pasteurii Can Be Used to Print a Layer of Calcium Carbonate. *Eng. Life Sci.* **2022**, *22*, 760–768. [CrossRef]
74. Hirsch, M.; Lucherini, L.; Zhao, R.; Clarà Saracho, A.; Amstad, E. 3D Printing of Living Structural Biocomposites. *Mater. Today* **2023**, *62*, 21–32. [CrossRef]
75. Tian, Z.; Tang, X.; Xiu, Z.; Xue, Z. The Spatial Distribution of Microbially Induced Carbonate Precipitation in Sand Column with Different Grouting Strategies. *J. Mater. Civ. Eng.* **2023**, *35*, 04022437. [CrossRef]
76. Van Paassen, L.A.; Ghose, R.; van der Linden, T.J.M.; van der Star, W.R.L.; van Loosdrecht, M.C.M. Quantifying Biomediated Ground Improvement by Ureolysis: Large-Scale Biogrout Experiment. *ASCE J. Soil. Mech. Found. Div.* **2010**, *136*, 1721–1728. [CrossRef]
77. Harkes, M.P.; van Paassen, L.A.; Booster, J.L.; Whiffin, V.S.; van Loosdrecht, M.C.M. Fixation and Distribution of Bacterial Activity in Sand to Induce Carbonate Precipitation for Ground Reinforcement. *Ecol. Eng.* **2010**, *36*, 112–117. [CrossRef]
78. Ginn, T.R.; Murphy, E.M.; Chilakapati, A.; Seeboonruang, U. Stochastic-Convective Transport with Nonlinear Reaction and Mixing: Application to Intermediate-Scale Experiments in Aerobic Biodegradation in Saturated Porous Media. *J. Contam. Hydrol.* **2001**, *48*, 121–149. [CrossRef] [PubMed]
79. Sun, X.; Miao, L.; Chen, R. The Application of Bio-Cementation for Improvement in Collapsibility of Loess. *Int. J. Environ. Sci. Technol.* **2021**, *18*, 2607–2618. [CrossRef]
80. Cheng, L.; Shahin, M.A.; Cord-Ruwisch, R. Surface Percolation for Soil Improvement by Biocementation Utilizing In Situ Enriched Indigenous Aerobic and Anaerobic Ureolytic Soil Microorganisms. *Geomicrobiol. J.* **2017**, *34*, 546–556. [CrossRef]
81. Yang, Y.; Chu, J.; Liu, H.; Cheng, L. Construction of Water Pond Using Bioslurry-Induced Biocementation. *J. Mater. Civ. Eng.* **2022**, *34*, 06021009. [CrossRef]
82. Sun, X.; Miao, L.; Wang, H.; Chen, R.; Wu, L. Bio-Cementation for the Mitigation of Surface Erosion in Loess Slopes Based on Simulation Experiment. *J. Soils Sediments* **2022**, *22*, 1804–1818. [CrossRef]
83. Cheng, L.; Cord-Ruwisch, R. Upscaling Effects of Soil Improvement by Microbially Induced Calcite Precipitation by Surface Percolation. *Geomicrobiol. J.* **2014**, *31*, 396–406. [CrossRef]
84. Lai, H.-J.; Cui, M.-J.; Chu, J. Effect of PH on Soil Improvement Using One-Phase-Low-PH MICP or EICP Biocementation Method. *Acta Geotech.* **2022**, *18*, 3259–3272. [CrossRef]
85. Joshi, S.; Goyal, S.; Reddy, M.S. Influence of Nutrient Components of Media on Structural Properties of Concrete during Biocementation. *Constr. Build. Mater.* **2018**, *158*, 601–613. [CrossRef]
86. Williams, S.L.; Kirisits, M.J.; Ferron, R.D. Optimization of Growth Medium for Sporosarcina pasteurii in Bio-Based Cement Pastes to Mitigate Delay in Hydration Kinetics. *J. Ind. Microbiol. Biotechnol.* **2016**, *43*, 567–575. [CrossRef]
87. Dhami, N.K.; Reddy, M.S.; Mukherjee, A. Significant Indicators for Biomineralisation in Sand of Varying Grain Sizes. *Constr. Build. Mater.* **2016**, *104*, 198–207. [CrossRef]

88. Heveran, C.M.; Liang, L.; Nagarajan, A.; Hubler, M.H.; Gill, R.; Cameron, J.C.; Cook, S.M.; Srubar, W.V. Engineered Ureolytic Microorganisms Can Tailor the Morphology and Nanomechanical Properties of Microbial-Precipitated Calcium Carbonate. *Sci. Rep.* **2019**, *9*, 14721. [CrossRef]
89. Wang, Y.; Wang, Y.; Soga, K.; DeJong, J.T.; Kabla, A.J. Microscale Investigations of Temperature-Dependent Microbially Induced Carbonate Precipitation (MICP) in the Temperature Range 4–50 °C. *Acta Geotech.* **2023**, *18*, 2239–2261. [CrossRef]
90. Peterson, M.E.; Daniel, R.M.; Danson, M.J.; Eisenthal, R. The Dependence of Enzyme Activity on Temperature: Determination and Validation of Parameters. *Biochem. J.* **2007**, *402*, 331–337. [CrossRef] [PubMed]
91. Aytekin, B.; Mardani, A.; Yazıcı, Ş. State-of-Art Review of Bacteria-Based Self-Healing Concrete: Biomineralization Process, Crack Healing, and Mechanical Properties. *Constr. Build. Mater.* **2023**, *378*, 131198. [CrossRef]
92. Joshi, K.A.; Kumthekar, M.B.; Ghodake, V.P. Bacillus Subtilis Bacteria Impregnation in Concrete for Enhancement in Compressive Strength. *Int. Res. J. Eng. Technol.* **2016**, *3*, 1229–1234.
93. De Leeuw, N.H.; Parker, S.C. Surface Structure and Morphology of Calcium Carbonate Polymorphs Calcite, Aragonite, and Vaterite: An Atomistic Approach. *J. Phys. Chem. B* **1998**, *102*, 2914–2922. [CrossRef]
94. Liu, L.; Liu, H.; Xiao, Y.; Chu, J.; Xiao, P.; Wang, Y. Biocementation of Calcareous Sand Using Soluble Calcium Derived from Calcareous Sand. *Bull. Eng. Geol. Env.* **2018**, *77*, 1781–1791. [CrossRef]
95. Choi, S.-G.; Park, S.-S.; Wu, S.; Chu, J. Methods for Calcium Carbonate Content Measurement of Biocemented Soils. *J. Mater. Civ. Eng.* **2017**, *29*, 06017015. [CrossRef]
96. *ASTM D2166/D2166M-16*; Standard Test Method for Unconfined Compressive Strength of Cohesive Soil. ASTM: West Conshohocken, PA, USA, 2016.
97. *ASTM D1883-21*; Standard Test Method for California Bearing Ratio (CBR) of Laboratory-Compacted Soils. ASTM: West Conshohocken, PA, USA, 2021.
98. *ASTM D3441-16*; Standard Test Method for Mechanical Cone Penetration Testing of Soils. ASTM: West Conshohocken, PA, USA, 2017.
99. *ASTM D2434-22*; Standard Test Methods for Measurement of Hydraulic Conductivity of Coarse-Grained Soils. ASTM: West Conshohocken, PA, USA, 2022.
100. *ASTM D2435/D2435M-11(2020)*; Standard Test Methods for One-Dimensional Consolidation Properties of Soils Using Incremental Loading. ASTM: West Conshohocken, PA, USA, 2020.
101. *ASTM C1585-20*; Standard Test Method for Measurement of Rate of Absorption of Water by Hydraulic-Cement Concretes. ASTM: West Conshohocken, PA, USA, 2020.
102. *ASTM D5856-15*; Standard Test Method for Measurement of Hydraulic Conductivity of Porous Material Using a Rigid-Wall, Compaction-Mold Permeameter. ASTM: West Conshohocken, PA, USA, 2016.
103. Xu, X.; Guo, H.; Li, M.; Deng, X. Bio-Cementation Improvement via CaCO3 Cementation Pattern and Crystal Polymorph: A Review. *Constr. Build. Mater.* **2021**, *297*, 123478. [CrossRef]
104. Cheng, L.; Cord-Ruwisch, R. In Situ Soil Cementation with Ureolytic Bacteria by Surface Percolation. *Ecol. Eng.* **2012**, *42*, 64–72. [CrossRef]
105. Terzis, D.; Laloui, L. Cell-Free Soil Bio-Cementation with Strength, Dilatancy and Fabric Characterization. *Acta Geotech.* **2019**, *14*, 639–656. [CrossRef]
106. Elmaloglou, A.; Terzis, D.; De Anna, P.; Laloui, L. Microfluidic Study in a Meter-Long Reactive Path Reveals How the Medium's Structural Heterogeneity Shapes MICP-Induced Biocementation. *Sci. Rep.* **2022**, *12*, 19553. [CrossRef]
107. Maleki-Kakelar, M.; Azarhoosh, M.J.; Golmohammadi Senji, S.; Aghaeinejad-Meybodi, A. Urease Production Using Corn Steep Liquor as a Low-Cost Nutrient Source by Sporosarcina pasteurii: Biocementation and Process Optimization via Artificial Intelligence Approaches. *Env. Sci. Pollut. Res.* **2022**, *29*, 13767–13781. [CrossRef]
108. Mortazavi Bak, H.; Kariminia, T.; Shahbodagh, B.; Rowshanzamir, M.A.; Khoshghalb, A. Application of Bio-Cementation to Enhance Shear Strength Parameters of Soil-Steel Interface. *Constr. Build. Mater.* **2021**, *294*, 123470. [CrossRef]
109. Yu, X.; Pan, X. Seawater Based Bio-Cementation for Calcareous Sand Improvement in Marine Environment. *Mar. Georesources Geotechnol.* **2023**, *41*, 1–10. [CrossRef]
110. Bhaduri, S.; Debnath, N.; Mitra, S.; Liu, Y.; Kumar, A. Microbiologically Induced Calcite Precipitation Mediated by Sporosarcina pasteurii. *J. Vis. Exp.* **2016**, *110*, e53253. [CrossRef]
111. Henze, J.; Randall, D.G. Microbial Induced Calcium Carbonate Precipitation at Elevated PH Values (>11) Using Sporosarcina pasteurii. *J. Environ. Chem. Eng.* **2018**, *6*, 5008–5013. [CrossRef]
112. Optimizing the Use of Sporosarcina pasteurii Bacteria for the Stiffening of Sand. Available online: http://www.envirobiotechjournals.com/article_abstract.php?aid=6965&iid=212&jid=1 (accessed on 4 August 2023).
113. Onal Okyay, T.; Frigi Rodrigues, D. Optimized Carbonate Micro-Particle Production by Sporosarcina pasteurii Using Response Surface Methodology. *Ecol. Eng.* **2014**, *62*, 168–174. [CrossRef]
114. Williams, S.L.; Kirisits, M.J.; Ferron, R.D. Influence of Concrete-Related Environmental Stressors on Biomineralizing Bacteria Used in Self-Healing Concrete. *Constr. Build. Mater.* **2017**, *139*, 611–618. [CrossRef]
115. Dosier, G.K. Production of Masonry with Bacteria. U.S. Patent 9796626B2, 24 October 2017.
116. Biolith®Tile from Biomason | Build with Biolith. Available online: https://biomason.com/biolith (accessed on 30 March 2023).
117. Lin, H.; Suleiman, M.T.; Brown, D.G.; Kavazanjian, E. Mechanical Behavior of Sands Treated by Microbially Induced Carbonate Precipitation. *J. Geotech. Geoenvir. Eng.* **2016**, *142*, 04015066. [CrossRef]

118. Filet, A.E.; Gadret, J.-P.; Loygue, M.; Borel, S. Biocalcis and Its Applications for the Consolidation of Sands. *Grouting Deep. Mix.* **2012**, *2012*, 1767–1780. [CrossRef]
119. Gomez, M.G.; Martinez, B.C.; DeJong, J.T.; Hunt, C.E.; deVlaming, L.A.; Major, D.W.; Dworatzek, S.M. Field-Scale Bio-Cementation Tests to Improve Sands. *Proc. Inst. Civ. Eng.—Ground Improv.* **2015**, *168*, 206–216. [CrossRef]
120. Phillips, A.J.; Troyer, E.; Hiebert, R.; Kirkland, C.; Gerlach, R.; Cunningham, A.B.; Spangler, L.; Kirksey, J.; Rowe, W.; Esposito, R. Enhancing Wellbore Cement Integrity with Microbially Induced Calcite Precipitation (MICP): A Field Scale Demonstration. *J. Pet. Sci. Eng.* **2018**, *171*, 1141–1148. [CrossRef]
121. Van Paassen, L.A. Biogrout, Ground Improvement by Microbial Induced Carbonate Precipitation. Ph.D. Thesis, Delft University of Technology (TU Delft), Delft, The Netherlands, 2009.
122. Graddy, C.M.R.; Gomez, M.G.; Kline, L.M.; Morrill, S.R.; DeJong, J.T.; Nelson, D.C. Diversity of *Sporosarcina*-like Bacterial Strains Obtained from Meter-Scale Augmented and Stimulated Biocementation Experiments. *Environ. Sci. Technol.* **2018**, *52*, 3997–4005. [CrossRef]
123. Lee, M.; Gomez, M.G.; El Kortbawi, M.; Ziotopoulou, K. Effect of Light Biocementation on the Liquefaction Triggering and Post-Triggering Behavior of Loose Sands. *J. Geotech. Geoenviron. Eng.* **2022**, *148*, 04021170. [CrossRef]
124. Gomez, M.G.; Anderson, C.M.; Graddy, C.M.R.; DeJong, J.T.; Nelson, D.C.; Ginn, T.R. Large-Scale Comparison of Bioaugmentation and Biostimulation Approaches for Biocementation of Sands. *J. Geotech. Geoenviron. Eng.* **2017**, *143*, 04016124. [CrossRef]
125. Hang, L.; Gao, Y.; van Paassen, L.A.; He, J.; Wang, L.; Li, C. Microbially Induced Carbonate Precipitation for Improving the Internal Stability of Silty Sand Slopes under Seepage Conditions. *Acta Geotech.* **2022**, *18*, 2719–2732. [CrossRef]
126. Lai, H.-J.; Cui, M.-J.; Wu, S.-F.; Yang, Y.; Chu, J. Retarding Effect of Concentration of Cementation Solution on Biocementation of Soil. *Acta Geotech.* **2021**, *16*, 1457–1472. [CrossRef]
127. Sohail, M.G.; Disi, Z.A.; Zouari, N.; Nuaimi, N.A.; Kahraman, R.; Gencturk, B.; Rodrigues, D.F.; Yildirim, Y. Bio Self-Healing Concrete Using MICP by an Indigenous Bacillus Cereus Strain Isolated from Qatari Soil. *Constr. Build. Mater.* **2022**, *328*, 126943. [CrossRef]
128. Achal, V.; Mukherjee, A.; Basu, P.C.; Reddy, M.S. Lactose Mother Liquor as an Alternative Nutrient Source for Microbial Concrete Production by Sporosarcina pasteurii. *J. Ind. Microbiol. Biotechnol.* **2009**, *36*, 433–438. [CrossRef]
129. Chidara, R.; Nagulagama, R.; Yadav, S. Achievement of Early Compressive Strength in Concrete Using *Sporosarcina pasteurii* Bacteria as an Admixture. *Adv. Civ. Eng.* **2014**, *2014*, e435948. [CrossRef]
130. Mirshahmohammad, M.; Rahmani, H.; Maleki-Kakelar, M.; Bahari, A. Performance of Biological Methods on Self-Healing and Mechanical Properties of Concrete Using S. pasteurii. *Env. Sci. Pollut. Res.* **2023**, *30*, 2128–2144. [CrossRef] [PubMed]
131. Kim, H.K.; Park, S.J.; Han, J.I.; Lee, H.K. Microbially Mediated Calcium Carbonate Precipitation on Normal and Lightweight Concrete. *Constr. Build. Mater.* **2013**, *38*, 1073–1082. [CrossRef]
132. Chen, H.-J.; Peng, C.-F.; Tang, C.-W.; Chen, Y.-T. Self-Healing Concrete by Biological Substrate. *Materials* **2019**, *12*, 4099. [CrossRef]
133. Chahal, N.; Siddique, R. Permeation Properties of Concrete Made with Fly Ash and Silica Fume: Influence of Ureolytic Bacteria. *Constr. Build. Mater.* **2013**, *49*, 161–174. [CrossRef]
134. Chahal, N.; Siddique, R.; Rajor, A. Influence of Bacteria on the Compressive Strength, Water Absorption and Rapid Chloride Permeability of Concrete Incorporating Silica Fume. *Constr. Build. Mater.* **2012**, *37*, 645–651. [CrossRef]
135. Zaerkabeh, R.; Sadeghi, A.M.; Afshin, H.; Majdani, R. Crack Healing and Mechanical Properties of Bacteria-Based Self-Healing Cement Mortar. *Period. Polytech. Civ. Eng.* **2022**, *66*, 581–592. [CrossRef]
136. Abo-El-Enein, S.A.; Ali, A.H.; Talkhan, F.N.; Abdel-Gawwad, H.A. Application of Microbial Biocementation to Improve the Physico-Mechanical Properties of Cement Mortar. *HBRC J.* **2013**, *9*, 36–40. [CrossRef]
137. Metwally, G.A.M.; Mahdy, M.; Abd El-Raheem, A.E.-R.H. Performance of Bio Concrete by Using Bacillus Pasteurii Bacteria. *Civ. Eng. J.* **2020**, *6*, 1443–1456. [CrossRef]
138. Nasser, A.A.; Sorour, N.M.; Saafan, M.A.; Abbas, R.N. Microbially-Induced-Calcite-Precipitation (MICP): A Biotechnological Approach to Enhance the Durability of Concrete Using Bacillus Pasteurii and Bacillus Sphaericus. *Heliyon* **2022**, *8*, e09879. [CrossRef] [PubMed]
139. Xu, H.; Lian, J.; Gao, M.; Fu, D.; Yan, Y. Self-Healing Concrete Using Rubber Particles to Immobilize Bacterial Spores. *Materials* **2019**, *12*, 2313. [CrossRef]
140. Qin, C.-Z.; Hassanizadeh, S.M.; Ebigbo, A. Pore-Scale Network Modeling of Microbially Induced Calcium Carbonate Precipitation: Insight into Scale Dependence of Biogeochemical Reaction Rates. *Water Resour. Res.* **2016**, *52*, 8794–8810. [CrossRef]

sustainability

Article

Use of Waste Tires as Transverse Reinforcement and External Confinement in Concrete Columns Subjected to Axial Loads

Mahesh Mahat [1], Mahesh Acharya [2,*], Manish Acharya [2] and Mustafa Mashal [2]

[1] Blue Planet Systems, Los Gatos, CA 95032, USA; maheshmahat@isu.edu
[2] Department of Civil & Environmental Engineering, Idaho State University, Pocatello, ID 83209, USA; manishacharya@isu.edu (M.A.); mashmust@isu.edu (M.M.)
* Correspondence: achamahe@isu.edu; Tel.: +1-(980)-339-0764

Abstract: Approximately 246 million waste tires are generated annually in the United States. That is roughly three tires per four individuals in the country. Most tires end up in landfills, adversely affecting the environment. In the last two decades, researchers have explored using tire chips in concrete to replace a portion of coarse aggregates. Past studies have indicated that up to 50% of coarse aggregates in concrete can be replaced with tire chips. This research proposes using recycled rubber tires and rubber chips in concrete columns. The tires are used as external transverse reinforcement in plain concrete columns. The tires function as formwork during the pour while providing confinement after curing. The concepts in this research can be used for retrofitting structures with inadequate foundations and constructing new structures. After analyzing the data from this research, the axial compressive test of confined columns was 50% greater on average than unconfined columns. The confinement effectiveness ratio for all confined specimens was greater than one.

Keywords: rubberized concrete; axial strength; confinement; rubber aggregate; energy dissipation

check for updates

Citation: Mahat, M.; Acharya, M.; Acharya, M.; Mashal, M. Use of Waste Tires as Transverse Reinforcement and External Confinement in Concrete Columns Subjected to Axial Loads. *Sustainability* **2023**, *15*, 11620. https://doi.org/10.3390/su151511620

Academic Editors: Mahdi Kioumarsi and Vagelis Plevris

Received: 1 June 2023
Revised: 30 June 2023
Accepted: 8 July 2023
Published: 27 July 2023

1. Introduction

Concrete is one of the world's oldest building materials [1]. The design community constantly offers new ideas and methods to find alternative materials to make concrete inexpensive or more sustainable than it already is by reducing its carbon footprint. New additives for concrete mixtures are also being researched and produced regularly. Cementitious materials such as fly ash, slag, silica fumes, and waste materials such as crumbled rubber are used as filler for cement or as a replacement for aggregate in concrete. In this research, waste tires have been recycled and used as reinforcement for the concrete column. Approximately 246 million waste tires are generated annually in the United States [2,3]. That is roughly three tires per four individuals in the country. Most tires end up in landfills, adversely affecting the environment. Using scrap tire rubber in making concrete has been proposed as an environmentally friendly alternative to disposing of such waste [4,5].

Studies have demonstrated that rubber tires contain elements that do not disintegrate, posing a threat to the environment [6]. The hazardous chemical produced by scrap tires at junkyards and landfills impact the environment [7]. Obtaining the most efficient methods of recycling old tires helps to conserve the environment [8]. Energy recovery is a widespread practice for managing waste tires. Entire or partial tires are an alternative fuel in power plants, paper mills, and cement kilns. According to the most recent statistical report on the management of end-of-life tires released by ETRMA, approximately 91% of waste tires in the nations of the European Union, as well as Norway, Serbia, Switzerland, and Turkey, were properly collected and managed in 2018 through material recycling (56.4%) and energy recovery (34.9%) [9]. As a result, studies have investigated techniques to recycle scrap tires that are both ecologically benign and cost-effective [3,4]. Because tire production is continually expanding, appropriate recycling of old tires is one of the most pressing concerns in the twenty-first century. Adequate tire management has necessitated the adoption of innovative technologies that improve recycling [8,10]. Improving recycling methods would aid in resolving the environmental issue posed by automobile tires.

1.1. Literature Review

Researchers have investigated the possibility of using rubber in various civil engineering projects for decades. Dabri et al. conducted an analytical investigation on the seismic behavior of reinforced concrete piers using elastomeric materials, while Jia et al. examined the seismic evaluation of precast bridge columns incorporating built-in elastomeric pads [11,12]. Additionally, the use of tire rubber in applications such as cement mixtures, road construction, and geotechnical work has been demonstrated to be effective in environmental protection and natural resource conservation, making it a suitable disposal method considering both environmental and economic factors [13,14]. Incorporating recycled tire aggregate in concrete enhances deformability, energy dissipation, and damping. However, an increased concentration of rubber in the concrete leads to reduced compressive and tensile strengths and a diminished Young's modulus of elasticity [15].

Consequently, incorporating rubber aggregates as partial replacements for fine or coarse aggregates primarily compromises the compressive strength of the mixture [2,3,5,6,15–19]. Studies have shown significant reductions in compressive strength, up to 60%, when replacing 50% of the coarse aggregate with rubber crumbs [20]. A study involving 70 concrete samples demonstrated that increasing rubber content decreased strength and modulus. Nevertheless, the addition of silica fume improved properties and mitigated strength loss. The research suggested that up to 25% rubber contents could produce rubberized concretes with compressive strengths ranging from 16 to 32 MPa [21].

This research investigates the utilization of 15% rubber aggregate by volume as a replacement for coarse aggregate and incorporating waste tires as transverse reinforcement in concrete columns. Columns play a critical role in ensuring the structural integrity of buildings by bearing significant loads and transmitting them to the foundation. Both strength and ductility are equally essential considerations in column design [22]. To enhance the characteristics of concrete structural elements, concrete confinement techniques have been employed in beams, slabs, columns, and walls to strengthen existing structures, adapt them, or construct new ones [23]. It has been demonstrated that confinement significantly improves concrete performance [24].

Researchers have explored various methods of confinement in concrete structures. Abdullah et al. conducted a study using four normal reinforced concrete (R.C.) beams, demonstrating that the application of post-tensioned metal straps (PTMS) around the beams' tensile zone increased their load-carrying capacity by 36% to 39% [25]. Garcia et al. proposed an innovative dual retrofitting system that combined Carbon Fiber Reinforced Polymers (CFRP) with PTMS applied to opposing frames of a building [26]. This retrofitting approach enabled the structure to withstand peak ground acceleration (PGA) up to four times higher than the original building [26]. Imjai et al. conducted a comprehensive analysis of multiple studies, concluding that PTMS effectively enhances the capacity and ductility of reinforced concrete elements under compression, shear, flexure, and bond-splitting-dominated conditions [27]. Ma et al. conducted tests on 24 cylinders under axial compression, revealing that the implementation of post-tensioned steel straps increased the strength of pre-damaged specimens by 313% and the axial deformability of pre-damaged cylinders by 2.64–2.95% [28]. Similarly, Imjai et al. performed an analytical and experimental study on four concrete beams to investigate the capacity enhancement of pre-cracked Side-Near Surface Mounted (SNSM) strengthened beams, resulting in a capacity increase of up to 55%. In contrast, beams strengthened with PTMS exhibited a capacity increase of only 8%, indicating that the former benefited from additional flexural reinforcement provided by CFRP bars [29].

The choice of confinement material significantly influences confined concrete's load paths and deformation behavior. Different materials, such as steel hoops or stirrups,

FRP, and GFRP, exhibit diverse load-deformation properties, leading to variations in concrete performance [30]. As a result, the selection of the confinement material has a substantial impact on the behavior of concrete. Mander et al. developed a unified stress-strain approach (Figure 1) specifically designed for circular transverse reinforcement in confined concrete [31].

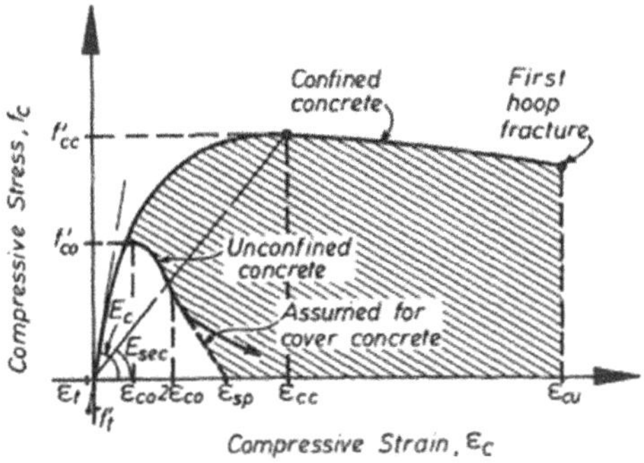

Figure 1. Stress-Strain Model Proposed for Monotonic Loading of Confined and Unconfined Concrete [31].

The confining material creates tensile hoop stress balanced by a uniform radial pressure, which responds to the concrete's lateral expansion as the axial stress grows, increasing the corresponding lateral strain [32,33]. Figure 2 shows the hoop tension developed by transverse steel reinforcements at yield and lateral stress on the concrete core. Equation (1) provides the expression for the confining pressure (f_l) at force equilibrium.

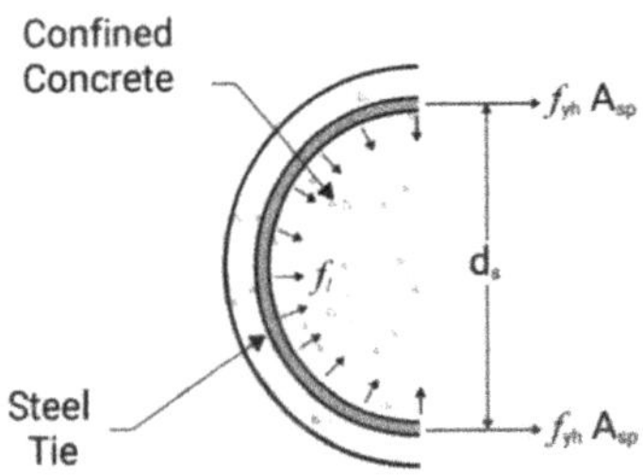

Figure 2. Stress State of Laterally Confined Concrete [34].

$$f_l = \left(2f_{yh}A_{sp}\right) / \left(d_s S_{sp}\right) \quad (1)$$

where A_{sp} is the cross-sectional area of the spirals, mm^2

f_{yh} is the yield stress of the confining spirals, MPa
S_{sp} is the pitch of the confining spirals, mm
d_s is the diameter of spirals between bar centers, mm.

Concrete expands laterally when an enclosed column is subjected to axial compression, but confining media constraints such expansion. Confinement effectiveness is a metric used to assess how successfully a material confines concrete, that is $f'_{cc}/'_{co}$, where f'_{cc} is the compressive strength of confined concrete and f'_{co} is the compressive strength of unconfined concrete.

The composition of tires can be seen in Figure 3, and tires are made using different materials in layers, as shown in Figure 4.

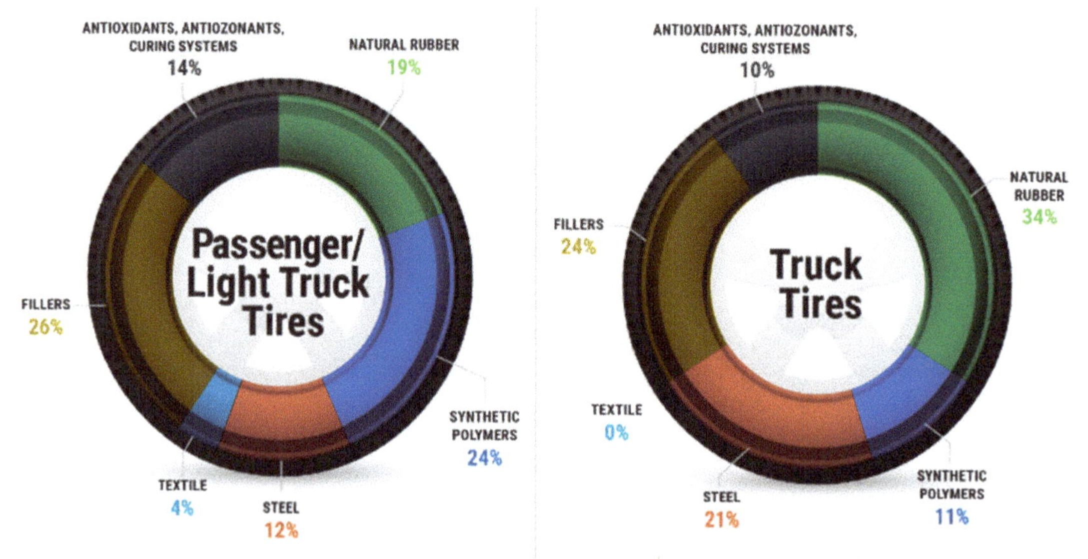

Figure 3. Reprinted with permission from ref. [35]. Copyright 2020 U.S. Tire Manufacturers Association.

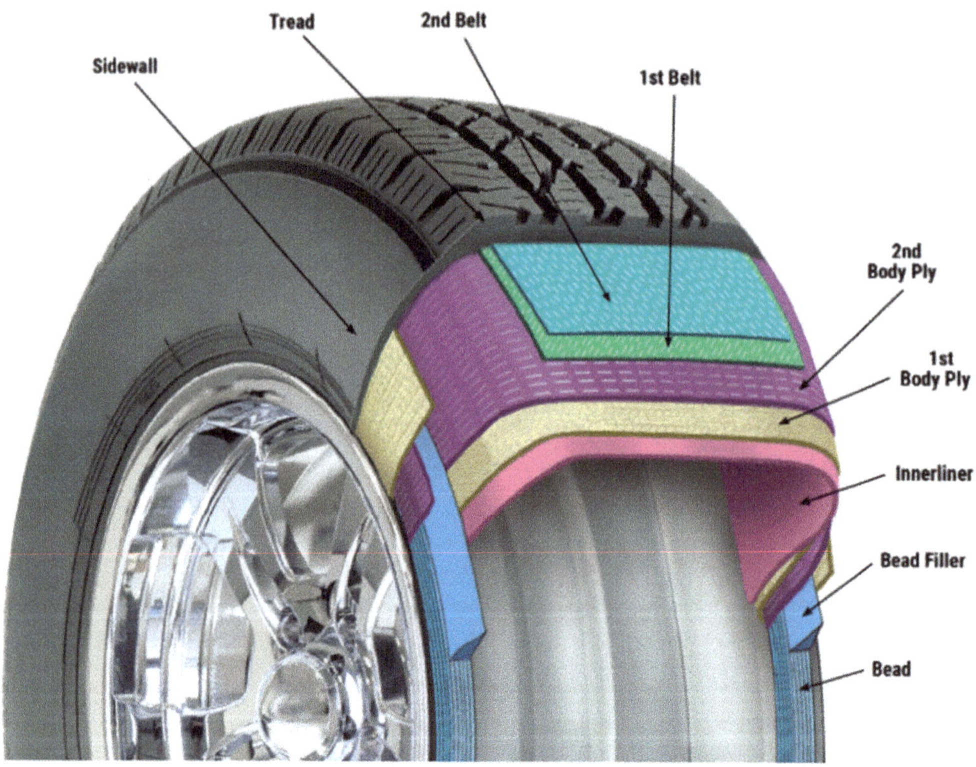

Figure 4. Reprinted with permission from ref. [35]. Copyright 2020 U.S. Tire Manufacturers Association.

Steel wire is used in tire belts, beads, and truck tire plies. The belts beneath the tread stiffen the tire casing, improving wear performance and handling. Polyester cord fabrics, rayon cord fabrics, nylon cord fabrics, and aramid cord fabrics are textiles used as reinforcing materials in tires. This steel wire and textiles in tires make them desirable for reinforcements. Tires can be used as external reinforcements around concrete columns and act as formworks for the columns. The reinforcement has been assumed to be similar to that created using circular hoops or stirrups. This assumption is made as an approximation to find out the confinement provided by tires. The axial compressive strength and confinement by tires were investigated in this study.

1.2. Problem Statement

The increasing production of vehicles has led to a rise in worn-out tires, posing significant risks to the environment and public health [8]. Improper recycling practices result in the loss of the economic value of tires and contribute to wasteful land usage. Moreover, disposing of waste tires often incurs additional charges. Therefore, it is crucial to understand the necessary steps for improving standards. As the demand for concrete reinforcement grows in the construction industry, tires can offer a valuable solution. By incorporating waste rubber scraps into concrete, both spaces can be saved, and the environment can be improved. Recycling vehicle tires through concrete reinforcement benefits the building sector financially, as materials that would have otherwise been discarded in landfills or junkyards are utilized. Utilizing scrap tire chunks or whole tires in construction promotes tire recycling. Rubber is more reusable than other building materials, creating a cleaner and more sustainable environment.

In this study, the authors propose an innovative approach of using recycled tires as external confinement to enhance concrete strength and deformation capacity under axial loads. The tires serve a similar function as stirrups or ties in concrete columns. To validate this concept, full-scale concrete pedestals are subjected to axial tests. These columns are constructed using plain rubber concrete with and without external tires. The results obtained from the confined columns are compared to those of unconfined (benchmark) specimens.

2. Experimental Works

2.1. Material and Specimens Specification

Two methods of utilizing waste tires have been used in this research. Method 1 involves utilizing waste tires, which have been shredded into small chunks and are used as a replacement for coarse aggregate. Method 2 reuses the whole tire with transverse reinforcement to provide confinement to the concrete columns.

It is important to note that specimens were not reinforced with steel bars. It is so because we are not investigating the effects of rebar in the concrete. The scope of the research is to see if tires can provide confinement without reinforcing. Adding reinforcing would bring other parameters into play (e.g., reinforcing ratio, volumetric transverse reinforcing, etc.). This would make the comparison more complicated in differentiating how much confinement each (tire and stirrups) can provide. Since the research is just looking at the compressive strength, smaller specimens would be sufficient; however, it is very difficult to obtain the smaller tires.

Twelve specimens of concrete columns were made; two with regular concrete, two with rubber concrete, four with regular concrete confined with tires, and four with rubber concrete confined with tires. The dimensions of the eight unconfined columns were 0.406 m (16 inches) in diameter and 0.914 m (36 inches) in height, and those for the confined columns were 0.762 ± 0.02 m (30 ± 0.75 inches) in diameter and 0.914 ± 0.006 m (36 ± 0.25 inches) in height. The tires utilized in this research were exclusively 16-inch rim diameter tires, readily accessible in tire shops near Idaho State University. The variation in diameter observed among the confined columns can be attributed to the utilization of different tires, each with its distinct level of wear and tear. The height measurement of each tire employed was 0.229 ± 0.013 m (9.0 ± 0.5 inches), as documented in Figure 5, which presents the tire dimensions. It is important to note that unconfined columns are not fabricated with the same diameter as the confined columns. If the amount of concrete in the unconfined columns is the same as in confined columns, then the comparison would not be easier. Also, the scrap tires have different diameters, so some were cut on one side along the height to have that consistent diameter for better comparison. Most tires used were classified as all-season tires, while a smaller portion consisted of winter and summer tires. Most tires featured steel cord belts with nylon ply as their casing material, while the sidewalls were reinforced with steel wire beads. Additionally, it is worth noting that the thread pattern varied among different tire brands.

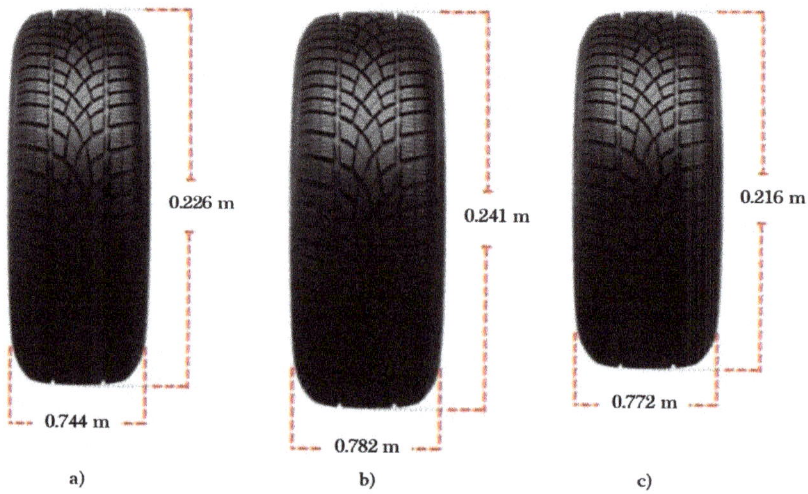

Figure 5. Dimensions of Tires Used: (**a**) 225/75R16, (**b**) 235/80R16, (**c**) 215/85R16.

For rubber concrete, 15% of the volume of course aggregate is replaced by rubber aggregate. To have a similar strength, the quantity of cement was increased to 520.6 kg./cu. m and fly ash was decreased to 130 kg./cu. m. The rubber aggregate used was 19 mm or smaller. The mix design for the normal and rubberized concrete is given in Table 1.

Table 1. Mix Design for Normal Concrete and Rubberized Concrete.

Normal Concrete		Rubberized Concrete	
Material	**Weight (kg/cu.m)**	**Material**	**Weight (kg/cu.m)**
Cement	352.4	Cement	520.6
Fly Ash	296.9	Fly Ash	130.4
Fine Aggregate	1210.9	Fine Aggregate	1210.9
Coarse Aggregate	576.7	Coarse Aggregate	490.2
Water	267.5	Rubber Aggregate	29.0
		Water	267.5

2.2. Construction

The unconfined columns (UC) were constructed using sonotubes, with wooden formwork employed to support them, as illustrated in Figure 6. To prevent any leakage during the pouring process, the lower ends of the sonotubes were caulked. Four tires were stacked together for the tire columns (TC), and three threaded rods were inserted into each tire to prevent collapse, as depicted in Figure 7. The top of each tire was drilled to avoid the entrapment of air bubbles. Initially, the tires were joined together using four three-inch screws, and the connection between the tires was sealed with epoxy to prevent any concrete leakage during pouring. The complete set of test specimens was prepared and poured exclusively within the research laboratory of ISU. The concrete mixer employed for this purpose was a 12-cubic-ft drum mixer. During the pouring process, the specimens underwent three layers of vibration. Once the columns were poured, steps were taken to level and smooth the top surface using a smooth trough. To ensure optimal curing conditions, the specimens were carefully wrapped with wet burlap and securely covered with a plastic tarp, preventing any loss of moisture. In accordance with the guidelines specified in ASTM C31-19 [36], three standard compression cylinders were randomly prepared for each test specimen. These cylinders were then placed in a concrete bath tank and underwent the curing process at laboratory temperature. After a curing period of 28 days, the compressive strength of the sample cylinders was evaluated through testing. The average compressive strength of each configuration specimen can be found in Table 2.

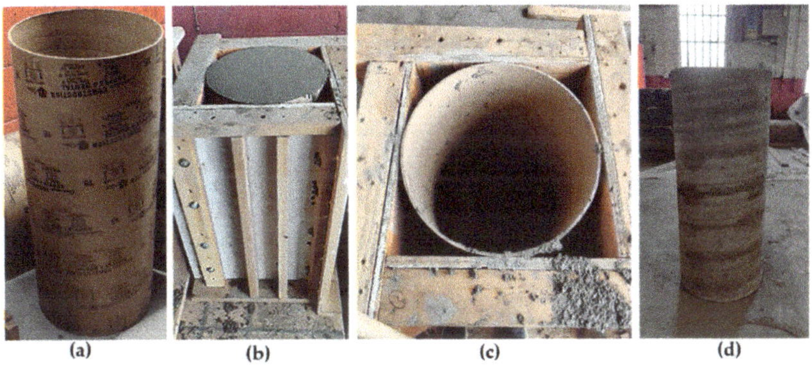

Figure 6. Preparation of UC Specimen: (**a**) sonotube, (**b**) specimen after pour, (c) sonatube in formwork, (**d**) cured specimen.

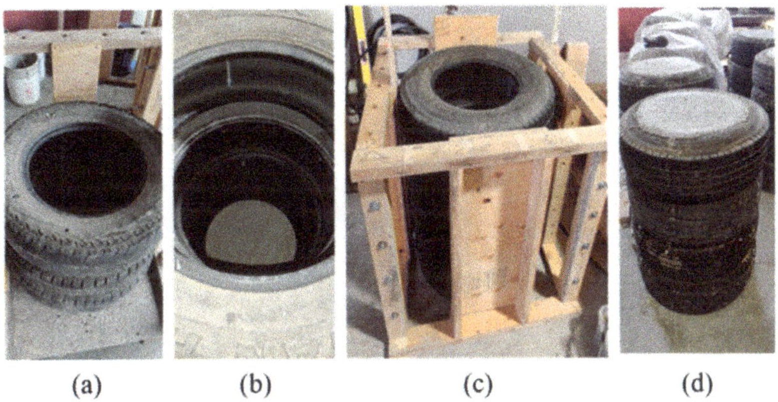

Figure 7. Preparation of TC Specimen: (**a**) tires stacked side view, (**b**) tires top view, (**c**) tires in formwork for pour, (**d**) cured specimen.

Table 2. Compressive strength of the sample cylinder.

Specimen	Average Compressive Strength (MPa)
Unconfined Normal Concrete	35.2
Confined Normal Concrete	34.8
Confined Rubberized Concrete	35.1
Unconfined Rubberized Concrete	34.9

2.3. Instrumentation and Test Setup

The Campbell Scientific data acquisition system (DAQ) was used to simultaneously monitor and record the loads and the axial deflections during the tests. The data collection rate for the test was set at two data points per second. The loads were recorded using a 2500-kips capacity load cell attached to the top of the isolation frame. The axial deflections of the specimen were measured using two 50-inch stroke string potentiometers placed at the centerline of the loading plates and attached at the edge. Four-string potentiometers were mounted on top of the isolation frame to measure the deflection of the top beam during the test. Eight potentiometers were used to measure the deflection of the vertical columns.

The monotonic compressive testing follows ASTM C39-21 [37], with a minor difference in that the height-to-diameter ratio is slightly greater than 2.2. That is due to the recyclable tires available affecting the height. The testing was performed using an isolation frame setup shown in Figure 8. The hydraulic jack can reach up to 8900 kN force with a maximum stroke of 4 inches. The load from the hydraulic jack was then transferred to a 16-inch diameter steel plate to distribute the axial load uniformly across the ends of the columns. The loading plates were used to have similar loading for all specimens. The loading apparatus was the same for tire columns. The tire columns had a diameter of around 30 inches, but the loading area was only 16 inches.

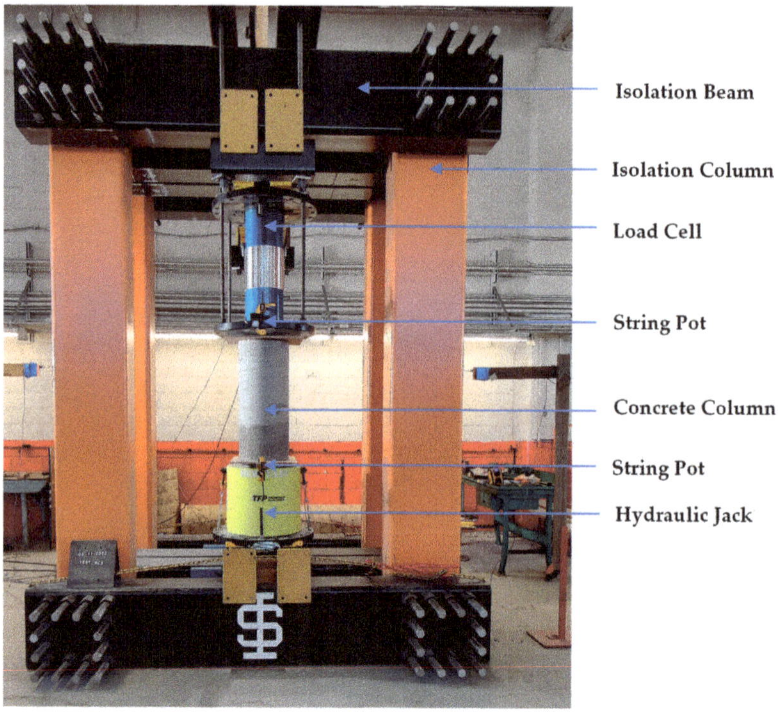

Figure 8. Test setup for the compression testing of column specimens.

A monotonic quasi-static loading protocol in accordance with ASTM C39-21 [37] was used to load the specimen for this test. The loading for the compression tests was applied at an increment of around 1500 pounds per second and continued until failure. For the unconfined columns, the tests were continued until the failure of the specimen. For tire columns, the loading continued until either the tires ruptured or there was a significant drop in axial load. The axial force and displacement were recorded for each specimen from the load cell and string potentiometers.

2.4. Experimental Results

For the test, six different configurations were used:

(a) Two unconfined columns with normal concrete (UCNCs),
(b) Two unconfined columns with rubberized concrete (UCRCs),
(c) Two confined columns with normal concrete (CCNCs),
(d) Two confined columns with rubberized concrete (CCRCs),
(e) Two confined columns with normal concrete with tires cut along the height (CCNCHs),
(f) Two confined columns with rubberized concrete with tires cut along the height (CCRCHs).

The results from each configuration have been compared with results from other configurations. The main parameters compared were the peak force, peak displacement, peak stress, peak strain, and energy dissipation. A force-controlled testing method was used in this study. The inconsistency in the graph is due to the force-controlled testing method. The data from the table for peak force and displacement are more relevant in this type of test. Other plots and comparisons, such as energy dissipated before failure, may change depending on the type of test, such as force based vs. displacement-based or monotonic vs. cyclic test. Figure 9 illustrates the axial force versus displacement plot for the tested specimens.

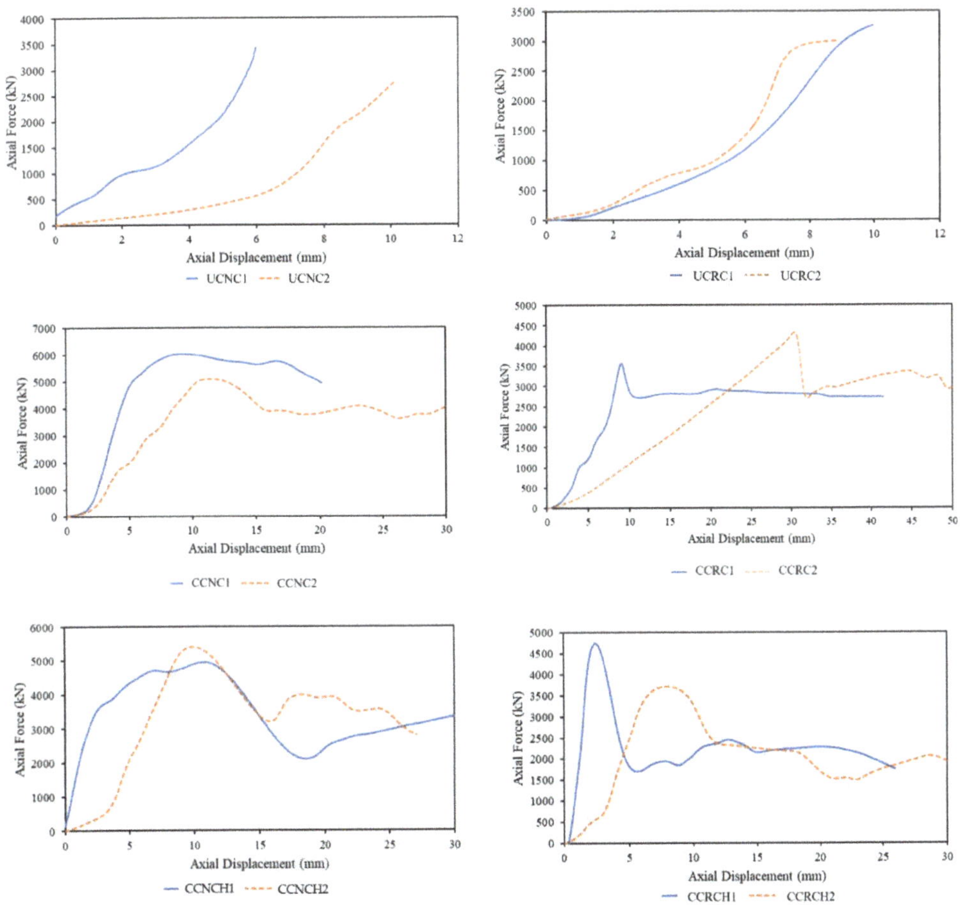

Figure 9. Axial force and displacement relationship.

The 28-day compressive strength of the unconfined normal concrete cylinder was 35.2 MPa, surpassing the strength of the UCNC1 and UCNC2 specimens. This suggests that the full-scale columns experienced non-uniform loading due to using steel plates instead of rubber or wood caps. Similarly, the UCRC1 and UCRC2 specimens displayed a lower cylinder compressive strength at 34.8 MPa, indicating non-uniform loading. Furthermore, all of these specimens exhibited a brittle failure pattern.

On the other hand, the CCNC, CCRC, CCNCH, and CCRCH columns were effectively confined using four tires with a diameter of 0.762 m and a limited loading surface area of 0.406 m. The 28-day compressive strengths of the concrete cylinder samples were measured at 35.1 kN for CCNC, 34.9 kN for CCRC, 35.1 kN for CCNCH, and 34.9 MPa for CCRCH. The increase in axial compressive strength suggests that the tires provided a level of confinement for the columns.

The compressive strength of CCNC1 was 32% higher than the sample, while CCNC2 showed a 12% increase. The columns experienced maximum force and subsequent decline, indicating concrete failure. Tire reinforcement became effective after concrete failure, with significant drops in axial force. In CCNC1, tire reinforcement started after a 7% force reduction, followed by a 2% increase and tire spitting. CCNC2 had a 23% force reduction, followed by force fluctuations until tire splitting. CCRC1 exhibited a 23% force drop, while CCRC2 had a 37% force reduction, with subsequent force fluctuations. CCNCH1 experienced a 57% force drop, picked up force by 58%, and fell again. CCNCH2 had a 37% force reduction, picking up force briefly before a significant dropping. CCRCH1 had a 64% force drop, followed by force fluctuations, while CCRCH2 had a 37% force reduction, a 22% drop, and significant force fluctuations. Concrete cracks produced audible popping sounds but remained hidden due to tire confinement.

After the failure of CCNC1, a 304 mm diagonal split occurred in the second tire from the top, with only a hairline crack visible in the exposed concrete. CCNC2 had an approximately 50.8 mm crack in the exposed concrete at the tire split, and the tire connection appeared lifted by the failed concrete. CCRC1 tilted to one side after failure, with a failed tire at the bottom and 50.8 mm cracks in the exposed concrete. CCRC2 exhibited a 100 mm crack on the exposed concrete at the tire split, and the top surface had a rupture on the tire sidewall, unique among columns. CCNCH1 had approximately 25.4 mm cracks in the exposed concrete, while CCNCH2 had a 12.7 mm crack at the tire split. Both failed from the section where the tire split. CCRCH1 had an 88.9 mm crack on the exposed concrete, while CCRCH2 had a 63.5 mm crack at the tire split. Popping sounds accompanied the concrete cracking, and force reduction was observed with each sound.

Figures 10–15 illustrate the failure modes of the UCNC, UCRC, CCNC, CCRC, CC-NCH, and CCRCH specimens. Similarly, all samples' performance factors (e.g., maximum axial force, corresponding displacement, maximum stress, corresponding strain, and energy dissipation) are summarized in Table 3.

(a) (b)

Figure 10. The failure mode of UCNC specimen: (**a**) columnar failure, (**b**) conical failure.

(a) (b)

Figure 11. The failure mode (columnar failure) of UCRC specimen: (**a**) UCRC1, (**b**) UCRC2.

(a) (b) (c) (d)

Figure 12. Failure mode: (**a**) CCNC1 specimen, (**b**) close-up view of a split in tire-CCNC1, (**c**) CCNC2 specimen, (**d**) close-up view of a split in tire-CCNC2.

(a) (b) (c) (d)

Figure 13. Failure mode: (**a**) CCRC1 specimen, (**b**) close-up view of a split in tire-CCRC1, (**c**) CCRC2 specimen, (**d**) close-up view of a split in tire-CCRC2.

Figure 14. Failure mode: (**a**) CCNCH1 specimen, (**b**) close-up view of a split in tire-CCNCH1, (**c**) CCNCH2 specimen, (**d**) close-up view of a split in tire-CCNCH2.

Figure 15. Failure mode: (**a**) CCRCH1 specimen, (**b**) close-up view of a split in tire-CCRCH1, (**c**) CCRCH2 specimen, (**d**) close-up view of a split in tire-CCRCH2.

Table 3. Summary of performance factors for all specimens.

Type: Unconfined	Maximum Axial Force (kN)	Corresponding Displacement (mm)	Maximum Stress (MPa)	Strain (mm/mm)	Energy Dissipation (kJ)
UCNC1	3426.9	5.99	26.4	0.00654	7.99
UCNC2	2783.9	10.13	21.4	0.00811	9.15
UCRC1	3258.5	9.96	25.1	0.01089	11.75
UCRC2	2998.7	8.89	23.1	0.00965	10.53
CCNC1	6031.7	10.1	46.5	0.01109	94.18
CCNC2	5093.2	11.4	39.2	0.01259	46.77
CCRC1	3567.0	9.1	27.5	0.00990	12.68
CCRC2	4317.8	30.2	33.3	0.03323	62.10
CCNCH1	4948.2	10.7	38.1	0.01177	47.03
CCNCH2	5412.2	9.9	41.7	0.01095	55.29
CCRCH1	4752.4	2.4	36.6	0.00259	9.59
CCRCH2	3714.7	7.9	28.6	0.00870	23.75

2.4.1. Unconfined Concrete Column: Normal Concrete vs. Rubberized Concrete

Four specimens with these configurations (i.e., two of each—UCNC and UCRC) were tested. The plot (Figure 16) shows the combined data from each configuration and graph. It is evident from the plot that the columns with rubber aggregate underwent greater axial deformation than the normal concrete columns. The axial strain of the UCRCs was 40% higher than that of the UCNCs on average. UCNC specimens failed in a brittle manner with visible cracks, indicating shear or conical failure. UCRC specimens also failed in a brittle manner, accompanied by a popping sound and columnar failure. Crack size was not measured for safety reasons. Thus, the failure modes for both configurations were similar: shear and columnar failure. The peak stress values were close to each other. The failure is brittle, but compared to this configuration, the rubber concrete had a higher strain at the point of failure. UCRCs can disperse more energy from monotonic compression loading than UCNCs. Based on these results, rubberized concrete will have less damage than columns with normal concrete under similar loading conditions.

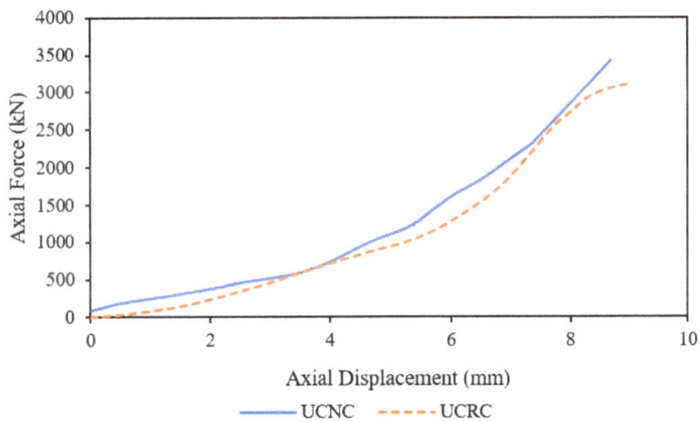

Figure 16. Unconfined concrete columns comparison.

2.4.2. Confined Concrete Column: Normal Concrete vs. Rubberized Concrete

Four specimens were tested from these configurations, two CCNC and CCRC specimens. The force vs. displacement graph in Figure 17 shows the combined data for each configuration. Each specimen was loaded until the tire failure was externally observed or until the specimen kept picking up force after a sudden drop in the force due to the failure of the concrete. Here, the concrete axial force that the column reached before the concrete failed was higher than that for the UCs. The columns reached a peak force. Then the force dropped by 15% on average in the CCNCs and around 30% in the CCRCs. Beyond this point, columns in both configurations showed strain-softening behavior. The CCRCs had 46% more axial strain on average but lower stress at the point of concrete failure. In both cases, the columns continued to pick up force until the sudden drop but never reached the previous peak force. As shown in the plot, the columns stopped picking up force after the tires ruptured. The average strain at the point of failure for a column with rubber aggregate was higher. The energy dissipation of the CCNCs was almost double that of the CCRCs. This means CCNCs can disperse more energy from monotonic compression loading than CCRCs. These results show that confined columns with normal concrete will have less damage than columns with rubberized concrete under similar loading conditions. The variation in energy dissipation might be due to different tires and different wear and tear on them. The sample size of columns tested in these configurations is small, so further investigation is needed to verify these results.

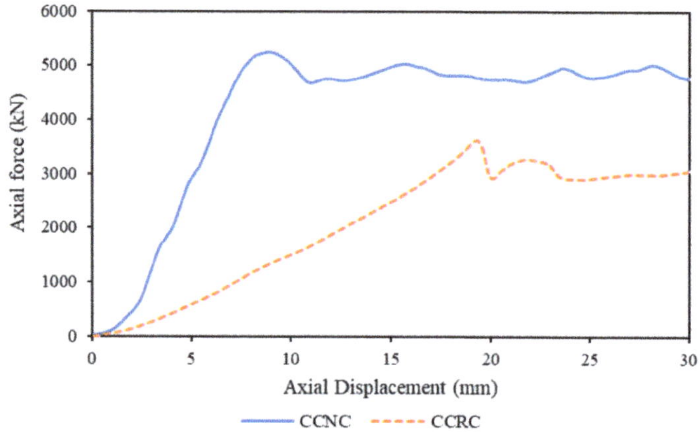

Figure 17. Confined concrete columns comparison.

2.4.3. Confined Concrete Columns with Tires Cut along the Height: Normal Concrete vs. Rubberized Concrete

Four specimens were tested from these configurations, two CCNCH and CCRCH specimens. The force vs. displacement graph is shown in Figure 18 for the combined data of each configuration. The pattern of the force and displacement is similar to confined. Still, the force reduction in these columns after concrete failure was considerably higher than in the confined column without tire cuts. The average drop in force for the CCNCHs was almost 47%, and that for the CCRCHs was around 50%. The peak stress and corresponding strain are significantly higher for columns with normal concrete. The average peak stress for the CCNCHs was 19% higher than that for the CCRCHs. The average corresponding strain for the CCNCHs was 50% higher than that for the CCRCHs. Variables such as different tires and varying wear and tear patterns might have influenced this. As mentioned earlier, the testing was force-controlled and is one of the reasons for the inconsistency in the plots. Axial deformation and energy dissipated before failure. CCNCH's energy dissipation is more than three times that of CCRCHs. This means CCNCH can disperse more energy from monotonic compression loading than CCRCH. Based on these results, CCNCHs will have less damage than CCRCHs under similar loading conditions. The sample size of columns tested in these configurations is small, so further investigation is needed to verify these results.

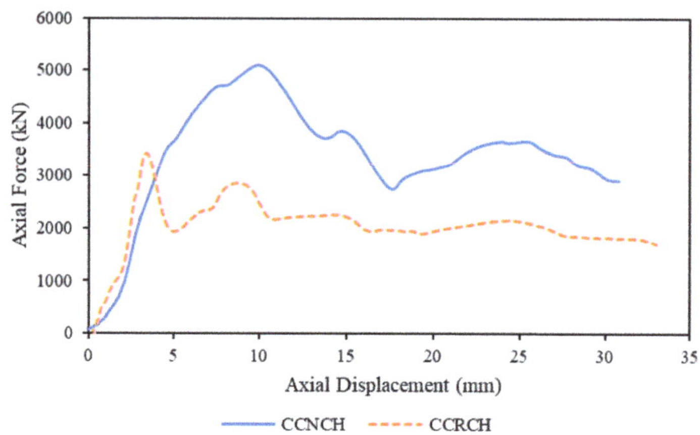

Figure 18. Confined concrete columns with tires cut along the height comparison.

2.4.4. Normal Concrete Columns Comparison

This section compares the concrete properties of normal concrete when it is unconfined, confined, and confined with a height cut. Before the concrete fails, there is a significant increase in axial force, nearly 80%, in the confined column compared to the unconfined column. The axial force capacities of CCNCs and CCNCHs are just 7% higher than that of UCNCs. There is a significant drop in force in CCNCHs once the concrete begins to fail. On average, the force drop is 15% for CCNCs, while that for CCNCHs is 47%. This shows that the tires reinforce both columns, but the tires in the CCNCHs were cut along the height, so the overall confinement effect is lower. The average rise in force capacity after the drop from peak force was 5% for CCNCs, and that for CCNCH was 35%. This shows that the reinforcement by tires starts after the concrete fails. The average axial strain corresponding to peak stress for the CCNCs was 39% higher than that for the UCNCs and 4% higher than that for the CCNCHs. Figure 19 shows the compressive force and displacement difference between confined and unconfined columns with normal concrete. The energy dissipation is more than eight times higher in CCNCs than UCNCs and almost six times higher in CCNCHs. The significant increase in energy dissipation shows that the tires acted as reinforcements for the columns. CCNCs can disperse more energy from monotonic compression loading than UCNCs. These results indicate that confined columns with normal concrete will undergo lesser damage than unconfined columns under similar loading conditions.

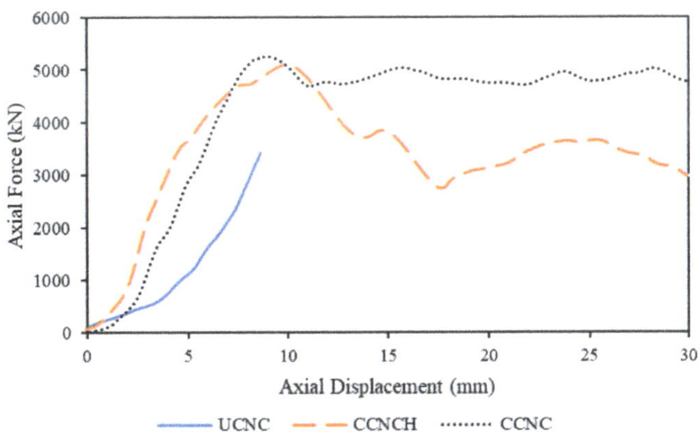

Figure 19. Normal concrete columns comparison.

2.4.5. Rubberized Concrete Columns Comparison

The concrete properties of rubberized concrete when it is unconfined, confined, and confined with a height cut have been compared in this section. Before concrete fails, there is a significant increase of nearly 25% in axial force in the confined column compared to the unconfined column. The axial force capacities of CCRCs and CCRCHs are similar. There is a significant drop in force in CCRCHs once the concrete begins to fail. On average, the force drop is 30% for CCRCs, while that for CCRCHs is 51%. This shows that the tires reinforced both columns, but for CCRCHs, the tires were cut along the height, so the overall confining effect is lower. The average rise in force capacity after the drop from the peak force was 13% for the CCRCs, and that for the CCRCH was 40%. This shows that the reinforcement effect of the tires engages after the concrete fails. The average axial strain on CCRCs corresponding to peak stress is twice that of UCRCs and is almost four times higher than that of CCRCHs. The concrete undergoes brittle failure when unconfined but undergoes significant deformation before failure when it is confined with tires. Figure 20 shows the compressive force and displacement difference between confined columns and unconfined columns with rubberized concrete. The energy dissipation is more than three times higher in CCRCs than UCNCs, while it is 50% higher than in CCRCHs. The significant increase in energy dissipation shows that the tires act as reinforcements for the columns. These results show that confined columns with rubberized concrete will have lesser damage than unconfined columns under similar loading conditions.

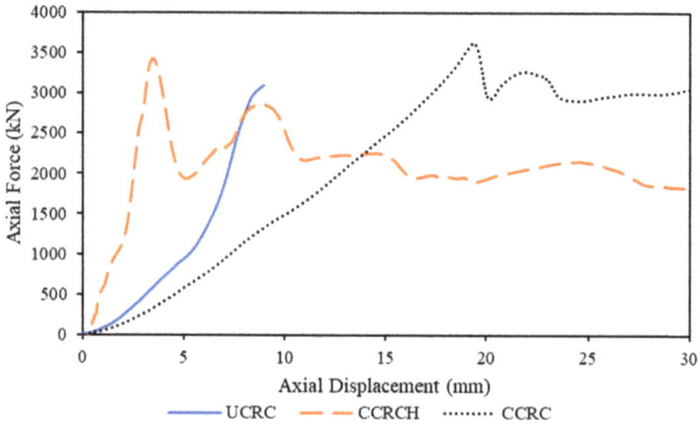

Figure 20. Rubberized concrete columns comparison.

2.5. *Tire Confinement Effect*

Tires act as transverse reinforcements when used as confining materials. There is a significant rise in the axial capacity of confined columns, which undergo considerable deformation before failure. The confinement effectiveness (k_l) is defined as the maximum compressive strength of confined concrete (f'_c) to that of unconfined concrete (f'_{co}). The average f'_{co} for UCNC is 23.9 MPa, and that for UCRC is 24.1 MPa. The confinement effectiveness is greater than 1. Table 4 shows the confinement effectiveness of confined columns based on the average compressive strength of unconfined columns. However, the confinement was observed to be passive. Confinement using tires is assumed to be similar to confinement using circular hoops. This assumption is made as an approximation to find out the confinement provided by tires. If the hoop tension from the tire at yield exerts uniform lateral stress on the concrete core, then the equilibrium of forces, as shown in Figure 21, is given by Equation (2).

$$f_l \times D \times S_{sp} = 2 \times f_{yh} \times A_t \quad (2)$$

Table 4. Confinement effectiveness of confined column.

Type	Average Maximum Compressive Stress (f'_{cc}) MPa	Average Confinement Effectiveness, *kl* (f'_{cc}/f'_{co})
CCNC	42.8	1.78
CCRC	30.4	1.26
CCNCH	39.9	1.67
CCRCH	36.6	1.35

The confinement pressure (f_l) of the tire is given by Equation (3),

$$f_l = \left(2f_{yh}A_t\right)/\left(DS_{sp}\right) \quad (3)$$

where f_{yh} is the yield stress of the confining tires, MPa

A_t is the cross-sectional area of the tire at the threaded wall, mm^2
D is the inside diameter of the column, mm

S_{sp} is the pitch between the tires, mm.

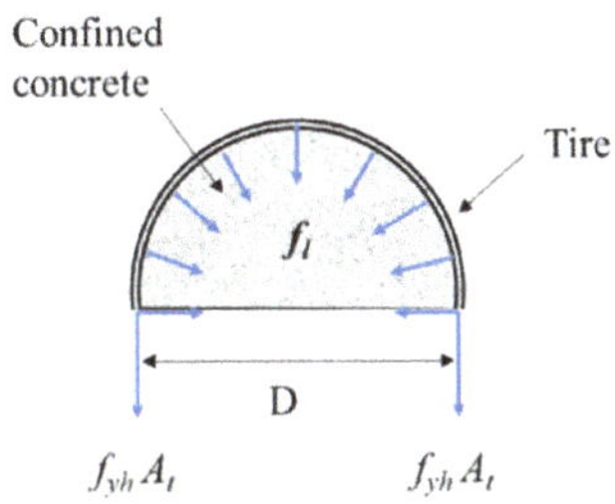

Figure 21. Confinement by tire.

Tires include rubber, synthetic polymers, steel, textile, carbon black or silica fillers, antioxidants, and curing systems such as sulfur and zinc oxide. Tires are multi-layered materials. Since the tire rubber would be the material with the least tensile strength, the tensile strength of tire rubber at yield (Equation (4)) is considered. The equilibrium between the hoop tension from the tires and the lateral stress on the concrete core has been assumed to be at the tensile limit of the tire's rubber.

$$f_l = (2f_tA_t)/(DS_{sp}) \tag{4}$$

where f_t is the tensile strength of a tire's rubber at yield, MPa.

The tensile strength of waste tire rubber is around 10.6 to 14.4 MPa [38]. The tensile strength of the rubber compound in tires is around 16.5 to 21.2 MPa [39]. The tensile strength considered in this study was 10.64 MPa, to be on the conservative side. The lateral stress of concrete under compression acts on the threaded section of the tires. The average thickness of the threaded section from the groove of the thread to the inner lining of the tire was 9.5 mm, and the average thread depth was 2.0 mm. The average pitch between tires was 25.4 mm. Table 5 shows the confining pressure of the tires based on different threaded widths and inner diameters using Equation (4).

Table 5. Confinement pressures of different tires.

Tire	Inner Diameter (mm)	Threaded Section Width (mm)	Confining Pressure, f_l (MPa)
225/75R16	712.5	200.7	2.24
235/80R16	750.6	210.8	2.24
215/85R16	740.4	190.5	2.05

The average confinement pressure of the tires is 2.2 MPa. The actual confinement is given by the ratio of confining pressure f_l and compressive strength of unconfined concrete f'_{co}. Table 6 shows the actual confinement values for the different specimens.

Table 6. Actual confinement in columns.

Type	Specimen	Actual Confinement (f_l/f'_{co})
CCNC	CCNC1	0.094092
	CCNC2	0.094092
CCRC	CCRC1	0.093553
	CCRC2	0.093553
CCNCH	CCNCH1	0.094092
	CCNCH2	0.094092
CCRCH	CCRCH1	0.093553
	CCRCH2	0.093553

The general equation used to calculate the capacity of concrete confined by tires is obtained using the graph of confinement effectiveness (f'_{cc}/f'_{co}) and the actual confinement (f_l/f'_{co}) of all specimens. The trendline of the data is given by Equation (5). Similarly, the compressive strength of confined concrete can be calculated using Equation (6).

$$\frac{f'_{cc}}{f'_{co}} = 774.58 \left(\frac{f'_l}{f'_{co}}\right) - 71.157 \quad (5)$$

$$f'_{cc} = 774.58\, f'_l - 71.157\, f'_{co} \quad (6)$$

In this preliminary research, an equation was developed for measuring the compressive strength of columns confined by tires used in this study. This equation can be used as an approximation to find out the confinement provided by tires. Detailed finite element analysis and further experimental investigations are required to get an accurate equation incorporating the connection between the tires, spacing between the tires, and varying diameter and thickness.

3. Discussion

This study explores and corroborates the use of waste tires as transverse reinforcements for concrete columns, a novel approach compared to previous studies involving rubber and tire fiber's use in concrete mixtures. Regardless of the concrete mix, using tires as reinforcements boosted the average axial capacity by approximately 50%. The compressive strength of the two concrete mixes used in the study was initially measured at 35.5 MPa. However, during the full-scale testing of the columns, the compressive strength decreased and averaged around 25.5 MPa. Despite the decrease in compressive strength observed in the unconfined specimens, the confined specimens displayed higher compressive strength, averaging around 5300 psi. Furthermore, this study provides equations for calculating compressive stress and the confinement pressure of confined concrete columns.

The force-controlled methodology used for testing this study and several variables, including different wear and tear levels in the tires, different brands, meaning there were different materials in the tires, and different tire diameters and heights, led to inconsistency in the plots, especially energy dissipated before failure. Strain-controlled testing of the sample is suggested for further investigations into this research topic. The test should be performed by changing the spacing between tires to understand how spacing or the connections between the tires affect the confinement.

Confining columns by tires could cause a lack of space between two columns which could be an issue in terms of architectural design but applications for something like this are not limited to buildings. It could support underground structures or civil works type applications as long as the axial load on the columns. This kind of system can be used for buildings, but then architecturally, this can be an issue; however, this can be used for other applications such as civil works and underground structures.

Based on the findings, tire reinforcement has a promising future as a component of civil infrastructure. An added advantage of utilizing easily available and cheap tires is their ability to act as formwork for structures and provide extra protection to the concrete against chemical exposure. These tire columns can be proposed to be used in underdeveloped nations where resources are limited and where houses have no foundation. To retrofit against earthquake/typhoon forces, new external tire columns can be introduced on the four sides of the building. A typical column can be constructed using recycled tires on each other and filled with rubber concrete. The base of the column extends can also be down the natural ground level to provide a nearly fixed connection. The concrete columns will restrain the sliding of the house under lateral loads. They will also help in reducing the moment demand on the walls. To provide gravity support, the soil under the four corners of the building can be excavated, and tire columns can be inserted. The pedestals resist gravity loads, prevent settlements, and act as dampeners under vertical excitation. This concept recycles the waste tire from landfill, so getting the same age, model, material properties, thread life, and tire size is impossible. But tires with similar wares, size, and thickness can be used to make such columns. The tires with tears are to be rejected and can be further scraped to get rubber aggregates or for other recycling/reuse applications.

Future research should also address the issues of the physical properties of waste tires to understand wear and tear in detail. Namely, tire aging is the very reason there are waste tires. Note that the lifespan of an average tire is usually six years, a maximum of 10 years, as the rubber compound hardens with time, and the tire is more likely to crack. Similarly, tire aging is related to oxidation, as the rubber dries out and becomes stiffer when exposed to oxygen. So, studying the rubber-concrete contact relationship on acceleration/deceleration of rubber aging would be very critical to understand. Further research is required to understand how tires function in concrete structures fully. The following are a few suggested research areas for future studies:

- Cyclical loading of tire columns using displacement-controlled testing methods to understand the confinement properties provided by tires fully.
- Rubber-concrete contact relationship with the tire aging.
- Testing the physical properties of waste tires to understand their wear and tear.
- Testing single tires filled with concrete to quantify the confinement of the tire.
- Testing tire columns and benchmark unconfined columns with the same dimensions.
- Testing tire columns with and without reinforcing rebar under flexure loading.

4. Conclusions

This study experimentally investigated the concept of using recycled rubber tires and rubber chips in concrete columns. According to experiments and their results, the following conclusions were made:

(a) Shear and columnar failure were common failure modes in unconfined normal concrete and rubber concrete specimens.
(b) In the case of rubber concrete, only the unconfined specimens exhibited a higher total energy dissipation under monotonic compression loading compared to normal concrete specimens. This variation in energy dissipation could be attributed to tire variations and their respective wear and tear.
(c) The confined columns, whether made of normal or rubber concrete, achieved a higher concrete axial force before failure than the unconfined columns.
(d) After concrete failure, confined normal concrete and rubber concrete columns displayed strain-softening behavior. However, the columns continued to pick up force after the drop but never reached the previous peak force.
(e) Confined concrete columns with tires cut along their height exhibited a significantly greater reduction in force after concrete failure compared to confined concrete columns without such cuts.
(f) Among columns with normal concrete, the highest energy dissipation under monotonic compression loading was observed in CCNCs, followed by CCNCHs and UCNCs.
(g) Among columns with rubber concrete, the greatest energy dissipation under monotonic compression loading was observed in CCRCs, followed by CCRCHs and UCRCs.

Author Contributions: Conceptualization, M.M. (Mustafa Mashal) and M.M. (Mahesh Mahat); methodology, M.M. (Mustafa Mashal) and M.M. (Mahesh Mahat); software, M.M. (Mahesh Mahat), and M.M. (Mahesh Acharya); validation, M.M. (Mustafa Mashal), M.M. (Mahesh Mahat), and M.M. (Mahesh Acharya); formal analysis, M.M. (Mustafa Mashal), M.M. (Mahesh Mahat), and M.M. (Mahesh Acharya); investigation, M.M. (Mahesh Mahat), M.M. (Mustafa Mashal), and M.A. (Mahesh Acharya); resources, M.M. (Mustafa Mashal), M.M. (Mahesh Mahat), and M.A. (Mahesh Acharya); data curation, M.M. (Mahesh Mahat), M.A. (Mahesh Acharya) and M.A. (Manish Acharya); writing—original draft preparation, M.M. (Mahesh Mahat), M.M. (Mustafa Mashal), M.A. (Mahesh Acharya), and M.A. (Manish Acharya); writing—review and editing, M.M. (Mahesh Mahat), M.M. (Mustafa Mashal), M.A. (Mahesh Acharya), and M.A. (Manish Acharya); visualization, M.M. (Mustafa Mashal), M.M. (Mahesh Mahat), and M.A. (Mahesh Acharya); supervision, M.M. (Mustafa Mashal);

project administration, M.M. (Mustafa Mashal); funding acquisition, M.M. (Mustafa Mashal). All authors have read and agreed to the published version of the manuscript.

Funding: This research received no external funding.

Institutional Review Board Statement: Not applicable.

Informed Consent Statement: Not applicable.

Data Availability Statement: Data is available upon request to the corresponding author.

Acknowledgments: The authors thank everyone who contributed to this research. The authors would like to express their gratitude to Jared Cantrell, Katie Hogarth, Jose Duran, Aashish Thapa, Bipin Aryal, Joshua Peck, Gage Cussin, and the entire structural team at Idaho State University for their assistance with this project. The authors would like to thank Jose Duran for proofreading the Journal article.

Conflicts of Interest: The authors declare no conflict of interest.

References

1. Gagg, C.R. Cement and concrete as an engineering material: An historic appraisal and case study analysis. *Eng. Fail. Anal.* **2014**, *40*, 114–140. [CrossRef]
2. Chen, M.; Si, H.; Fan, X.; Xuan, Y.; Zhang, M. Dynamic compressive behaviour of recycled tyre steel fibre reinforced concrete. *Constr. Build. Mater.* **2022**, *316*, 125896. [CrossRef]
3. Siddique, R.; Naik, T.R. Properties of concrete containing scrap-tire rubber–an overview. *Waste Manag.* **2004**, *24*, 563–569. [CrossRef]
4. Formela, K. Sustainable development of waste tires recycling technologies–recent advances, challenges and future trends. *Adv. Ind. Eng. Polym. Res.* **2021**, *4*, 209–222.
5. Ahmad, I.; Iqbal, M.; Abbas, A.; Badrashi, Y.I.; Jamal, A.; Ullah, S.; Yosri, A.M.; Hamad, M. Enhancement of Confinement in Scaled RC Columns using Steel Fibers Extracted from Scrap Tyres. *Materials* **2022**, *15*, 3219. [CrossRef]
6. Wang, J.; Dai, Q.; Si, R.; Guo, S. Mechanical, durability, and microstructural properties of macro synthetic polypropylene (PP) fiber-reinforced rubber concrete. *J. Clean. Prod.* **2019**, *234*, 1351–1364. [CrossRef]
7. Halsband-Lenk, C.; Sørensen, L.; Booth, A.; Herzke, D. Car tire crumb rubber: Does leaching produce a toxic chemical cocktail in coastal marine systems? *Front. Environ. Sci.* **2020**, *8*, 125. [CrossRef]
8. Sitepu, M.H.; Matondang, A.R.; Sembiring, M.T. Used tires recycle management and processing: A review. In Proceedings of the IOP Conference Series: Materials Science and Engineering. *IOP Publ.* **2020**, *801*, 012116.
9. European Tire and Rubber Manufacturers Association. ETRMA. Europe—91% of All End of Life Tyres Collected and Treated in 2018. Brussels, Belgium. Available online: https://www.etrma.org/library/europe-91-of-all-end-of-life-tyres-collected-and-treated-in-2018/ (accessed on 24 July 2022).
10. Huang, W. Sustainable management of different systems for recycling end-of-life tyres in China. *Waste Manag. Res.* **2021**, *39*, 966–974. [CrossRef]
11. Dabiri, H.; Kheyroddin, A. An analytical study into the seismic behavior of RC pier with elastomeric materials. *Asian J. Civ. Eng. (Bhrc)* **2017**, *18*, 1183–1193.
12. Jia, J.; Zhang, K.; Saiidi, M.S.; Guo, Y.; Wu, S.; Bi, K.; Du, X. Seismic evaluation of precast bridge columns with built-in elastomeric pads. *Soil Dyn. Earthq. Eng.* **2020**, *128*, 105868.
13. Pitilakis, K.; Karapetrou, S.; Tsagdi, K. Numerical investigation of the seismic response of RC buildings on soil replaced with rubber–sand mixtures. *Soil Dyn. Earthq. Eng.* **2015**, *79*, 237–252. [CrossRef]
14. Oikonomou, N.; Mavridou, S. The use of waste tyre rubber in civil engineering works. In *Sustainability of Construction Materials*; Elsevier: Amsterdam, The Netherlands, 2009; pp. 213–238.
15. Strukar, K.; Šipoš, T.K.; Miličević, I.; Bušić, R. Potential use of rubber as aggregate in structural reinforced concrete element—A review. *Eng. Struct.* **2019**, *188*, 452–468. [CrossRef]
16. Venu, M.; Rao, P. Study of rubber aggregates in concrete: An experimental investigation. *Int. J. Civ. Eng. Technol.* **2010**, *1*, 15–26.
17. Siddiqui, M.M.A. Study of rubber aggregates in concrete an experimental investigation. *Int. J. Latest Res. Eng. Technol.* **2016**, *2*, 36–57.
18. Batayneh, M.; Marie, I.; Asi, I. Use of selected waste materials in concrete mixes. *Waste Manag.* **2007**, *27*, 1870–1876. [CrossRef] [PubMed]
19. Liu, L.; Cai, G.; Liu, S. Compression properties and micro-mechanisms of rubber-sand particle mixtures considering grain breakage. *Constr. Build. Mater.* **2018**, *187*, 1061–1072. [CrossRef]
20. Batayneh, M.K.; Marie, I.; Asi, I. Promoting the Use of Crumb Rubber Concrete in Developing Countries. *Waste Manag.* **2008**, *28*, 2171–2176. [CrossRef]
21. Güneyisi, E.; Gesoğlu, M.; Özturan, T. Properties of rubberized concretes containing silica fume. *Cem. Concr. Res.* **2004**, *34*, 2309–2317. [CrossRef]

22. Razvi, S.W.N.; Shaikh, M. Effect of confinement on behavior of short concrete column. *Procedia Manuf.* **2018**, *20*, 563–570. [CrossRef]
23. Pham, T.M.; Hao, H. Confinement efficiency of concrete cylinders wrapped with different types of FRP under impact loads. In Proceedings of the Mechanics of Structures and Materials: Advancements and Challenges-Proceedings of the 24th Australasian Conference on the Mechanics of Structures and Materials, Perth, Australia, 6–9 December 2016; Volume 2017, pp. 515–520.
24. Wang, W.; Martin, P.R.; Sheikh, M.N.; Hadi, M.N. Eccentrically loaded FRP confined concrete with different wrapping schemes. *J. Compos. Constr.* **2018**, *22*, 04018056. [CrossRef]
25. Abdullah, W.; Omar, A.; Rafiq, S.K. Experimental Work on Using Fully Wrapped Post-Tensioned Metal Straps Around Normal Reinforced Concrete Beams to Increase Flexural Strength of RC Beams. *Sulaimania J. Eng. Sci.* **2021**, *8*. [CrossRef]
26. Garcia, R.; Pilakoutas, K.; Hajirasouliha, I.; Guadagnini, M.; Kyriakides, N.; Ciupala, M.A. Seismic retrofitting of RC buildings using CFRP and post-tensioned metal straps: Shake table tests. *Bull. Earthq. Eng.* **2017**, *15*, 3321–3347.
27. Imjai, T.; Chaisakulkiet, U.; Garcia, R.; Pilakoutas, K. Strengthening of RC members using Post-Tensioned Metal Straps: State of the research. *Lowl. Technol. Int.* **2018**, *20*, 109–118.
28. Ma, C.K.; Garcia, R.; Yung, S.C.S.; Awang, A.Z.; Omar, W.; Pilakoutas, K. Strengthening of pre-damaged concrete cylinders using post-tensioned steel straps. *Proc. Inst. Civ. Eng.-Struct. Build.* **2019**, *172*, 703–711. [CrossRef]
29. Imjai, T.; Setkit, M.; Garcia, R.; Figueiredo, F.P. Strengthening of damaged low strength concrete beams using PTMS or NSM techniques. *Case Stud. Constr. Mater.* **2020**, *13*, e00403. [CrossRef]
30. Rubel, S.N.R.; Rahman, M.; Saju, J. Review Study on Different Concrete Confinement Materials and Mechanisms. In Proceedings of the 2nd International Conference on Research and Innovation in Civil Engineering, Chittagong, Bangladesh, 11 January 2020.
31. Mander, J.B.; Priestley, M.J.; Park, R. Theoretical stress-strain model for confined concrete. *J. Struct. Eng.* **1988**, *114*, 1804–1826. [CrossRef]
32. De Lorenzis, L.; Tepfers, R. Comparative study of models on confinement of concrete cylinders with fiber-reinforce polymer composites. *J. Compos. Constr.* **2003**, *7*, 219–237. [CrossRef]
33. Teng, J.G.; Chen, J.F.; Smith, S.T.; Lam, L. *FRP: Strengthened RC Structures*; Wiley: Hoboken, NJ, USA, 2002.
34. Abdelrahman, K.; El-Hacha, R. Analytical prediction model for circular SMA-confined reinforced concrete columns. *Eng. Struct.* **2020**, *213*, 110547. [CrossRef]
35. U.S Tire Manufacturers Association. WHAT'S IN A TIRE. Washington, DC., USA. Available online: https://www.ustires.org/whats-tire-0 (accessed on 24 July 2022).
36. *ASTM C31/C31M-22*; Standard Practice for Making and Curing Concrete Test Specimens in the Field. ASTM International: West Conshohocken, PA, USA, 2019.
37. *ASTM C39/C39M-21*; Standard Test Method for Compressive Strength of Cylindrical Concrete Specimens. ASTM International: West Conshohocken, PA, USA, 2021.
38. Faizah, R.; Priyosulistyo, H.; Aminullah, A. The Properties of Waste Rubber Tires in Increasing the Damping of Masonry Wall Structure. In Proceedings of the IOP Conference Series: Materials Science and Engineering. *IOP Publ.* **2019**, *650*, 012041.
39. Bijarimi, M.; Zulkafli, H.; Beg, M.D.H. Mechanical properties of industrial tyre rubber compounds. *J. Appl. Sci. (Faisalabad)* **2010**, *10*, 1345–1348. [CrossRef]

sustainability

Article

Impact of Fracture–Seepage–Stress Coupling on the Sustainability and Durability of Concrete: A Triaxial Seepage and Mechanical Strength Analysis

Zhuolin Shi [1,2], Chengle Wu [1,2], Furong Wang [1,2], Xuehua Li [1,2,*], Changhao Shan [1,2] and Yingnan Xu [1,2]

1 Key Laboratory of Deep Coal Resource Mining, Ministry of Education, School of Mines, China University of Mining and Technology, Xuzhou 221008, China
2 School of Mines, China University of Mining and Technology, Xuzhou 221116, China
* Correspondence: xuehua_cumt@163.com

Abstract: As an indispensable material in construction and engineering, concrete's mechanical properties and permeability are crucial for structures' stability and durability. In order to reasonably assess and improve the durability of fracture-containing concrete structures and to enhance the sustainable working life of concrete structures, this research investigated the seepage characteristics of fracture-containing concrete and its mechanical property deterioration under fracture–seepage coupling by testing the permeability and strength of concrete samples before and after water penetration using triaxial seepage test and mechanical strength test. The results show that the fracture–seepage coupling action significantly affects the permeability characteristics and mechanical strength of fracture-containing concrete. In particular, the strength of concrete samples containing a single fracture decreased with increased fracture angle, with a maximum decrease of 32.8%. The fracture–seepage–stress coupling significantly reduced the strength of the fracture-containing concrete samples, which was about twice as much as the strength of the no-fracture concrete samples. Different fracture angles affect the mode of fracture expansion and damage (The fracture angle varies from small to large, and the damage form of concrete changes from tensile damage to tensile–shear composite damage). Moreover, the coupling effect of fracture–seepage–stress will further increase fracture-containing concrete's fragmentation in the damage process. Therefore, improving the seepage and fracture resistance of concrete plays a vital role in improving the sustainable working life of concrete structures.

Keywords: mechanical properties; durability; fracture-containing concrete; seepage characteristics; fracture–seepage coupling

Citation: Shi, Z.; Wu, C.; Wang, F.; Li, X.; Shan, C.; Xu, Y. Impact of Fracture–Seepage–Stress Coupling on the Sustainability and Durability of Concrete: A Triaxial Seepage and Mechanical Strength Analysis. *Sustainability* **2024**, *16*, 1187. https://doi.org/10.3390/su16031187

Academic Editors: Mahdi Kioumarsi and Vagelis Plevris

Received: 2 January 2024
Revised: 25 January 2024
Accepted: 28 January 2024
Published: 31 January 2024

1. Introduction

1.1. Importance of Concrete and Its Strength: Seepage Characteristics

Concrete, as an important material in construction and engineering, is widely used in the building industry and underground engineering. Investigating the permeability and mechanical properties of concrete with fracture is essential to improve the sustainable use of concrete in buildings and extend its service life. An in-depth understanding of the effect of fracture on the permeability and strength of concrete can help design structures with better durability, which can effectively prevent internal corrosion and deterioration, thus significantly improving the safety and stability of buildings. Such research can also facilitate the development of new durable concrete materials, further promoting concrete development for buildings towards sustainability and long life. While in working conditions, concrete is mainly subjected to pressure, so compressive strength becomes an important consideration in engineering design. With the complexity of the stress situation of concrete structure, in the actual project, concrete components will be affected by many uncertain factors (such as impact loading, temperature, construction technology, etc.) and produce different forms of fracture defects inside. Concrete structures are in long-term working conditions with internal fractures. These fractures will further develop and expand under external stresses. This will reduce the durability and bearing capacity of the structure and accelerate its destruction. For example, in 2007, in Minneapolis, Minnesota, United States of America, the Highway 35 bridge collapsed as a result of a severe weakening of the concrete's load-bearing capacity due to long-accumulated cracks. This catastrophic event resulted in 13 deaths and some 145 injuries. The accident not only caused significant casualties and property damage but also raised widespread concerns about the safety and durability of infrastructure. Fractures in the concrete, in turn, create a large number of water-conducting channels. Underground structures such as tunnels, mining tunnels, and other critical large-scale projects such as dams, water surges in tunnels, tunnel collapses, and landslides are the leading causes of engineering safety accidents. Water penetrates the structure through the concrete matrix and internal fracture to damage the integrity and durability of the structure. In 2017, for example, a severe dam failure at the Oroville Dam in California, USA, was caused by seepage and erosion problems resulting from cracks in the dam's concrete structure. This resulted in the partial collapse of the dam's spillway and the emergency evacuation of nearly 188,000 residents, resulting in significant economic and social impacts.

In summary, the research on the degradation effect of fractures in concrete on the mechanical properties of concrete and the influence of concrete seepage characteristics is significant.

1.2. Progress of Research on the Effect of Fracture in Rock on the Strength of Concrete

Some scholars [1–6] investigated single prefabricated fracture expansion patterns in rock materials under uniaxial loading. It was shown that multiple forms of secondary cracks would be generated at the tip of the prefabricated fracture in chronological order, and the expansion direction of the cracks was in the direction of the maximum principal stress. Yang et al. [7] obtained that fracture length and angle under axial pressure are the key influencing factors on the strength and deformation properties of sandstone samples by axial compression tests on single-fracture brittle sandstones. Qian et al. [8] analyzed the effect of fracture on the strength of rocks by axial compression tests on rock specimens. They concluded that the sensitivity of the factors influencing the strength of rocks is fracture filling, fracture angle, and fracture length, in descending order of magnitude. Chen et al. [9,10] investigated the effect of curing factors on the strength of concrete and the effect of chemical corrosion on the mechanical properties of concrete by triaxial mechanical tests. Xie et al. [11,12] used energy release and dissipation to determine the critical stress of the abrupt structural damage of rock samples under different stress states. Combined with CT scanning and acoustic emission, damage theory was introduced into analyzing damage mechanisms. Li et al. [13] investigated the crack extension damage evolution process by uniaxial compression test and CT scanning test of rocks. They concluded that the crack tip and edge are the cracking starting points of the fracture of rock samples. Lin et al. [14–16] investigated the expansion law of single fracture in mortar materials at different inclination angles. The results showed that when the angle between the fracture and the axial pressure direction was large, the cracks expanded unstably, and the samples were prone to cracking.

1.3. Progress of Seepage Characteristics of Fracture-Bearing Rock

Regarding the research on the effect of fracture on the seepage characteristics of materials in rock materials, Bian et al. [17] investigated the evolution law of seepage properties in the water storage space of fractured sandstone in combination with in situ CT scanning technology. By using three-dimensional digital image technology, Chen et al. [18] investigated the mechanical model and the seepage characteristics of fracture in sandstone samples under different seepage pressures.

In general, although there has been extensive research on the mechanism of fracture extension, mechanical characteristic changes, and seepage evolution law of fracture-containing rock materials, there is less research on the related properties of concrete materials. The effects of fractures in concrete and their comparison with the effects of fractures in rock materials are significantly different. In terms of differences in material properties, concrete is a composite material consisting of cement, sand, gravel, and water. This makes its internal structure and properties more complex than those of a single-component rock. Fracture extension in concrete is influenced by the interface between the aggregate and the cement paste. In terms of fracture generation and expansion, the formation of concrete fractures is influenced by various factors, such as loading, temperature changes, shrinkage, and corrosion. In contrast, rock fractures are usually caused by geological stresses, hydrogeological action, or external forces. In addition, fractures in concrete may extend gradually from microscopic defects. In contrast, macroscopic stress fields may affect rock fractures more directly. In terms of permeability differences, fractures in concrete usually affect its permeability, which is critical to the durability of the structure. The permeability of rock, on the other hand, is more influenced by the fracture system, and its permeation paths and rates differ significantly from those of concrete. The mechanical properties and seepage characteristics of concrete with fracture in actual engineering often determine the durability of concrete structures.

In this research, the permeability of concrete samples containing prefabricated fracture was obtained by conducting percolation tests on concrete samples under different triaxial stresses. In this way, the effects of different fracture morphologies (single fracture, Y-fracture, and X-fracture), fracture angle, and stress conditions on the percolation pattern of concrete samples were investigated. The strength decay characteristics of concrete strength under fracture–seepage coupling under different fracture morphology, fracture angle, and other conditions were investigated by conducting strength tests on fracture-containing concrete samples before and after seepage. Through this research, we can predict and respond more accurately to the effects of fracture on concrete performance so that measures can be taken at the design and construction stages to enhance the resistance to fracture and seepage of concrete structures. This is of great significance in improving the overall performance of buildings, extending their service life, and reducing the need for maintenance and repair at a later stage. Therefore, this research significantly impacts the promotion of concrete structures to be more sustainable and durable.

2. Experiment Preparation and Experimental Program Design

2.1. Sample Preparation

According to the sample size criteria recommended by the International Society for Rock Mechanics (ISRM) [19], cylindrical samples with dimensions ϕ = 50 mm and H = 100 mm will be used in this research. This sample size is characterized by good uniformity, easy handling, and uniform stress distribution. Moreover, this sample size is consistent with the sample size in the international test method, and the experimental results obtained are also convenient for academic research and communication. The prepared fracture-containing concrete samples were made from concrete poured in advance and finished curing. The ratio of cement:sand:water:coarse sand in the concrete samples was 1:2:0.6:2, in which natural river sand was chosen as the concrete aggregate. Because of its uniform particle size and hardness, it can ensure the strength and compactness of the concrete sample. They were pouring concrete samples through molds. The concrete was vibrated using a vibrating table to remove internal gases and to compact the concrete. The poured concrete samples need to be cured at a temperature of about 24 °C and in 80–90% humidity. The concrete under cover should also be covered to prevent it from drying out too quickly. After 28 days of curing, the concrete samples with prefabricated single fracture, prefabricated Y-fracture, and prefabricated X-fracture were obtained by wire cutting. The sample preparation is shown in Figure 1. The concrete samples with a precast fracture are shown in Figure 2.

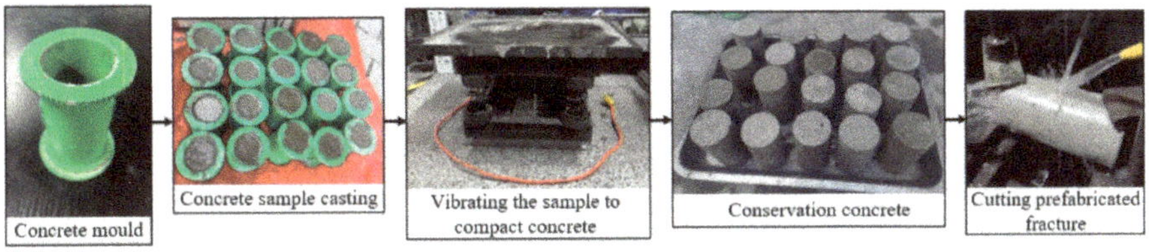

Figure 1. Sample preparation.

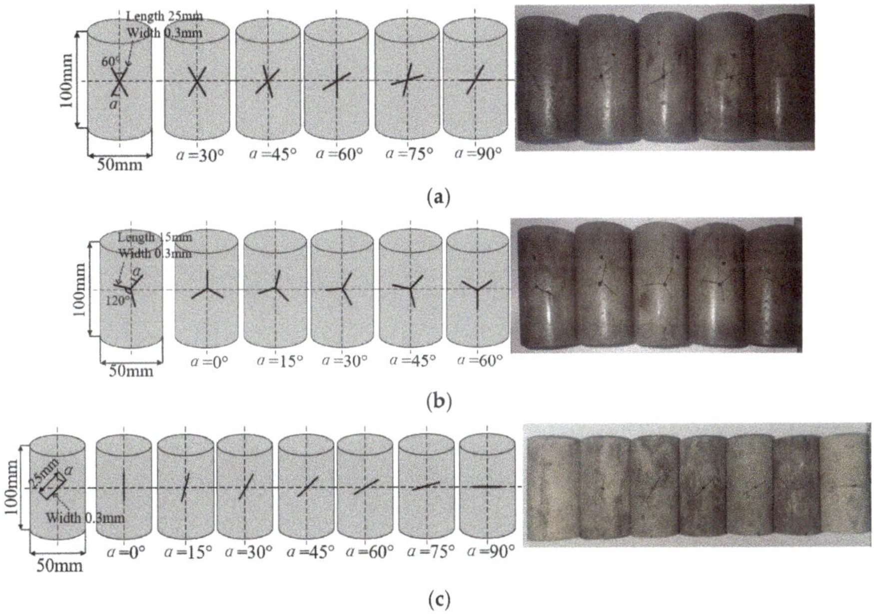

Figure 2. Concrete samples containing precast fractures: (**a**) X-fracture, (**b**) Y-fracture, and (**c**) single fracture.

2.2. Experimental Equipment

MTS electro-hydraulic servo testing machine was selected for mechanical characterization of concrete samples with fracture. The stress–strain curves and strengths of concrete samples before and after seepage under different fracture morphologies obtained from experiments can be analyzed to obtain the strength decay characteristics of fracture-containing concrete samples under fracture–seepage coupling.

The triaxial-loading rock-seepage test platform was selected for the test of seepage characteristics of concrete samples with fracture. The mass of outflow water per unit of time under different stress conditions can be obtained by applying different stresses to fracture-containing concrete samples using the seepage experimental platform. The permeability of the concrete samples can then be calculated to characterize the seepage properties of the concrete samples under different fracture patterns and stress conditions. The test platform consists of a load-bearing system, a loading system, a temperature control system, a seepage system, a measurement system, and an automatic acquisition and control system. In the loading system, the axial load $\sigma_{1,\max}$ = 300 kN, radial load $\sigma_{2,\max} = \sigma_{3,\max}$ = 25 MPa, and the upstream pressure of liquid seepage in the seepage system $p_{w,\max}$ = 20 MPa. The experimental equipment is shown in Figure 3. Among them, the principle of the triaxial-loading rock-seepage test platform is shown in Figure 4.

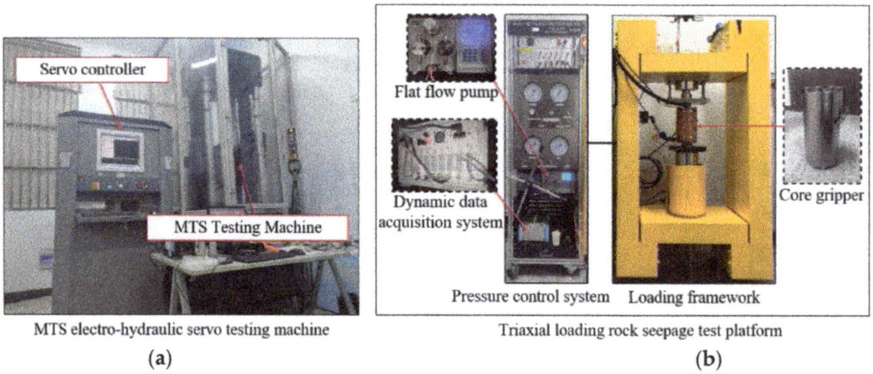

Figure 3. Experimental equipment: (**a**) MTS electro-hydraulic servo testing machine and (**b**) triaxial-loading rock-seepage test platform.

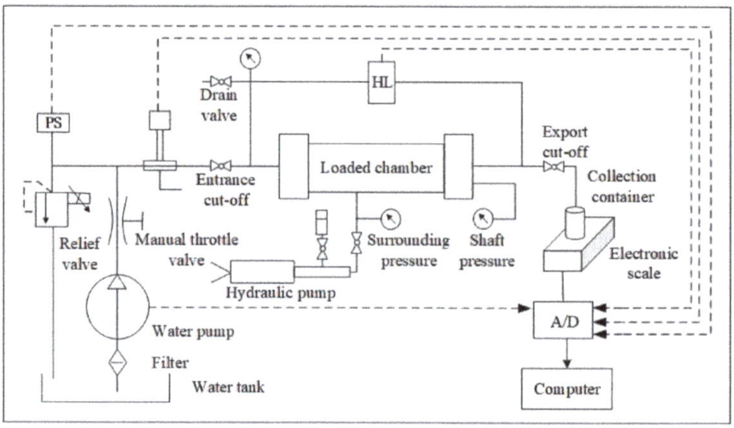

Figure 4. Principle of the triaxial-loading rock-seepage test platform.

2.3. *Experimental Program*

In natural rock and artificial concrete structures, fracture patterns are very complex. However, most complex fractures can be regarded as a combination of single fracture and cross-fracture. In order to investigate the effects of different fracture morphologies on the seepage characteristics and mechanical properties of concrete, single fracture, Y-fracture, and X-fracture are selected as the fracture morphologies to be studied in this research. In order to investigate the effect of concrete fracture on the seepage characteristics of concrete, seepage experiments were carried out on fracture-containing concrete samples using a triaxial-loading rock-seepage testing platform. The specific parameters of the fracture are shown in Figure 2. The experiments were conducted using concrete sample fracture morphology, fracture inclination, and external loading applied to the sample as the research variables. The experiment was conducted using stress-controlled loading, with axial stress σ_1 and radial stress σ_2 (σ_3) applied at a rate of 1 MPa/min for preloading. Once the surrounding pressure reached the predetermined value of the experimental program, the surrounding pressure was kept constant, and the axial pressure was applied to the predetermined value of the experimental program. Keeping the axial stress σ_1, radial stress σ_2, and σ_3 stable, the percolation pressure P_w was applied through an advection pump, and percolation was carried out using a constant flow pattern of 5 mL/min. Based on the strength of the concrete samples in the pre-test and the equipment parameters, the experimental program in this research is shown in Table 1.

Table 1. Loading scheme for seepage test of concrete with fracture.

No.	Fracture Morphology	Fracture Angle/°	Surrounding Stress σ_2 (σ_3)/MPa	Axial Stress σ_1/MPa	Seepage Stress P_w/MPa
1	Without fracture	/	4, 5, 6	6, 7, 8	3
2	Single fracture	0, 15, 30, 45, 60, 75, 90			
3	Y-fracture	0, 15, 30, 45, 60			
4	X-fracture	30, 45, 60, 75, 90			

In the research, the effect of prefabricated fractures on the mechanical strength of concrete can be obtained by conducting compressive strength tests on concrete samples with single fracture, "Y" fracture, and "X" fracture with different inclination angles. The effect of fracture–seepage–stress coupling on the deterioration in the mechanical properties of concrete was investigated by conducting compressive strength tests on concrete samples after seepage tests and without seepage tests, respectively.

Under different seepage pressures and external stresses, fractured concrete's fracture and microcrack morphology will change. Seepage water and external stresses also affect the microstructure of concrete and, hence, its water absorption. Therefore, concrete samples' water absorption must be tested before starting the subsequent study. The saturated water content of the concrete samples before and after the percolation test was obtained by immersion testing of dry concrete samples to investigate the water absorption characteristics of the fractured concrete samples under seepage–stress coupling.

3. Characteristics of Seepage in Fracture-Containing Concrete

3.1. Investigation of Water Absorption Characteristics of Fracture-Containing Concrete Samples under Seepage–Stress Coupling

A non-destructive water immersion device was used to saturate the samples. The water content of the concrete sample at the current moment was calculated by Equation (1).

$$w_t = \frac{m_t - m_s}{m_s} \times 100\% \tag{1}$$

where w_t is the moisture content of the sample after immersion time t expressed as %; m_t is the total mass of the sample after immersion time in g; and m_s is the mass of the sample when dry in g.

The average water content–time relationship of the fracture-containing concrete samples before and after the percolation experiment is shown in Figure 5.

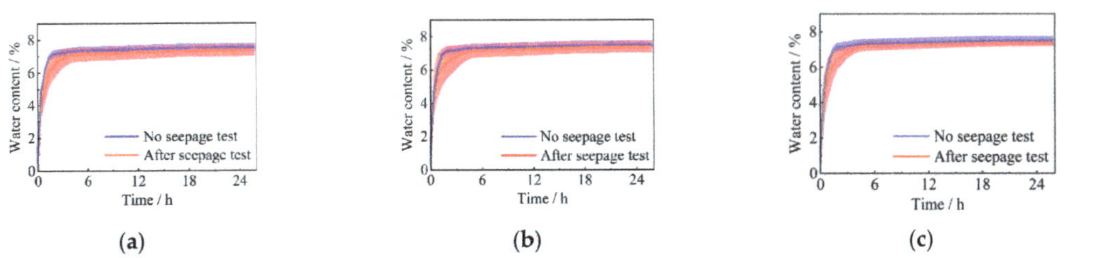

Figure 5. Mean moisture content–time relationship for concrete samples with fractures: (**a**) single fracture, (**b**) Y-fracture, and (**c**) X-fracture.

According to Figure 5, the average saturated water content of concrete samples without the percolation experiment is about 7.54%—the average saturated water content of concrete samples after the reduced percolation experiment. The average saturated water content of concrete samples with a single fracture is about 7.37%, about 2.25% lower. The average saturated water content of concrete samples with Y-fracture is about 7.40%, about 1.86% lower. The average saturated water content of concrete samples with X-fracture is about 7.38 percent, about 2.12 percent lower. Concrete samples without percolation experiments reach 95% of the saturated water content in about 2 h. The time to reach saturated water content was prolonged for all the concrete samples containing fractures after the percolation experiment (single fracture, 4 h; Y-fracture, 4.5 h; and X-fracture, 4.5 h). From the microstructural aspect of the concrete samples, it is possible that the pore structure of the concrete was adjusted by the coupling of stress and seepage. The internal pores of concrete are reduced due to external stresses and seepage stresses. The capillary action of the concrete pores is reduced so that the water absorption efficiency of the sample is reduced, the saturated water content is reduced, and the time to reach the saturated state is prolonged. The presence of fracture reduces the saturated water content of concrete. This also provides additional space and channels for seepage, allowing water to penetrate deeper into the concrete. Moreover, due to the fracture in the sample, the stress concentration phenomenon occurs near the fracture of the sample when the external stress is applied. This stress concentration phenomenon causes the fracture to expand and change the water absorption channel, thus affecting the water absorption characteristics of concrete containing fractures. The prolonged percolation may also lead to changes in the internal microstructure of the concrete, such as stripping the aggregate–cement paste interface. These changes may also further affect the water absorption properties of concrete.

From the distribution of error bands in Figure 5, it can be seen that with the complexity of the geometrical characteristics of the fracture in the concrete sample (single fracture–Y-fracture–X-fracture), the water absorption characteristics of the concrete sample tend to stabilize (the width of the error band of the average water content gradually decreases). Comparing the error bands of concrete samples with the same fracture morphology before and after percolation, it can be seen that the water absorption characteristics of concrete changed significantly after percolation, and there are apparent differences in the water content state between samples at different times. The significant difference in the error bands of concrete samples with different fracture patterns before and after percolation also reflects the significant effect of fracture–seepage coupling on the water absorption of concrete. The range of the error bands can also reflect the permeability changes in concrete samples with different fracture patterns to a certain extent.

3.2. *Seepage Characteristics of Concrete Samples Containing Fractures*

Further investigation of the seepage characteristics of concrete samples can better reveal how water is transported along the fracture in fracture-containing concrete and how this process is affected by the external stress conditions of the concrete. Combined with the results of its water absorption characteristics, it can better help us to analyze the mechanical characteristics of fracture-containing concrete under complex environmental conditions. Fracture size, fracture angle, and fracture morphology are important fracture parameters that affect the seepage characteristics of concrete. Among them, previous research has explored the effect of rock fracture size on the seepage characteristics of rock and found that the rough fracture seepage of rock has an apparent size effect [20]. Next, the effect of fracture angle and fracture morphology on the seepage characteristics of fracture-containing concrete samples will be investigated. According to Darcy's law [21], a linear relationship exists between the seepage velocity of the fluid and the hydraulic gradient, as shown in Equation (2).

$$Q = k\frac{\Delta p A}{\mu l} \tag{2}$$

where Q is the fluid flow rate through the sample per unit time in cm^3/s; k is the permeability of the sample in m^2; A is the cross-sectional area of the fluid through the sample in cm^2; μ is the viscosity of the fluid in Pa·s; l is the length of the seepage path in cm; and Δp is the pressure difference before and after the passage of the fluid through the sample in MPa.

The permeability k of the sample can be obtained according to Equation (3).

$$k = \frac{q\mu l}{\Delta p} \tag{3}$$

where q is the seepage velocity in cm/s.

The permeabilities of single-fracture concrete samples with different fracture angles under different stresses can be obtained from the seepage test of fracture-containing concrete samples, as shown in Figure 6. With the gradual increase in the angle between the fracture and the direction of seepage in the concrete samples, the permeability of concrete samples under each stress decreases. Among them, the permeability change was most evident in the fracture angle from 0° to 30°. As an example, when the surrounding pressure is 4 MPa and the axial pressure is 6 MPa, the permeability of the concrete sample decreases from 6.46×10^{-16} m^2 to 2.87×10^{-16} m^2, which is a decrease of about 55.57% when the fracture angle in the concrete is increased from 0° to 30°. Throughout the process of increasing the fracture angle from 0° to 90°, the permeability of the concrete samples decreased from 6.46×10^{-16} m^2 to 1.73×10^{-16} m^2, a decrease of about 73.22%. It can be seen that the permeability of concrete samples containing a single fracture is more significantly affected by the angle between the single fracture and the direction of seepage. The existence of fracture in the concrete sample shortens the distance of water seepage in its interior, so the sample's permeability is more significant when the angle between the single fracture and the seepage direction is minor. According to Figure 6, it can be seen that the permeability of concrete samples containing fractures is mainly affected by the stress perpendicular to the seepage direction. Due to the effect of stress perpendicular to the direction of seepage, the internal fracture and pore space of concrete is closed by pressure so that the seepage channel is reduced, which reduces the permeability of concrete samples. Moreover, this effect is more significant at low stress. Taking σ_1 = 6 MPa as an example, when σ_2 (σ_3) increased from 4 MPa to 6 MPa, the permeability of concrete samples decreased by about 34.84% on average. Moreover, when σ_1 = 8 MPa, the average decrease in permeability of concrete samples under the same condition is about 29.13%.

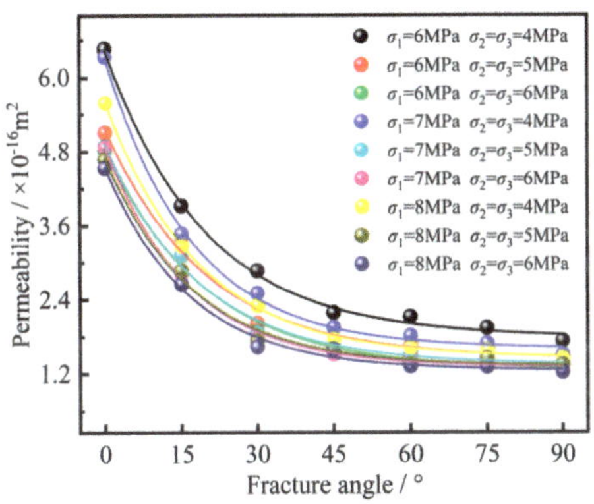

Figure 6. Permeabilities of single-fracture concrete samples with different fracture angles.

The permeabilities of Y-fracture and X-fracture concrete samples at different stresses are shown in Figure 7. Compared with the permeability change in concrete samples containing a single fracture at different fracture angles in Figure 6, the fracture angle does not significantly affect the permeability of concrete samples with Y-fracture and X-fracture. Taking the average permeability of concrete samples at each stage during the seepage experiments as an example, the average permeability of single-fracture concrete samples was reduced from 3.03×10^{-16} m^2 to 2.04×10^{-16} m^2, a reduction of about 32.67%, during the process of increasing the stresses applied to the samples from σ_1 = 6 MPa and σ_2 (σ_3) = 4 MPa to

σ_1 = 8 MPa and σ_2 (σ_3) = 6 MPa. The average permeability of Y-fracture concrete samples was reduced from 6.24 × 10^{-16} m^2 to 5.75 × 10^{-16} m^2, about 7.85%. The average permeability of X-fracture concrete samples was reduced from 6.73 × 10^{-16} m^2 to 5.65 × 10^{-16} m^2, about 16.05%. From the above, it can be seen that with the increase in the axial pressure and the surrounding pressure, the permeability of Y-fracture and X-fracture with different fracture angles shows an overall decreasing trend. From the analysis of fracture morphology, it can be seen that the more complex the fracture morphology is under the same stress condition, the smaller the effect of stress on the permeability of concrete samples. This suggests that a complex fracture structure may produce more seepage paths, thus reducing the effect of stress on permeability. In the seepage experiments, the average permeability of the concrete samples containing a single fracture was 2.40 × 10^{-16} m^2, the average permeability of the concrete samples containing a Y-fracture was 5.93 × 10^{-16} m^2, and the average permeability of the concrete samples containing an X-fracture was 6.16 × 10^{-16} m^2. The more complex the fracture morphology of the concrete samples, the greater the permeability of the concrete samples. This also indicates that the complex fracture structure provides more channels for water seepage and increases the permeability of the concrete samples.

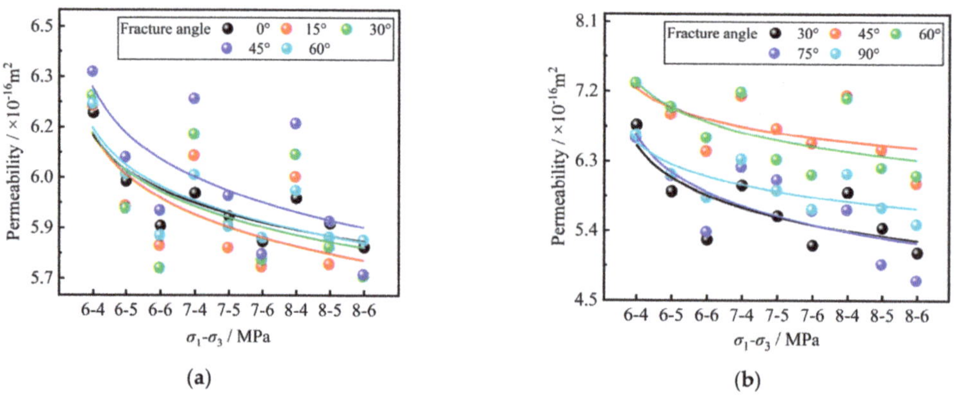

Figure 7. Permeabilities of concrete samples with complex fractures at different fracture angles: (**a**) Y-fracture and (**b**) X-fracture.

4. Degradation of Mechanical Properties of Fracture-Containing Concrete

By exploring the effect of fracture–seepage–stress coupling on the strength of concrete, it is possible to reveal the specific effect of its complex coupling on the bearing capacity of concrete. Secondly, by investigating the damage characteristics of concrete, a more comprehensive understanding of the effects of different conditions on concrete can be obtained. The above research can provide a theoretical basis for better predicting and improving concrete performance and durability.

4.1. Effect of Fracture–Seepage–Stress Coupling on the Strength of Concrete

Formation and development of fracture are the primary manifestations of concrete deterioration [22–24]. Fractures increase the stress concentration in concrete. The local stress concentration due to stress redistribution, in turn, promotes the generation of new fractures or the extension of existing ones, which reduces the overall load-bearing capacity of the concrete structure and reduces its strength [25]. Seepage of water and other fluids through concrete fracture may lead to chemical erosion or physical stripping of materials around concrete fracture [26–28]. This may lead to further expansion of the fracture in the concrete. At the same time, seepage's physical erosion effect may change the concrete's microstructure. This will increase the brittleness and reduce the strength of concrete to some extent. Different stress states may change the opening and direction of expansion and development of concrete fracture, thus affecting the permeability and mechanical properties of concrete. It can be seen that stress redistribution, seepage erosion, and stress action generated by fracture have apparent effects on the mechanical properties of concrete. However, there is less research on coupling fracture, seepage, and stress. In general, the research on the effect of fracture–seepage–stress coupling on the strength of concrete is of great significance.

4.1.1. Deterioration in Concrete Strength by Fractures in Concrete

The stress–strain curves of the fracture-containing concrete samples without the seepage test are shown in Figure 8. The strength of the concrete sample without fracture is about 14.15 MPa. The strength of the concrete sample with a single fracture is about 11.16 MPa, which is 21.13% lower than that of the concrete sample without fracture. The strength of the concrete sample with Y-fracture is about 9.01 MPa, which is 36.33% lower than that of the concrete sample without fracture. The strength of the concrete samples containing X-fracture is about 9.48 MPa, which is 33.00% lower than those without fracture. It can be seen that fracture in concrete has a more significant deteriorating effect on the strength of concrete. The existence of the fracture leads to the stress concentration phenomenon in concrete, which makes the local area of concrete reach the destructive strength earlier. The Y- and X-fractures, due to their more complex structure, produce a more obvious stress concentration at the tip of the fracture, leading to further degradation of the strength of the concrete samples. From the figure, it can also be found that the peak strain of the concrete samples decreases under the influence of the deteriorating effect of the fracture. The more complex the fracture morphology, the smaller the peak strain of concrete samples. During the compression process of the sample, there is a prominent stage of fracture closure in the concrete samples containing fracture, which is manifested by the rapid change in the stress–strain curve of the sample on the strain axis within a short period. The more complex the fracture morphology in the sample, the earlier the fracture closure phenomenon occurs. This also indicates that the deterioration effect of cracks in concrete samples is more apparent when the fracture morphology is more complex.

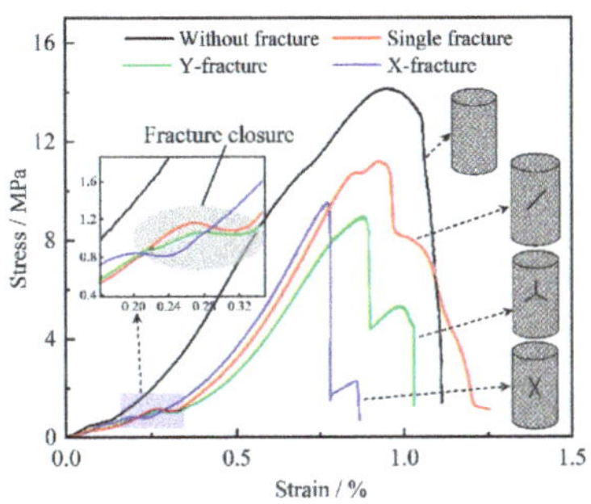

Figure 8. Stress–strain curves of fracture-containing concrete samples without seepage test.

4.1.2. Deterioration in Concrete Strength by Fracture–Seepage–Stress Coupling

The strengths of fracture-containing concrete samples after the seepage experiment are shown in Table 2. The relationship between the strength attenuation in concrete samples under the coupled action of fracture–seepage–stress and the angle of the fracture is shown in Figure 9. Table 2 and Figure 9 show that different fracture morphologies will cause the strength of concrete samples to be significantly attenuated after the combined action of seepage and stress. Among them, after the same seepage pressure and external stress, the average attenuation in the strength of concrete samples with a single fracture is 26.06%, the average attenuation in the strength of concrete samples with Y-fracture is 32.06%, and the average attenuation in the strength of concrete samples with X-fracture is 37.83%. The more complex the fracture morphology, the stronger the deterioration effect on the concrete strength under the combined effect of fracture–seepage–stress. From Figure 9, it can be seen that there is a positive correlation between the strength attenuation in concrete samples containing a single fracture and the single-fracture angle. This may be because the increase in the single-fracture angle leads to an increase in the area of the fracture in the direction perpendicular to the loading direction. Under the same external stress, the effective bearing area of the concrete samples decreases, which leads to different degrees of attenuation in the strength of the concrete samples. In contrast, in relatively complex fracture morphology (X and Y shapes), there is no significant relationship between the strength attenuation in concrete samples and the fracture angle.

Table 2. Strengths of fracture-containing concrete samples after seepage experiments.

No.	Fracture Morphology	Fracture Angle/°	Strength Average/MPa	Strength Decay/%
A1	Without fracture	/	13.02	/
B1	Single fracture	0	10.89	16.36
B2		15	10.29	20.97
B3		30	9.87	24.19
B4		45	9.50	27.04
B5		60	9.18	29.49
B6		75	8.91	31.57
B7		90	8.75	32.80
C1	Y-fracture	0	8.06	38.10
C2		15	9.31	28.49
C3		30	9.41	27.73
C4		45	9.23	29.11
C5		60	8.22	36.87
D1	X-fracture	30	8.19	37.10
D2		45	8.27	36.48
D3		60	7.83	39.86
D4		75	7.93	39.09
D5		90	8.25	36.64

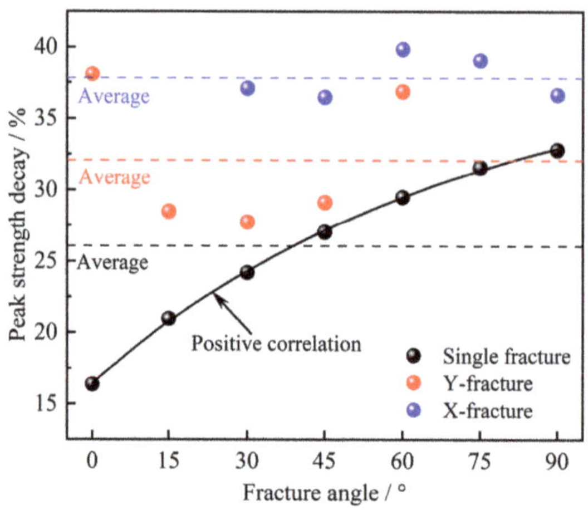

Figure 9. Strength decay of concrete samples under the coupled action of fracture–seepage–stress in relation to the angle of the fracture.

The average strengths of concrete samples before and after the seepage experiment are shown in Table 3. According to the data in the table, it can be seen that after the seepage experiment, the strength of concrete samples has different degrees of decline due to the joint effect of seepage and stress. Among them, the strength of concrete samples with fracture decreased significantly compared with that of concrete samples without fracture. Under the influence of seepage and stress, new microcracks may be generated around the cracks in the concrete samples, exacerbating the cracks' extension. Concrete samples with complex fractures may have more complex seepage channels, which may exacerbate the reduction in the load-carrying capacity of the samples after the seepage experiments. When there are fractures in concrete, the seepage medium (water) seeps through these channels. This seepage not only causes changes in pressure within the fracture. It also accelerates erosion within the concrete when the seepage medium contains chemicals that can react with the concrete components. All of these effects affect the development and expansion of microfractures in concrete. The presence of fracture reduces the overall continuity of concrete, making it more prone to fracture and damage under external stress. When the original fracture, seepage, and stress coupling will accelerate the expansion of secondary cracks, thus reducing the bearing capacity and durability of concrete. In general, the mechanical properties of concrete samples containing fractures will produce a more obvious deterioration phenomenon under fracture–seepage–stress coupling.

Table 3. Average strengths of concrete samples before and after seepage experiments.

Fracture Morphology	Average Strengths of Concrete Samples/MPa		Strength Decay/%
	No Seepage Experiment	After the Seepage Experiment	
Without fracture	14.15	13.02	7.99
Single fracture	11.16	9.50	14.87
Y-fracture	9.01	8.06	10.54
X-fracture	9.48	8.19	13.61

4.2. Damage Characteristics of Fracture-Containing Concrete under Fracture–Seepage–Stress Coupling

4.2.1. Characteristics of Crack Extension in Concrete with Fractures under Compression

In order to further clarify the damage characteristics of fracture-containing concrete, the crack extension characteristics of concrete samples with different fracture angles under compression after seepage tests are investigated. Currently, researchers have identified the types of cracks that can be observed in rock-like materials under uniaxial compression [29,30]. Under compression, wing cracks (smaller, spiky lateral cracks that accompany the main crack on either side) and secondary cracks (cracks initiated by extension of the main crack or stress concentration at the crack end) occur in rock-like materials [31]. Among them, wing cracks start to expand from the tip of the original fracture, which appears earlier. Moreover, it expands in a curvilinear path with increasing load. Wing cracks are a type of tensile crack, and the direction of initiation is approximately perpendicular to the direction of the original fracture. Secondary cracks appear later, usually in the form of shear cracks in the same plane of the original fracture expansion. According to the characterization of the crack surface, the crack form can be distinguished: wing cracks and tensile cracks have smooth surfaces. Secondary shear cracks (cracks that are perpendicular to the extension direction of the main crack and are in a shear state) have a rough surface accompanied by a large amount of rock particle debris. The shear strength can be characterized by the degree of debris fragmentation. A schematic diagram of the crack extension morphology and types of fracture-containing rock materials under compression is shown in Figure 10.

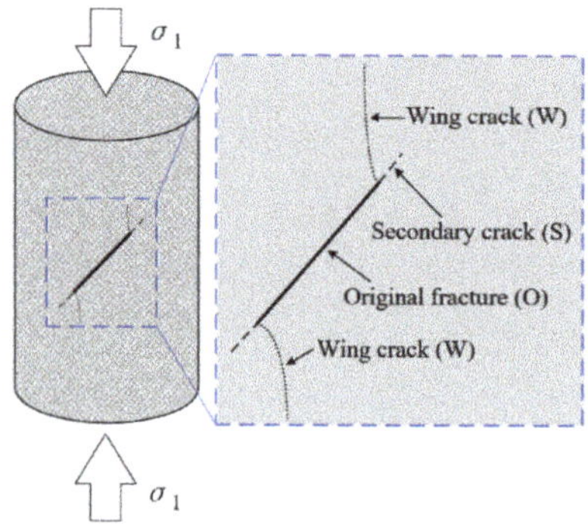

Figure 10. Morphology and type of crack extension after compression of fracture-containing rock-like materials.

Macroscopic crack expansion on the surface of concrete samples in various stages of compression was recorded by high-definition images. Now, the crack distribution images of each sample at the time of damage are taken to analyze the crack extension characteristics of the concrete containing fracture under compression. The crack expansion of concrete samples after compression damage is shown in Figure 11. In the compression process of the sample, the elastic deformation of concrete under compression led to the expansion of cracks. Moreover, the through fracture formed by the crack expansion led to the loss of bearing capacity and damage to the concrete sample. Wing cracks and secondary shear cracks mainly control the damage form of concrete samples. Among them, wing cracks initiate at the tip of the original fracture. When it expands to a certain extent, it will be connected to the secondary shear cracks generated from the end face of the sample, thus forming a through fracture between the end face of the sample and the original fracture, leading to the destruction of the sample.

According to the crack distribution of concrete samples under different fracture angles, it can be seen that the original fracture angle has different degrees of influence on the type of crack extension and final damage morphology of the samples. In concrete samples containing a single fracture, the macroscopic crack extension is more evident with increased fracture angle. Among them, when the fracture angle is slight, the damage form of the sample is dominated by tensile damage. When the fracture angle is larger, the damage form of the sample is dominated by tensile–shear composite damage. In the concrete samples containing Y-fracture or X-fracture, the influence of the fracture angle on the crack extension characteristics and the damage form of the sample is minor. According to the expansion of cracks in concrete samples with different fracture morphology, it can be seen that the more complex the original fracture morphology is, the more the number of wing cracks and secondary shear cracks produced by compression damage.

4.2.2. Damage Fractals of Concrete Containing Fracture under Compression

The broken mass of concrete samples after destabilization and damage directly results from the extension and penetration of fractures and cracks, which has the value of fractal research [32]. The fractal research used the particle size–mass classification method to fractalize the broken blocks produced by compressing concrete samples containing fracture [33]. The broken masses of concrete samples were sieved using a particle-size sieve. The sieved particle size ranges were larger than 25 mm, 25–20 mm, 20–15 mm, 15–10 mm, 10–5 mm, 5–2.5 mm, and less than 2.5 mm, respectively. The sieved fractions of broken blocks of some concrete samples are shown in Figure 12.

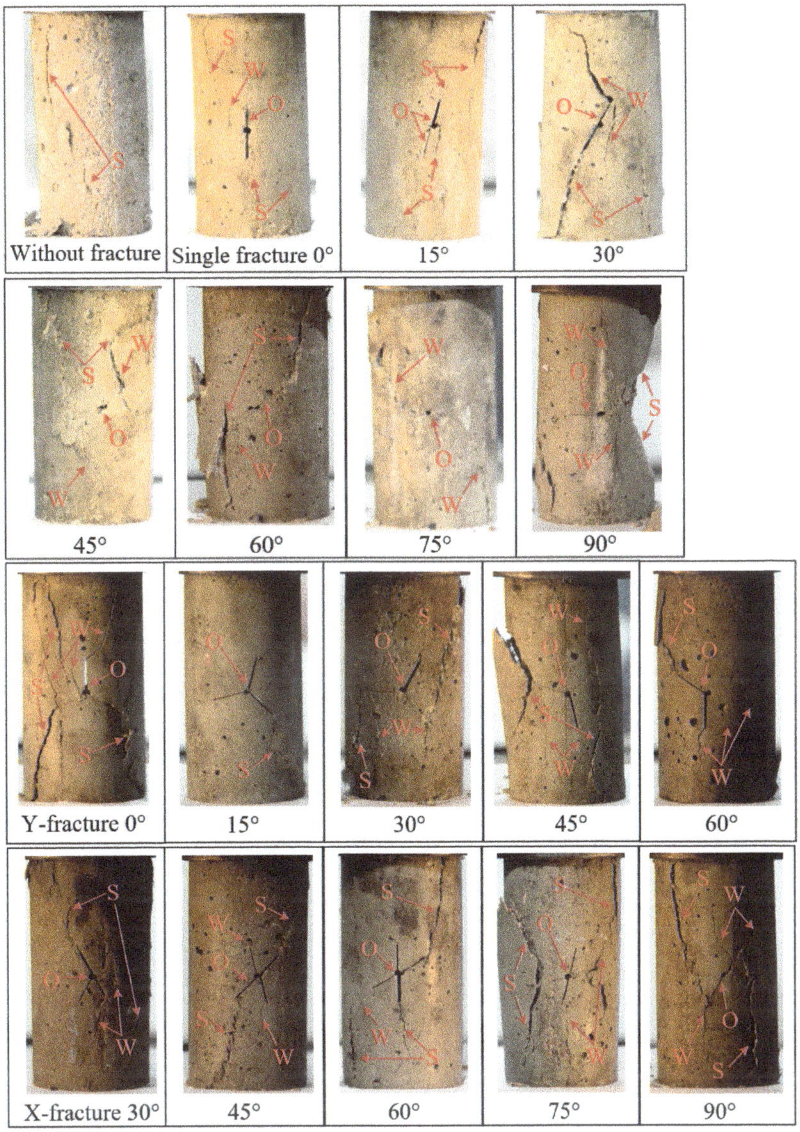

Figure 11. Expansion of cracks in concrete samples after damage under compression.

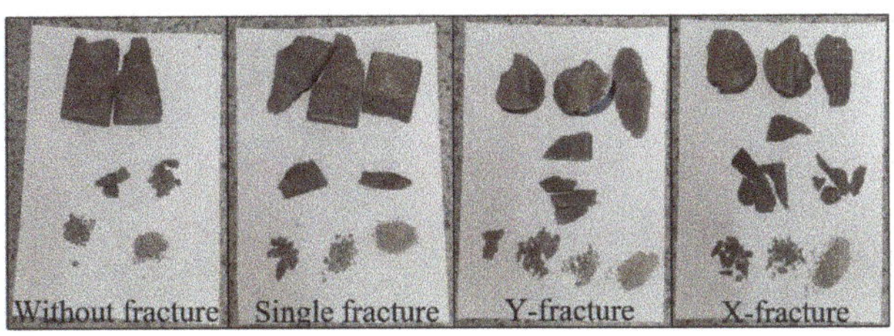

Figure 12. Sieving of broken masses of some concrete samples.

In this research, the mass of the broken mass M and the equivalent side length of the broken mass L_{eq} were used for statistical analysis to calculate the fractal dimension (a mathematical concept used to describe the complexity and irregularity of a rock's surface or internal structure) of the broken mass of the concrete sample D. The equations are as follows:

$$L_{eqi} = 1000 \cdot \sqrt[3]{\frac{m_i}{\rho \cdot 10^6}} \tag{4}$$

where L_{eqi} is the equivalent side length of the broken mass of the i sample in mm; ρ is the average density of the concrete sample in kg/m^3; and m_i is the mass of the broken mass of the i sample in g.

$$\alpha = \frac{\lg(M_{L_{eq}}/M)}{\lg L_{eq}} \tag{5}$$

where α is the slope of $M_{L_{eq}}/M{-}L_{eq}$ in double logarithmic coordinates; $M_{L_{eq}}/M$ is the cumulative percentage of fragments with equivalent side lengths less than L_{eq}; $M_{L_{eq}}$ is the mass of fragments corresponding to an equivalent side length of L_{eq} in g; and M is the total mass of fragments in the computational scale in g.

$$D = 3 - \alpha \tag{6}$$

The concrete sample broken-block mass–equivalent edge length relationships are shown in Figure 13. From the figure, it can be seen that the concrete sample broken-block mass–equivalent edge length has good linear correlation and self-similarity. It conforms to the fractal statistical distribution.

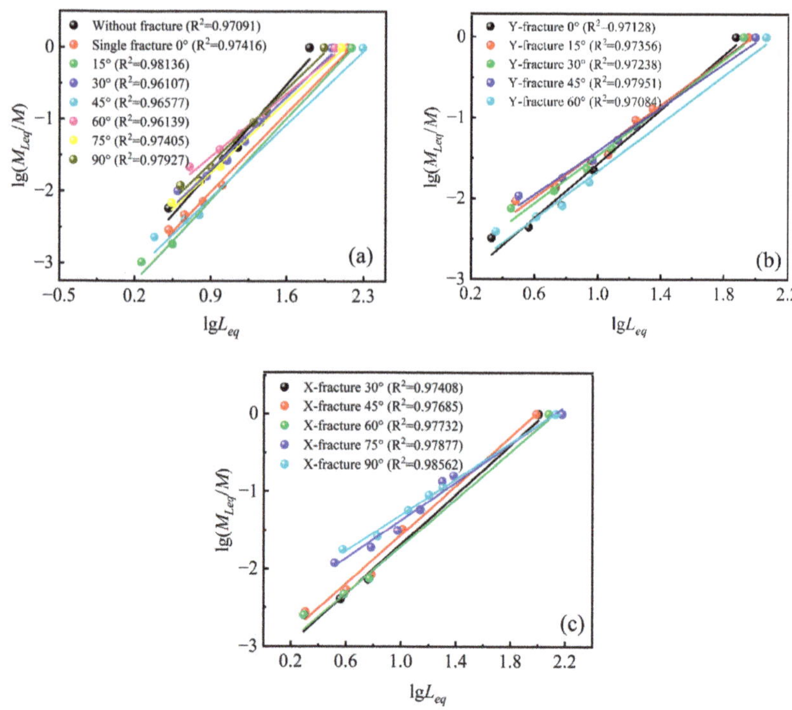

Figure 13. Mass–equivalent edge length relationships for broken blocks of concrete samples: (**a**) single fracture, (**b**) Y-fracture, and (**c**) X-fracture.

Statistically, the distribution of fractal dimensions of broken blocks of concrete samples in this research is shown in Figure 14. Concrete samples without fracture have the lowest fractal dimension, indicating that the structure of their broken blocks is relatively simple. The fractal dimension of concrete samples with a single fracture increased compared to the fractal dimension of concrete samples without fracture, indicating the effect of fracture on the structural complexity of the broken blocks of concrete samples. The fractal dimension of broken blocks of concrete samples further increased with the complexity of fracture morphology (Y-fracture and X-fracture). It can be seen that the presence and expansion of the fracture add new interfaces and discontinuities to the concrete sample on a macroscopic level. These fractures make the concrete sample more fragmented during compressive instability damage, producing broken blocks with smaller dimensions and less mass. A comparison of the fractal dimension of the broken blocks of concrete samples before and after the seepage experiments also shows a slight increase in the fractal dimension of the broken blocks of concrete samples after the seepage experiments. This may be related to the dynamic effect of fluid percolation. Under seepage and external stress, fracture-containing concrete's internal pore structure, microcrack, and stress state are redistributed. Moreover, this coupling effect will change the bond state inside the concrete, which reduces the strength of the concrete structure and causes more block fragmentation after damage.

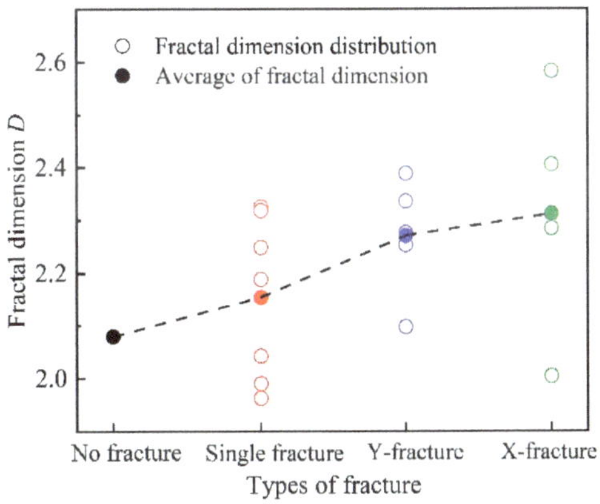

Figure 14. Distribution of fractal dimensions of broken blocks of concrete samples.

5. Conclusions

This research mainly carried out the seepage experiment and uniaxial compression test of concrete samples containing different fractures. It includes investigating the water absorption characteristics of concrete samples containing fractures, the change in permeability of samples under different stresses, and the mechanical strength and damage characteristics of samples with different fracture morphologies before and after the seepage experiments. Finally, the seepage characteristics of fracture-containing concrete under different stresses and the deterioration characteristics of mechanical properties of fracture-containing concrete under the action of fracture–seepage coupling were obtained. The main conclusions are as follows:

The action of seepage and external stresses can reduce the saturated water content of concrete and prolong the time it takes for the concrete to reach the saturated water content. The saturated water content is reduced by about 2.1%. In fracture-containing concrete, the stress concentration near the fracture will cause the fracture to expand and change the water-conducting channel. Therefore, the more complex the geometrical characteristics of the fracture in the concrete, the less time it takes for each concrete sample to reach saturation versus the change in saturated water content.

The permeability of concrete samples containing a single fracture decreases and stabilizes with the increase in the angle between the single fracture and the direction of seepage. Under the stress conditions of this research, the permeability of the concrete samples was reduced by about 73.2% when the pinch angle was 90° at maximum. However, in concrete with more complex fractures, the angle between the fracture and the seepage direction has no noticeable effect on the permeability of the concrete. The more complex fracture morphology can reduce the effect of stress on concrete permeability.

A single fracture in concrete samples reduces their strength with increasing angle between the fracture and the stress direction. More complex fracture morphology correlates with lower strength and peak strain. Seepage–stress coupling significantly affects strength, with fracture-containing samples experiencing double the strength decay compared to samples without fracture. Concrete samples with different fracture types also showed notable strength reductions, averaging 26.06% for single fracture, 32.06% for Y-fracture, and 37.83% for X-fracture.

The elastic deformation of concrete under compression is the leading cause of crack expansion in concrete samples, and crack development eventually leads to sample damage. The original fracture angle significantly affects the crack expansion and damage pattern. Smaller fracture angles in concrete samples with single fracture lead to tensile damage, and larger fracture angles lead to tensile–shear composite damage. The higher the complexity of the fracture morphology, the higher the number of wing cracks and secondary shear cracks generated during concrete damage, resulting in a more complex damage pattern. The fractal dimension of broken blocks of fracture-containing concrete samples increases with the complexity of fracture morphology. The coupling effect of fracture, seepage, and stress can further increase the fractal dimension of fracture-containing concrete broken blocks.

From the research on the effect of fracture–seepage–stress coupling on the strength and damage characteristics of concrete, it can be seen that the existence of fracture and its coupling with seepage and stress significantly reduces the bearing capacity and durability of concrete, which is manifested in the apparent deterioration in the mechanical properties of fracture-containing concrete after the occurrence of seepage phenomenon.

Author Contributions: Z.S., C.W. and X.L. conceived and designed the experimental program; Z.S., C.W. and F.W. derived the theoretical equations; Z.S., C.W., C.S. and Y.X. conducted the laboratory experiments; Z.S. and F.W. collated and analyzed the experimental data; Z.S. wrote the paper and X.L. helped Z.S. revise the manuscript. All authors have read and agreed to the published version of the manuscript.

Funding: This research was funded by the Science and Technology Innovation Special Fund of Jiangsu Provincial Science and Technology Department, approval No. BK20220024, project No. 2022-12460.

Data Availability Statement: No data were used to support this study.

Conflicts of Interest: The authors declare no conflicts of interest.

References

1. Wei, C.; Zhang, B.; Zhu, W.; Wang, S.; Li, J.; Yang, L.; Lin, C. Fracture Propagation of Rock like Material with a Fluid-Infiltrated Pre-existing Flaw Under Uniaxial Compression. *Rock Mech. Rock Eng.* **2021**, *54*, 875–891. [CrossRef]
2. Nolen-Hoeksema, R.C.; Gordon, R.B. Gordon Optical detection of crack patterns in the opening-mode fracture of marble. *Int. J. Rock Mech. Min. Sci. Geomech. Abstr.* **1987**, *24*, 135–144. [CrossRef]
3. Van Mier, J.; Man, H.K. Some Notes on Microcracking, Softening, Localization, and Size Effects. *Int. J. Damage Mech.* **2009**, *18*, 283–309. [CrossRef]
4. Lajtai, E.Z. Unconfined Shear Strength of Discontinuous Rocks. In Proceedings of the 2nd ISRM Congress, Belgrade, Yugoslavia, 21–26 September 1970.
5. Feng, G.; Zhu, C.; Wang, X.; Tang, S. Thermal effects on prediction accuracy of dense granite mechanical behaviors using modified maximum tangential stress criterion. *J. Rock Mech. Geotech.* **2023**, *15*, 1734–1748. [CrossRef]

6. Feng, G.; Wang, X.C.; Kang, Y.; Zhang, Z.T. Effect of thermal cycling-dependent cracks on physical and mechanical properties of granite for enhanced geothermal system. *Int. J. Rock Mech. Min.* **2020**, *134*, 104476. [CrossRef]
7. Wong, R.; Lin, P.; Tang, C.A. Experimental and numerical study on splitting failure of brittle solids containing single pore under uniaxial compression. *Mech. Mater.* **2006**, *38*, 142–159. [CrossRef]
8. Qianqian, D.; Guowei, M.; Qiusheng, W.; Mingjie, X.; Qiaoyan, L. Different damage parameters on failure properties with non-straight marble under uniaxial compression. *Electron. J. Geotech. Eng.* **2015**, *20*, 6151–6167.
9. Chen, D.; Yu, X.; Liu, R.; Li, S.; Zhang, Y. Triaxial mechanical behavior of early age concrete: Experimental and modelling research. *Cem. Concr. Res.* **2019**, *115*, 433–444. [CrossRef]
10. Chen, D.; Yu, X.T.; Guo, M.Y.; Liao, Y.D.; Ouyang, F. Study on the mechanical properties of the mortars exposed to the sulfate attack of different concentrations under the triaxial compression with constant confining pressure. *Constr. Build. Mater.* **2017**, *146*, 445–454. [CrossRef]
11. Xie, H.P.; Yang, J.U. Criteria for Strength and Structural Failure of Rocks Based on Energy Dissipation and Energy Release Principles. *Chin. J. Rock Mech. Eng.* **2005**, *24*, 3003–3010.
12. Yuangao, L.; Weiyuan, Z.; Jidong, Z.; Qiang, Y. Discontinuous Bifurcation Model of Damage Localization for Jointed Rocks and its Application. *Acta Mech. Sin. Chin. Ed.* **2003**, *35*, 411–418.
13. Li, J.C.; Li, T.C.; Wang, G.; Bai, S.W. CT scanning test of built-in fracture extension under uniaxial compression. *J. Rock Mech. Eng.* **2007**, *26*, 484–492.
14. Wang, W.H.; Wang, X.J.; Jiang, H.T.; Yan, Z.; Li, S.R. Mechanical properties of rock-like samples with different inclination fracture under uniaxial compression. *Sci. Technol. Bull.* **2014**, *32*, 48–53.
15. Lin, P.; Huang, K.J.; Wang, R.K.; Zhou, W.Y. Crack extension and damage behaviour of samples with single crack defects at different angles. *J. Rock Mech. Eng.* **2005**, 5652–5657. Available online: https://www.researchgate.net/publication/294743512_Crack_growth_mechanism_and_failure_behavior_of_specimen_containing_single_flaw_with_different_angles (accessed on 9 December 2023).
16. Xie, Q.T.; Guo, J.Z.; Wang, J.L.; Chen, Y.G. Crack extension measurement in sandstone specimens with inclined single cracks under uniaxial compression. *Geotechnics* **2011**, *32*, 2917–2921.
17. Bian, H.; Qin, X.W.; Luo, W.J.; Ma, C.; Zhu, J.; Lu, C.; Zhou, Y.F. Evolution of hydrate habit and formation properties evolution during hydrate phase transition in fractured-porous medium. *Fuel* **2022**, *324*, 124436. [CrossRef]
18. Chen, Y.; Xu, J.; Peng, S.J.; Zhang, Q.W.; Chen, C.C. Strain localisation and seepage characteristics of rock under triaxial compression by 3D digital image correlation. *Int. J. Rock Mech. Min.* **2022**, *152*, 105064. [CrossRef]
19. Hudson, J.A.; Cornet, F.H.; Christiansson, R. ISRM Suggested Methods for rock stress estimation—Part 1: Strategy for rock stress estimation. *Int. J. Rock Mech. Min.* **2003**, *40*, 991–998. [CrossRef]
20. Shi, Z.; Yao, Q.; Wang, W.; Su, F.; Li, X.; Zhu, L.; Wu, C. Size Effects of Rough Fracture Seepage in Rocks of Different Scales. *Water-Sui* **2023**, *15*, 1912. [CrossRef]
21. Gray, W.G.; Miller, C.T. Examination of Darcy's law for flow in porous media with variable porosity. *Environ. Sci. Technol.* **2004**, *38*, 5895–5901. [CrossRef]
22. Zhang, D.; Yao, H. Disease Analysis and solutions of the prestressed concrete bridge cracks. In Proceedings of the 2nd International Conference on Civil Engineering, Architecture and Building Materials (CEABM 2012), Yantai, China, 25–27 May 2012; Volume 178–181, pp. 2398–2400.
23. Shaikh, F.U. A Effect of Cracking on Corrosion of Steel in Concrete. *Int. J. Concr. Struct. Mater.* **2018**, *12*, 3. [CrossRef]
24. Combrinck, R.; Steyl, L.; Boshoff, W.P. Influence of concrete depth and surface finishing on the cracking of plastic concrete. *Constr. Build. Mater.* **2018**, *175*, 621–628. [CrossRef]
25. Yang, J.Y.; Guo, Y.C.; Tan, J.J.; Shen, A.Q.; Wu, H.; Li, Y.; Lyu, Z.H.; Wang, L.S. Strength deterioration and crack dilation behavior of BFRC under dynamic fatigue loading. *Case Stud. Constr. Mater.* **2022**, *16*, e01051. [CrossRef]
26. Ting, S.; Yang, L.; IOP. Chloride ion erosion experiment research in cracked concrete. In Proceedings of the 2nd International Conference on Materials Science, Energy Technology and Environmental Engineering (MSETEE), Zhuhai, China, 28–30 April 2017; Volume 81.
27. Gao, Y.; Zou, C. CT study on meso-crack propagation of gradient composite concrete subjected to sulfate erosion. *Mag. Concr. Res.* **2015**, *67*, 1127–1134. [CrossRef]
28. Guo, J.; Wang, K.; Guo, T.; Yang, Z.; Zhang, P. Effect of Dry-Wet Ratio on Properties of Concrete Under Sulfate Attack. *Materials* **2019**, *12*, 2755. [CrossRef] [PubMed]
29. Lee, J.; Hong, J.W. Morphological aspects of crack growth in rock materials with various flaws. *Int. J. Numer. Anal. Met.* **2019**, *43*, 1854–1866. [CrossRef]
30. Silva, B.G.D.; Einstein, H.H. Modeling of crack initiation, propagation and coalescence in rocks. *Int. J. Fract.* **2013**, *182*, 167–186. [CrossRef]
31. Rhc, W.; Kt, C. Crack coalescence in a rock-like material containing two cracks. *Int. J. Rock Mech. Min.* **1998**, *35*, 147–164.

32. Xie, H.P.; Gao, F.; Zhou, H.W.; Zuo, J.P. Fractal study of rock fracture and fragmentation. *J. Disaster Prev. Mitig. Eng.* **2003**, 1–9. Available online: https://kns.cnki.net/KXReader/Detail?invoice=Ytr/JrSeUr7dE1unRvbIcxIZDvXD4kYuSm/DvinUguvfj3BIAvt3a8sFLwvQnkEQlJeL2HXfbfl5H4XGk/IHLW7aV2tVZZbJt9oq/pDGqplpJVrSJfRtwpK8L5bFcpXF9KqXdRZ72naQQLGthNT5tNW2UwoXm24v+u7bTY77hhM=&DBCODE=CJFQ&FileName=DZXK200304000&TABLEName=cjfd2003&nonce=3AA62C61C3594E6A8D149487A5BD609B&TIMESTAMP=1706602578386&uid= (accessed on 9 December 2023).
33. He, M.C.; Yang, G.X.; Miao, J.L.; Jia, X.N.; Jiang, T.T. Classification of debris in rock explosion experiments and its research method. *J. Rock Mech. Eng.* **2009**, *28*, 1521–1529.

sustainability

MDPI

Article

The Relationship between Concrete Strength and Classes of Resistance against Corrosion Induced by Carbonation: A Proposal for the Design of Extremely Durable Structures in Accordance with Eurocode 2

Luisa María Gil-Martín, Luisa Hdz-Gil, Emilio Molero and Enrique Hernández-Montes *

School of Civil Engineers, University of Granada, 18012 Granada, Spain; mlgil@ugr.es (L.M.G.-M.); luisahernandez@correo.ugr.es (L.H.-G.); emiliomolero@ugr.es (E.M.)
* Correspondence: emontes@ugr.es

Abstract: The new Eurocode 2 provides valuable information on the required concrete cover to protect reinforcement against corrosion induced by carbonation, for two design service life values of 50 and 100 years. However, to design structures with an even longer service life and assess existing ones, additional tools are necessary. The 'square root of time' relationship is a well-established method for estimating the penetration of the carbonation front, making it useful for long-term design and assessment purposes. In this article, we propose a new function that adjusts the evolution of the carbonation front to the Eurocode 2 values. This function is a powerful tool for designing extremely durable structures and assessing existing ones. To demonstrate its effectiveness, we provide two examples of its application.

Keywords: carbonation front; minimum cover; reinforced concrete structures

check for updates

Citation: Gil-Martín, L.M.; Hdz-Gil, L.; Molero, E.; Hernández-Montes, E. The Relationship between Concrete Strength and Classes of Resistance against Corrosion Induced by Carbonation: A Proposal for the Design of Extremely Durable Structures in Accordance with Eurocode 2. *Sustainability* **2023**, *15*, 7976. https://doi.org/10.3390/su15107976

Academic Editors: Mahdi Kioumarsi and Vagelis Plevris

Received: 12 April 2023
Revised: 10 May 2023
Accepted: 10 May 2023
Published: 13 May 2023

1. Introduction

There are several factors that cause damage to reinforced concrete (RC in what follows) structures, and the damage induced by the corrosion of reinforcement has a serious effect on durability [1–4] and on the loading capacity [5,6]. Corrosion is mainly caused by the action of chloride ions [7,8] and the penetration of carbon dioxide [9,10]. So, when the service life of RC structures is being considered [11], requirements such as cover thickness, concrete strength, type of cement, water–cement ratio, and environmental exposure must be taken into consideration [8]. Once corrosion starts, repairs are costly and difficult [12,13].

The increase in urban pollution [14] is just one of the factors that affects the durability of RC structures. The increase in carbon dioxide (CO_2) emissions into the atmosphere affects carbonation depth, making concrete carbonation a serious problem [15]. The service life of a structure is the period of time between the construction of the structure and the moment in which its performance fails to meet the requirements of the users [16]. According to [7], service life is defined as the sum of the corrosion initiation time (or the time the pollutant takes to affect the thickness of the whole concrete cover), and the time needed for the propagation of the corrosion (which is the time between the initiation period and the moment when the level of corrosion degradation is unacceptable). However, regarding carbonation, damage usually occurs at the instant when the carbonation front comes into contact with the reinforcement [17,18]. So, when assessing service life, the period of time corresponding to the propagation of corrosion is not considered. This is supported by the fact that when the carbonation front reaches the reinforcing bar, the damage to the structure is of a significant level.

On site, carbonation depths are measured using the phenolphthalein method or other more sophisticated techniques such as thermogravimetric analysis. In the literature, there are several models that can be used to estimate carbonation depth, and these models can be used to make service life predictions [15,19–21], which is a major advance in how RC structures exposed to CO_2 over time can be evaluated.

Some of the models used for determining carbonation depth [18,19,22] consider the influence of the different parameters involved in the carbonation process. These models can be difficult to apply in practice as obtaining all the input parameters [8] can be both time consuming and expensive.

Contrary to the analysis of chloride ingress in concrete structures [23], where simplified models can be inappropriate [24], there are models based on simplified equations [7] widely used to determine the carbonation depth [9].

Carbonation is mainly affected by environmental conditions (CO_2 content, relative humidity, and temperature), exposure conditions (protection from the rain), concrete material (water/binder ratio (w/b), cement type, and content), the amount of CO_2 in the environment, and pore size distribution [18,19,25–27].

The process of CO_2 diffusion is usually modeled using Fick's first law. So, the carbonation coefficient or carbonation rate constant, K, (mm/year^{0,5}) is calculated using the 'square root of time' relationship, i.e., as the ratio between the attack penetration depth (mm) and the square root of the exposure time to CO_2 (year) [7,28]. The square root model has been verified both in the laboratory and in the field [29].

Some authors [30] have indicated that it is impossible to consider all the parameters involved in the carbonation process as there are so many of them [28]. However, factors affecting carbonation can be intercorrelated. Statistical studies have proved that the water/binder ratio and the quantity of clinker in the binder have a direct relationship with the compressive strength of concrete [15]. In fact, a higher water-cement ratio makes the diffusion of CO_2 into concrete easier, while, for a fixed water–cement ratio, the use of a higher cement content in the concrete mix improves concrete durability [31]. Moreover, compressive strength influences the porous structure of concrete, and so it is a key factor in the diffusion of carbon dioxide through the material, which also makes it a key factor in carbonation. Generally, a high compressive strength is associated with more compact concrete mixes and with concretes with a higher clinker content. Additionally, due to the chemical reactions during carbonation, the higher the clinker content, the slower the carbonation coefficient is [15].

However, according to [21], the factors which have the most significant effect on the carbonation coefficient K are: the exposure conditions, cement type, and the compressive strength of the concrete, and there is a strong correlation between compressive concrete strength and carbonation resistance [28,32].

Current standards (e.g., [33]) have established minimum cover requirements $c_{min,dur}$ to protect reinforcement from corrosion. These cover values depend on the different classes of exposure and are defined for 50 and 100 years of design service life [32,34,35].

In the CEB Bulletin 34 [36], and in the Spanish and Portuguese standards [28,32,35], the minimum cover is obtained from the 'square root of time', with the time exposure to CO_2 (year) or the time when corrosion starts. In the Spanish Structural Code [35] the apparent carbonation coefficient, $K_{app,carb}$, is obtained as a function of the mean compressive strength of concrete and other factors, such as the amount of entrained air and the type of binder.

In prEN 1992 [34], the value of $c_{min,dur}$ is tabulated by classifying a structure in two ways: exposure class (EC) and exposure resistance class (ERC). The minimum levels of concrete cover for durability needed that can withstand carbonation are presented in a tabular format for 50 and 100 years of design service life.

In this paper, a least square technique is used to adjust a square root model to the data proposed by prEN 1992 [34] for the minimum levels of concrete cover needed to withstand carbonation. The main assumption of the presented proposal is that the minimum cover value is presented as a function of compressive concrete strength for each of the exposure conditions considered. The adjustment is based on the formulation proposed in [35].

2. Materials and Methods

Steel is stable within concrete thanks to the fact that, when hydrated, cement generates an alkaline environment, pH > 12, in which the metal generates a protective film against corrosion which is called the passive film. This film is very thin, very stable—as long as the alkalinity is maintained—and very impermeable to metal ions. Thanks to the passive film, the steel is effectively protected against corrosion [37–39].

The passive film can be damaged or even disappear either because the alkalinity of the concrete is not maintained(pH < 9), mainly induced by the carbonation of the cover, or because of the presence of a high concentration of chlorides in the cover [37]. The time between concreting and the disappearance or damage of the passive film is called the initiation time, see Figure 1.

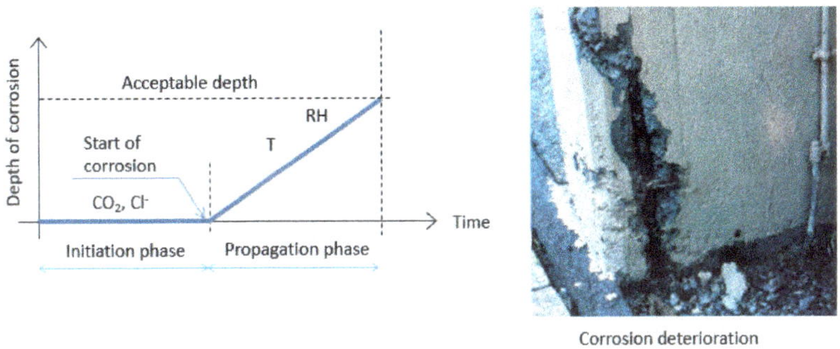

Figure 1. Conceptual scheme of a reinforcement corrosion process as a function of time, adapted from [7].

2.1. *Carbonation Front, a Deterministic Point of View*

Carbonation is the process through which the concrete cover loses its alkalinity, generating a front (carbonation front) that penetrates the RC element inwards.

The moderate solubility of calcium hydroxide (portlandite, $Ca(OH)_2$) formed in the hydration process of cement generates an alkaline environment as a result of the OH^- ions that it generates in the presence of water [40]. Due to the porous nature of concrete, atmospheric CO_2 slowly penetrates through the pores of the concrete and comes into contact with calcium hydroxide, with which it reacts and forms calcium carbonate ($CaCO_3$), which causes a drop in the concrete pH. This reaction causes the high and low pH zones to be divided by the carbonation front (see Figure 2).

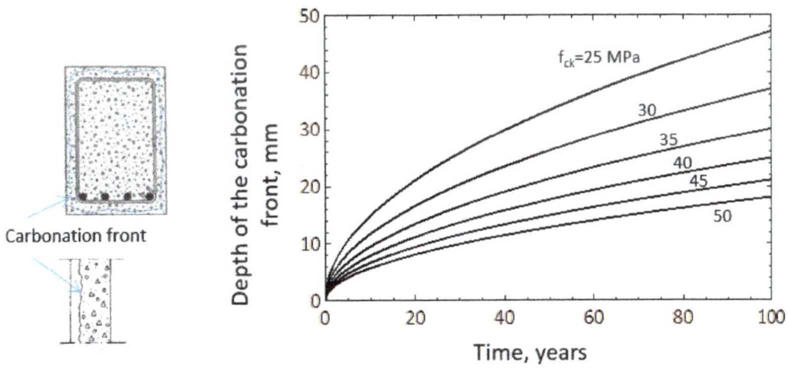

Figure 2. Depth of the carbonation front as a function of time.

The depth of the carbonation front (i.e., the zone affected by low pH) can be approximated by [7]:

$$x = V_{CO_2}\sqrt{t} \tag{1}$$

where x is the depth of the carbonation front, t is the time, and V_{CO_2} is the carbonation rate of the concrete, also called K. Coefficient K depends on environmental conditions (CO_2 concentration, humidity, etc.) and concrete characteristics (w/c ratio, type of cement, etc.) and it varies significantly from one structure to another [28]. In [28], the mean compressive strength is used to estimate K using a regression analysis.

As alkalinity decreases, the carbonation front advances. When the carbonation front reaches the reinforcement, it makes the passive film disappear, and so the reinforcing steel is left unprotected against corrosion. In the carbonation process, the passive film completely disappears, and the probability of corrosion reactions occurring is the same at every point on the metal surface, which leads to uniform or generalized corrosion.

As in almost every chemical reaction, moisture must be present for carbonation to occur. However, when the pores are saturated with water, it is difficult for CO_2 to penetrate the concrete as it is a gas. Therefore, the maximum carbonation rate is found at levels of intermediate humidity, and the most unfavorable scenarios are the wet–dry cycles. It is interesting to note that carbonation hardly occurs in either very dry concrete (because of the absence of water) or in completely saturated concrete (as CO_2 cannot move through the pores that are full of water).

Several mathematical models have been created to describe the evolution of the carbonation front, such as [19]. The following expression (2), extracted from Annex 12 of the Spanish Structural Code [35], provides mean values of the evolution of the depth of the carbonation front over time:

$$\begin{gathered} x(t) = k_{ap,carb}\sqrt{t} \\ k_{ap,carb} = c_{env}\, c_{air}\, a\, (f_{ck} + 8)^{b} \end{gathered} \tag{2}$$

where $k_{ap,carb}$ is the apparent carbonation coefficient in mm/year$^{0.5}$ so x(t) is obtained in mm, c_{env} is a coefficient that depends on environmental conditions (see Table 1). c_{air} is 1.0 if air-entrained concrete with less than 4.5% of entrained air (of the volume of the concrete) is used. If the amount of entrained air is greater than or equal to 4.5%, then c_{air} = 0.7. In Equation (2), a and b are dimensionless parameters that depend on the type of binder (see Table 2).

Table 1. Type of environment.

Environment	c_{env}
Sheltered from the rain	1.0
Exposed to rain	0.5
Buried elements, above the water table	0.3
Buried elements, below the water table	0.2

Table 2. Type of binder.

Binder	a	b
Portland cement	1800	−1.7
Portland cement +28% fly ash	360	−1.2
Portland cement + 9% silica fume	400	−1.2

Figure 3 shows the evolution of the carbonation front for Portland cement and for three types of environmental conditions defined by c_{env} (see Table 1) according to [35], Equation (2). In Figure 3, c_{air} = 1 is considered.

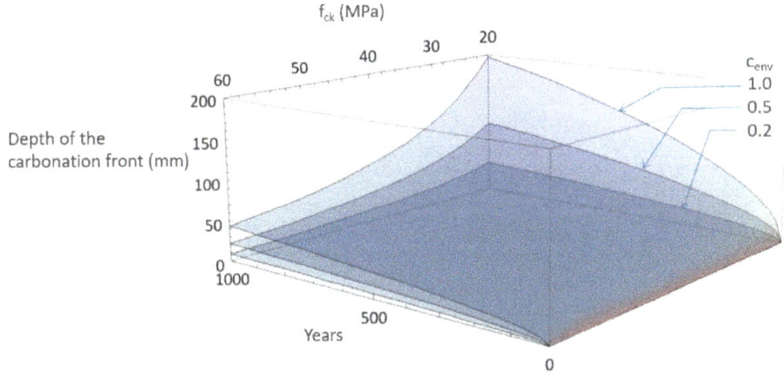

Figure 3. Evolution of the carbonation front for Portland cement according to [35].

Figure 3 shows that the compressive strength of concrete has a considerable influence on the depth of the carbonation front. From a comparison of the three surfaces in Figure 3, it is evident that when there is a higher risk of corrosion (i.e., for the higher value of c_{env}), the influence of concrete compressive strength is greater.

In Figure 4, Equation (2) for c_{env} = 1.0 (structure sheltered from the rain), c_{air} = 1, Portland cement, and f_{ck} = 50 MPa is plotted, together with the carbonation depth obtained from the Portuguese standard [28] for equivalent conditions. Figure 4 corresponds to: c = 400 ppm of the environmental carbon dioxide concentration (adopted as a value of reference in prEN 1992 [34]), f_{ck} = 50 MPa, f_{cm} = (f_{ck} + 8) MPa, k_0 = 3 (standard test conditions), k_2 = 1 (standard curing), and exposure class XC3 (moderate humidity, which corresponds to the external concrete sheltered from rain in prEN 1992 [34]). According to Eurocode 2, f_{ck} is the characteristic concrete cylinder compressive strength and f_{cm} is the mean concrete cylinder compressive strength.

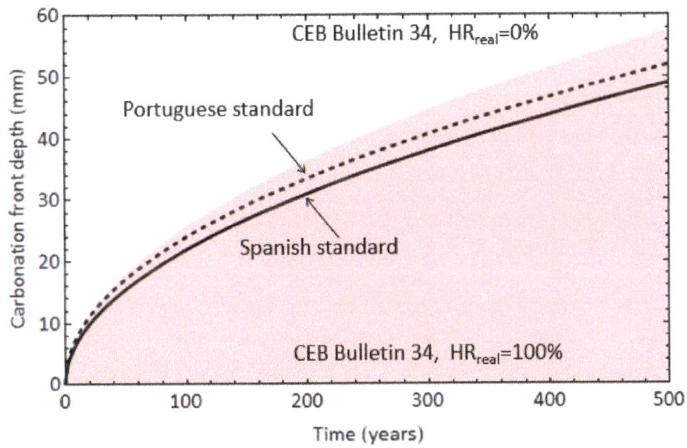

Figure 4. Comparison of the evolution of the carbonation front for Portland cement according to the Spanish Structural Code [35] and the Portuguese standard LNEC E-465 [28] for XC3 exposure class, c_{air} = 1.0, and f_{ck} = 50 MPa. The range of values of the carbonation depth according to Annex C of CEB Bulletin 34 [36] for h_{Nd} = 182 days, p_{sR} = 0 for HR_{real} ranging between 0 and 100% is shown as a shaded area.

In Annex C of the CEB Bulletin 34 [36], the carbonation depth in uncracked concrete is presented as a function of the design service life (time), and three climatic parameters: the real value of the relative humidity of the carbonated layer (HR_{real}), the average number of rainy days per year with a volume of precipitation water that is over 25 mm (h_{Nd}), and the probability of driving rain obtained from the average distribution of the wind direction during rain events (p_{sR}). The two last parameters should be obtained from weather station data. In Figure 4 (shaded area), the range of values of the carbonation depth according to [36] when HR_{real} varies between 0 and 100% has also been represented for h_{Nd} = 182 days (half a year) and p_{sR} = 0 (horizontal structural element).

As can be seen in Figure 4, using both the Spanish and Portuguese standards leads to similar values of the evolution of the carbonation depth over time for the most common exposure class [28], and the values given by the Portuguese standard are slightly higher than those estimated by the Spanish Structural Code [35] for each period of time. Moreover, the carbonation depths obtained from both standards lie within the range defined by the CEB Bulletin 34 [36].

2.2. *Minimum Cover Required for Protection against Carbonation in prEN 1992*

In prEN 1992 [34], the minimum concrete cover ($c_{min,dur}$) for protection of the reinforcement against corrosion induced by carbonation depends on the design service life, the exposure class, and the exposure resistance class (ERC). Each ERC is identified as XRC followed by a number that corresponds to the carbonation rate (see Equation (1)) related to the 90% fractile of the depth of the carbonation front (in mm) after 50 years and under the following reference conditions: 400 ppm CO_2 in a constant 65%-RH environment and at 20 °C.

Two design service life values are considered in prEN 1992 [34], 50 and 100 years, and the values of $c_{min,dur}$ are presented in a tabular format. The values given by prEN 1992 are shown in Figure 5, where XC is the exposure class for carbonation and XRC is the resistance class against corrosion caused by carbonation. To facilitate the identification of the different exposure classes and service design lives, discrete values given by prEN 1992 [34] have been connected with lines in Figure 5.

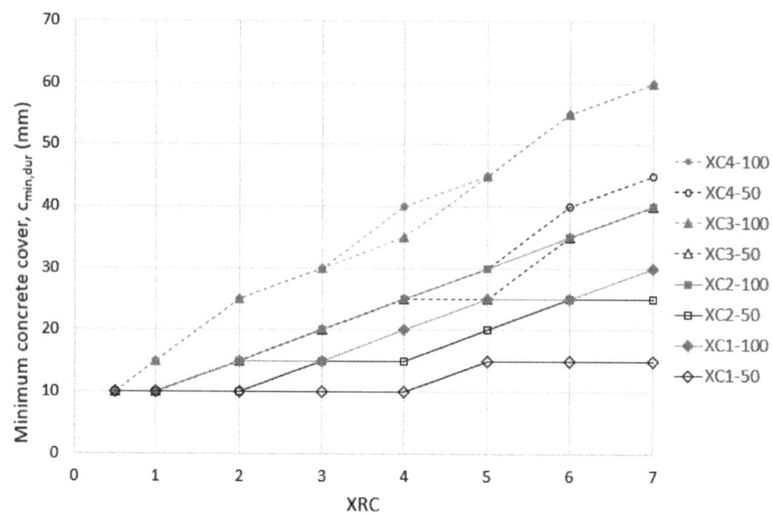

Figure 5. Minimum concrete cover for protection against oxidation induced by carbonation (XC1 = dry, XC2 = wet, or permanent high humidity, XC3 = moderate humidity, XC4 = cyclic wet and dry).

3. Results

As can be seen in Figure 5, the minimum concrete cover for durability proposed by the European standard prEN 1992 [34] depends on the resistance class against corrosion induced by carbonation (XRC). However, the regulation does not explicitly define the recommended values for XRC.

Moreover, in order to carry out long-term structural designs (e.g., design for a service life of five centuries) or to evaluate the remaining design service life of existing structures, it would be useful to have an expression for the depth of the carbonation front as a function of time, as in [28,32,35].

3.1. Continuous Formulation of the Carbonation front Based on Eurocode 2

This paper proposes, as a first approximation, a function similar to the one included in the Spanish Structural Code [35] and given in Equation (2) (Equation (3)). The expression proposed (Equation (3)) must give values that are as close as possible to those given by prEN 1992 [34] (Figure 5).

$$\begin{gathered} x(t) = k^{*}_{ap,carb}\sqrt{t} \\ k^{*}_{ap,carb} = c^{*}_{env}\, c_{air}\, a\, (f_{ck} + 1.8)^{b} \end{gathered} \quad (3)$$

As can be seen in Equation (2), the term in brackets is the approximate mean compressive concrete strength for cases when not enough statistical data are available:

$$f_{cm} = f_{ck} + 8\ [\mathrm{MPa}] \quad (4)$$

with f_{ck} in [35] as the characteristic concrete strength, which corresponds to a confidence level of 95%. Equation (4) is also proposed by the European standard [33,41,42]. So, the first change to make in the new expression Equation (3) to fit prEN 1992 [34], is to obtain the compressive concrete strength corresponding to the 90% fractile. Assuming that the compressive strength of concrete follows a standard normal distribution, this value can be obtained by using basic statistical concepts from the following system of equations:

$$\left.\begin{array}{l} \mathrm{CDF}[\mathrm{NormalDistribution}[(f_{ck}+8),\ \sigma],\ f_{ck}] = 0.05 \\ \mathrm{CDF}[\mathrm{NormalDistribution}[(f_{ck}+8),\ \sigma],\ f_{ck}+\delta] = 0.01 \end{array}\right\} \mathrm{P} \begin{array}{l} \sigma = 4.9 \\ \delta = 1.8 \end{array} \quad (5)$$

with CDF as the cumulative distribution function and σ as a standard deviation of the standard normal distribution followed by the concrete compressive strength. Solving the system of equations Equation (5), the 90% fractile of the compressive concrete strength is f_{ck} + 1.8, which is the value in brackets in Equation (3).

Using least square regression, the parameters c^{*}_{env} and f_{ck} in Equation (3) have been adjusted so that they match the values given by prEN 1992 [34], Figure 5. In this work, it has been assumed that each resistance class against corrosion induced by carbonation (XRC) corresponds to a value of the characteristic concrete cylinder compressive strength (f_{ck}) for each type of binder and entrained air content. By using this approach, the corrected environmental coefficient (c^{*}_{env}, see Table 1) in Equation (3) has been obtained from a least square regression for the case corresponding to the most common type of cement and with no entrained air (i.e.,: Portland cement and air-entrained concrete with less than 4.5% of entrained air, in volume of the concrete). So, according to [35] the following values for the parameters in Equation (3) are considered as: c_{air} = 1.0, a = 1800, and b = −1.7; see Table 3.

Table 3. Environmental coefficients adjusted (c^{*}_{env}) to fit the minimum cover for carbonation in prEN 1992 [34].

Adjusted Exposure Classes for Carbonation	XC1	XC2	XC3	XC4
c^{*}_{env}	0.45	0.60	0.90	0.95

The adjusted values of c^{*}_{env} for each exposure class in prEN 1992 [34] (XC) are summarized in Table 3.

Once the coefficients in Table 2 are known, each XRC class defined in prEN 1992 [34] can be associated with a value of characteristic concrete strength (f_{ck}). The corresponding values of characteristic concrete strength (f_{ck}), which are the unknown of the adjustment, are adjusted for each type of binder (i.e.,: Portland, Portland +28% fly ash, and Portland + 9% silica fume, see Table 2), and for the two cases related to the entrained air content (under or over 4.5% of entrained air of the volume of the concrete) in [35]. A least square regression has been used for the adjustment.

The adjustment has been carried out by simultaneously considering both of the design service life values in prEN 1992, 50 and 100 years, and the results are summarized in Table 4.

Table 4. Relationship between the resistance class against corrosion induced by carbonation (XRC) and the characteristic concrete cylinder compressive strength (f_{ck} [MPa]) for each type of binder and volume of entrained air content.

	Type of Binder	Portland		Portland + 28% Fly Ash		Portland + 9% Silica Fume	
	Entrained Air	>4.5%	<4.5%	>4.5%	<4.5%	>4.5%	<4.5%
ERC in prEN 1992 [34]	XRC 0.5	63	80	94	128	103	139
	XRC 1	49	60	66	90	73	98
	XRC 2	36	45	44	59	48	65
	XRC 3	32	40	36	49	39	54
	XRC 4	28	35	30	41	33	45
	XRC 5	25	31	26	35	28	38
	XRC 6	22	27	21	29	23	32
	XRC 7	21	26	20	27	21	29

Figure 6 shows the results of the adjustment carried out for Portland cement with an entrained air volume of under 4.5% of the volume of the concrete. For the sake of comparison, the values of the minimum cover for carbonation ($c_{min,dur}$) for the four exposure classes, XC1 to XC4, given by prEN 1992 [34] (see Figure 5) are also represented, and they are linked by dashed straight lines in Figure 6, where the term "approx" identifies the adjusted curves. Figure 6a,b represents the design service lives of 50 and 100 years, respectively.

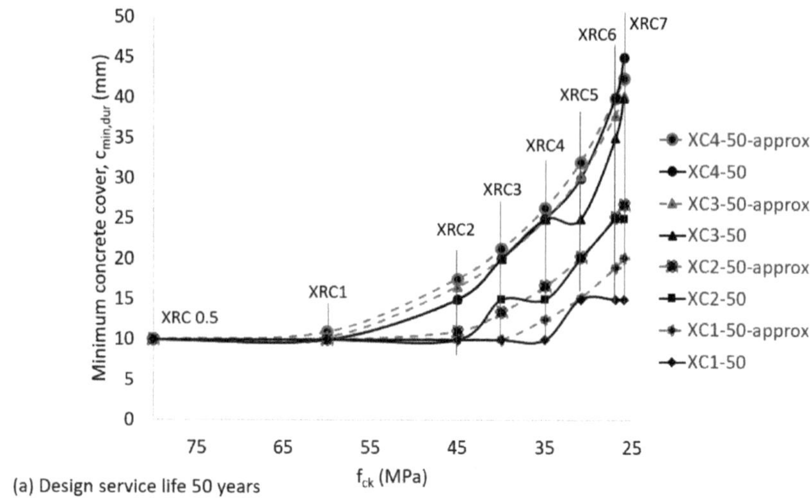

Figure 6. *Cont.*

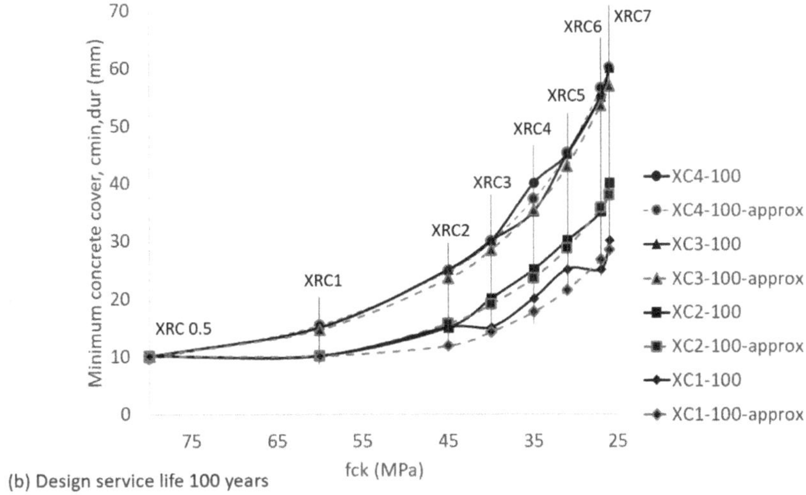

Figure 6. Relationship between the XRC class and the characteristic concrete strength for Portland cement with a volume of entrained air of under 4.5% of the volume of the concrete. Design service lives of (**a**) 50 and (**b**) 100 years.

Figure 6 shows that the adjustment proposed is a good fit for the values proposed by prEN 1992 [34], especially for the two higher XC classes.

As can be seen in Figure 6, the concrete characteristic compressive strength and the minimum cover for carbonation proposed by prEN 1992 [34] correlate well, which confirms that, as proposed in this work, the value of f_{ck} could be suitable for obtaining a first approximation of the XRC class, as proposed in Table 4.

Other conditions have to be added to the results in Figure 6 if the limitations of the indicative strength classes stated in EN 206 [41] for each exposure class are considered; see Table 5.

Table 5. Indicative strength class for corrosion induced by carbonation for each exposure class according to EN 206 [41].

Exposure Class	XC1	XC2	XC3/XC4
Strength class	≥C20/25	≥C25/30	≥C30/37

Moreover, the 'square root of time' expression proposed, Equation (3), enables other design service lives (not only 50 and 100 years) to be studied, which is very useful, as the case studies below show.

3.2. Case Studies

In order to demonstrate the usefulness of the proposal, it will be applied to two case studies.

3.2.1. The Green Tunnel on the London–Birmingham High Speed Railway Line

The prerequisites for concrete strength and nominal cover given by the authority for the pre-cast tunnel known as the Green Tunnel for the high-speed railway line between London and Birmingham are summarized in Table 6.

Table 6. Prerequisites for the design of the Green Tunnel.

Element	Precast/Cast In Situ	Nominal Cover *	
		Inner Face	Outer Face
Arch, Arch wall, Side Wall, Footing	Precast (C50/60)	45 mm	55 mm
Central Wall	Precast (C50/60)	45 mm	45 mm
Invert/Base Slab	Cast-in situ (C50/60)	60 mm	70 mm

* Clear cover to main reinforcement = nominal cover + stirrup diameter.

As Figure 6 shows, there is almost no difference between the minimum cover against carbonation given by prEN 1992 [34] for classes XC3 and XC4. In this example, both exposure classes are considered by using Equation (3) with $c_{air} = 1.0$ and $c^{*}_{env} = 0.90$ for XC3 class and $c^{*}_{env} = 0.95$ for XC4 (see Table 3).

Figure 7 represents the required minimum cover (i.e., the depth of the carbonation front) as a function of the design service life (time) for $f_{ck} = 50$ MPa and for these two exposure classes. Figure 7 shows that, as a first approximation, the life expectancy of the Green Tunnel is about 500 years. This result is based on the continuous approximation proposed in this work (Equation (3)), which is based on prEN1992 [34].

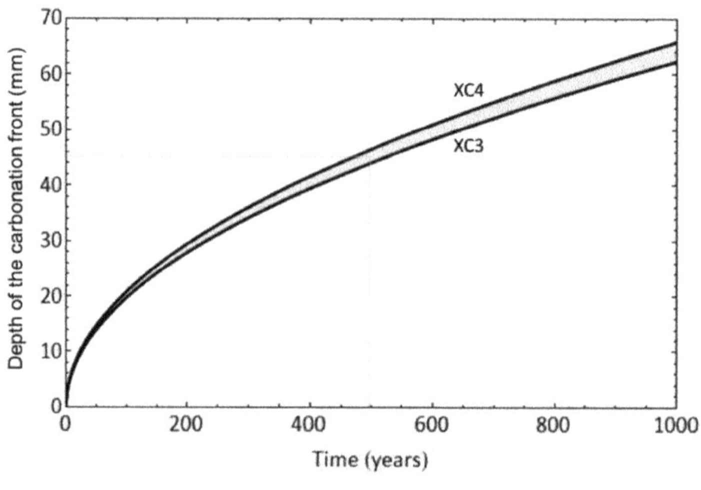

Figure 7. Expected service life of the Green Tunnel.

3.2.2. Camino de Ronda Street Buildings (Granada, Spain)

Camino de Ronda is the longest street in Granada, and it is also one of the most densely populated areas of the city. Along this street, reinforced concrete buildings of various ages can be found. As a preliminary approach, based on the regulations in force during the construction of the buildings, we have considered, as a working hypothesis, that the characteristic strength of concrete is 25 MPa in concrete manufactured up until 1970, and 30 MPa from 1970 onwards. The concrete cover in the exposed areas have been assumed to be 30 mm. The concrete cover in the internal areas have been assumed to be 15 mm for buildings from before 1970 and 25 mm for later ones. According to Equation (3), the initiation time is 25 years for pre-1970 buildings and 44 years for post-1970 buildings, see Figure 8. These premises are the worst-case scenario, given that a concrete structure is always protected from environmental factors, by plaster coatings, paint, bricks, etc.

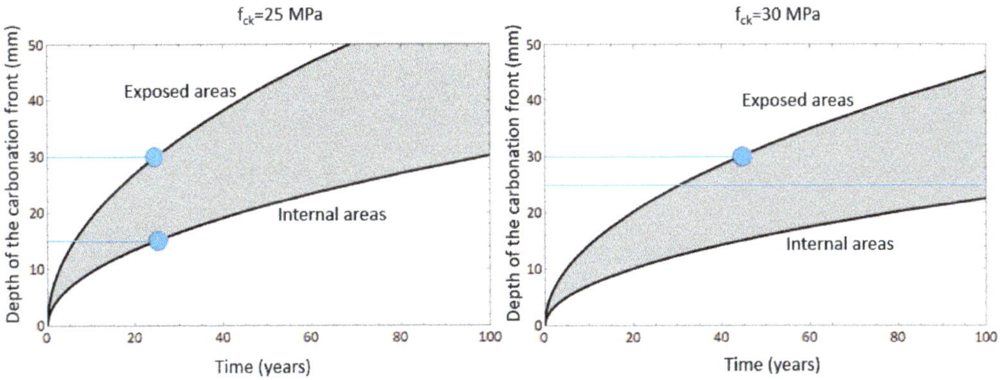

Figure 8. Evolution of the carbonation front in case study 2, Equation (3).

After the initiation time, the remaining service life depends on the result of periodic inspections. Note that the appearance of cracks parallel to the reinforcement requires an exhaustive structural analysis to determine the bearing capacity of the structure.

The results obtained have been summarized in Figure 9, which indicate when periodic inspections for the buildings of Camino de Ronda are necessary based on the estimated end of the initiation times.

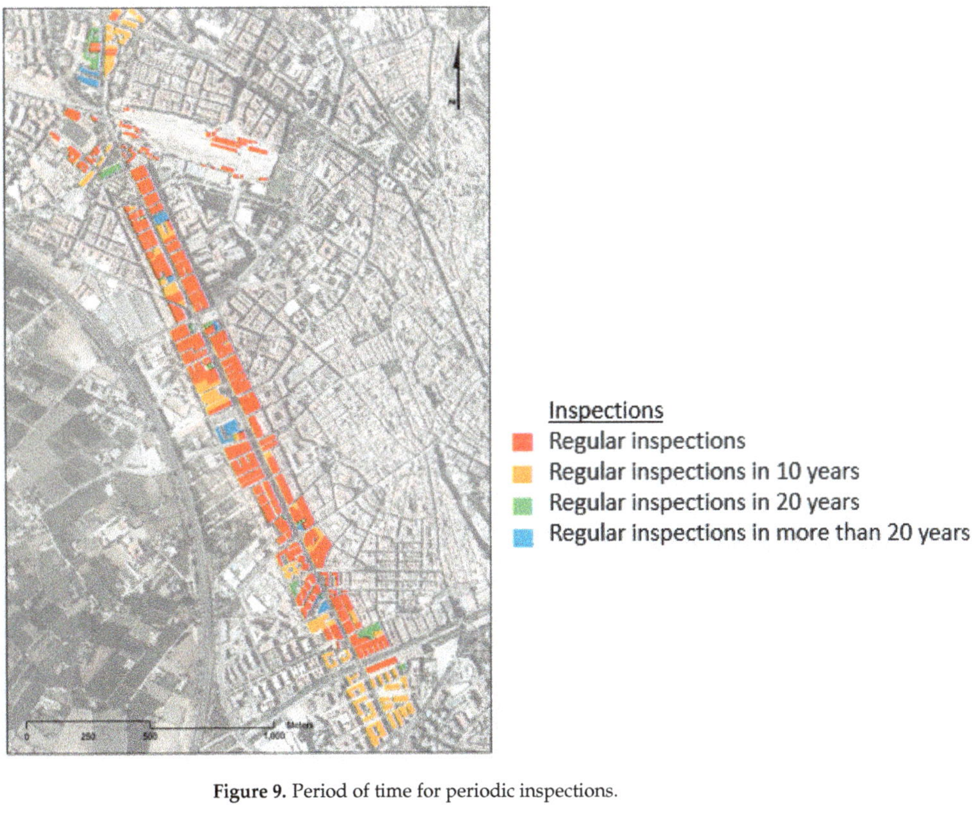

Figure 9. Period of time for periodic inspections.

4. Limitations of the Study

The aim of the present study is to establish the relationship between XRC classes and concrete strength for regular concrete, and not to replace the XRC classes introduced by the new version of Eurocode 2. It should be noted that the presented formulation may not be applicable if impermeability of the concrete or carbonation resistance is achieved through means other than increasing the concrete strength.

5. Conclusions

The most significant finding of this paper is that the evolution of the carbonation front and the expected service life of a structure can be approximated using the new Eurocode 2, without relying on the resistance class against carbonation-induced corrosion. Instead, only the strength resistance of the concrete and the exposure class are used.

Additionally, the following conclusions are addressed:

- The presented study proposes a new continuous formulation for the carbonation front.
- The formulation is based on the 'square root of time' expression given by the relevant literature and is in accordance with the minimum cover proposed by prEN 1992.
- The minimum cover required to protect against carbonation can be determined from the proposed expression, which is formulated as a function of the compressive strength of concrete.
- The proposed expression allows for the indicative strength classes against corrosion induced by carbonation proposed by prEN 1992 to be considered.
- The new expression is shown to be a useful tool for the design of extremely durable structures.

Author Contributions: Conceptualization, E.H.-M. methodology, E.H.-M. and L.M.G.-M.; formal analysis, L.H.-G. and L.M.G.-M.; investigation, L.H.-G.; data curation, E.M.; writing—original draft, L.H.-G.; writing—review and editing, L.H.-G. and L.M.G.-M.; visualization, E.M.; supervision, E.H.-M. project administration, E.H.-M. funding acquisition, E.H.-M. All authors have read and agreed to the published version of the manuscript.

Funding: This work is part of the HYPERION project (https://www.hyperion-project.eu/, accessed on 1 April 2003). HYPERION has received funding from the European Union's Framework Programme for Research and Innovation (Horizon 2020) under grant agreement no. 821054. This work also was supported by the Spanish Government (Ministerio de Ciencia, Innovación y Universidades) as part of the Research Projects RTI2018-101841-B-C21 and RTI2018-101841-B-C22 (MINECO/FEDER). The content of this publication is the sole responsibility of UGR and does not necessarily reflect the opinion of the European Union.

Institutional Review Board Statement: Not applicable.

Informed Consent Statement: Not applicable.

Data Availability Statement: Data are available upon reasonable request to the authors.

Conflicts of Interest: The authors declare no conflict of interest.

References

1. Sánchez, J.; Torres, J.E.; Rebolledo, N. Brittle Behavior of Corroded Rebars. *Hormigón Acero* **2020**, 1–7. [CrossRef]
2. Williamson, S.J.; Clark, L.A. Effect of Corrosion and Load on Reinforcement Bond Strength. *Struct. Eng. Int.* **2002**, *12*, 117–122. [CrossRef]
3. Val, D.V.; Stewart, M.G. Reliability Assessment of Ageing Reinforced Concrete Structures—Current Situation and Future Challenges. *Struct. Eng. Int.* **2009**, *19*, 211–219. [CrossRef]
4. de Boer, A.; Gulikers, J. Effect of Reinforcement Geometry on the Probability of Corrosion Initiation in Reinforced Concrete Structures. *Struct. Eng. Int.* **2009**, *19*, 198–202. [CrossRef]
5. Carbonell-Márquez, J.F.; Gil-Martín, L.M.; Fernández-Ruiz, M.A.; Hernández-Montes, E. Procedure for the Assessment of the Residual Capacity of Corroded B-Regions in RC Structures. *Constr. Build. Mater.* **2016**, *121*, 519–534. [CrossRef]
6. Hernández-Montes, E.; Fernández-Ruiz, M.A.; Carbonell-Márquez, J.F.; Gil-Martín, L.M. Residual Capacity Assessment of Reinforced Concrete D-Regions Affected by Corrosion. *Constr. Build. Mater.* **2020**, *263*, 120228. [CrossRef]

7. Tuutti, K. Corrosion of Steel in Concrete. Ph.D. Thesis, Swedish Cement and Concrete Research Institute, Stockholm, Sweden, 1982.
8. Possan, E.; Dal Molin, D.C.C.; Andrade, J.J.O. A Conceptual Framework for Service Life Prediction of Reinforced Concrete Structures. *J. Build. Pathol. Rehabil.* **2018**, *3*, 2. [CrossRef]
9. Andrade, C. Future Trends in Research on Reinforcement Corrosion. In *Corrosion of Steel in Concrete Structures*; Poursaee, A., Ed.; Woodhead Publishing: Oxford, UK, 2016; pp. 269–288. ISBN 978-1-78242-381-2.
10. Kouassi, F.A.; Dauxois, J.-Y.; Duprat, F.; De Larrard, T.; Deby, F. Engineering Statistical Models for Carbonation Depth. *Eur. J. Environ. Civ. Eng.* **2022**, *27*, 215–238. [CrossRef]
11. Isaksson, R.; Rosvall, M.; Espuny, M.; Nunhes, T.V.; de Oliveira, O.J. How Is Building Sustainability Understood?—A Study of Research Papers and Sustainability Reports. *Sustainability* **2022**, *14*, 12430. [CrossRef]
12. Andrade, M.C.; Castro-Borges, P.; Pazini, E. Detection of Corrosion Risk besides Patch Repairs. *Hormigón Acero* **2021**, *72*, 41–58.
13. Chinchón, S.; Sánchez, J.; Ortega, L.M.; García, J.; Rebolledo, N.; Torres, J.E.; Perilbáñez, J. Nueva Metodología Para La Delimitación de Las Zonas a Reparar de Un Hormigón Afectado Por Corrosión. *Hormigón Acero* **2022**, *73*, 51–57. [CrossRef]
14. Ali, M.E.; Hasan, M.F.; Siddiqa, S.; Molla, M.M.; Nasrin Akhter, M. FVM-RANS Modeling of Air Pollutants Dispersion and Traffic Emission in Dhaka City on a Suburb Scale. *Sustainability* **2023**, *15*, 673. [CrossRef]
15. Silva, A.; Neves, R.; De Brito, J. Statistical Modelling of Carbonation in Reinforced Concrete. *Cem. Concr. Compos.* **2014**, *50*, 73–81. [CrossRef]
16. Andrade, C.; Alonso, C.; Gonzalez, C. *Remaining Service Life of Corroding Structures*; International Association for Bridge and Structural Engineering: Zurich, Switzerland, 2020.
17. Keršner, Z.; Teplý, B.; Novak, D. Uncertainty in Service Life Prediction Based on Carbonation of Concrete. In *Durability of Building Materials and Components 7*; Sjostrom, C., Ed.; Routledge: London, UK, 1996; pp. 1–8. ISBN 9781315025025.
18. Saetta, A.V.; Vitaliani, R.V. Experimental Investigation and Numerical Modeling of Carbonation Process in Reinforced Concrete Structures Part I: Theoretical Formulation. *Cem. Concr. Res.* **2004**, *34*, 571–579. [CrossRef]
19. Papadakis, V.G.; Vayenas, C.G.; Fardis, M.N. Experimental Investigation and Mathematical Modeling of the Concrete Carbonation Problem. *Chem. Eng. Sci.* **1991**, *46*, 1333–1338. [CrossRef]
20. Papadakis, V.G.; Vayenas, C.G.; Fardis, M.N. Fundamental Modeling and Experimental Investigation of Concrete Carbonation. *ACI Mater. J.* **1991**, *88*, 363–373. [CrossRef]
21. Hills, T.P.; Gordon, F.; Florin, N.H.; Fennell, P.S. Statistical Analysis of the Carbonation Rate of Concrete. *Cem. Concr. Res.* **2015**, *72*, 98–107. [CrossRef]
22. Saetta, A.V.; Vitaliani, R.V. Experimental Investigation and Numerical Modeling of Carbonation Process in Reinforced Concrete Structures: Part II. Practical Applications. *Cem. Concr. Res.* **2005**, *35*, 958–967. [CrossRef]
23. Endale, S.A.; Taffese, W.Z.; Vo, D.H.; Yehualaw, M.D. Rice Husk Ash in Concrete. *Sustainability* **2023**, *15*, 137. [CrossRef]
24. Marchand, J.; Samson, E. Predicting the Service-Life of Concrete Structures—Limitations of Simplified Models. *Cem. Concr. Compos.* **2009**, *31*, 515–521. [CrossRef]
25. Chen, Y.; Liu, P.; Yu, Z. Effects of Environmental Factors on Concrete Carbonation Depth and Compressive Strength. *Materials* **2018**, *11*, 2167. [CrossRef] [PubMed]
26. Ekolu, S.O. Model for Practical Prediction of Natural Carbonation in Reinforced Concrete: Part 1-Formulation. *Cem. Concr. Compos.* **2018**, *86*, 40–56. [CrossRef]
27. Zhu, J.; Zhang, R.; Zhang, Y.; He, F. The Fractal Characteristics of Pore Size Distribution in Cement-Based Materials and Its Effect on Gas Permeability. *Sci. Rep.* **2019**, *9*, 17191. [CrossRef]
28. Monteiro, I.; Branco, F.A.; De Brito, J.; Neves, R. Statistical Analysis of the Carbonation Coefficient in Open Air Concrete Structures. *Constr. Build. Mater.* **2012**, *29*, 263–269. [CrossRef]
29. You, X.; Hu, X.; He, P.; Liu, J.; Shi, C. A Review on the Modelling of Carbonation of Hardened and Fresh Cement-Based Materials. *Cem. Concr. Compos.* **2022**, *125*, 104315. [CrossRef]
30. Bakker, R. *Corrosion of Steel in Concrete, "Initiation Period"*; Schiessl, P., Ed.; RILEM; Chapman & Hall: London, UK, 1988.
31. Ghanooni-Bagha, M.; YekeFallah, M.R.; Shayanfar, M.A. Durability of RC Structures against Carbonation-Induced Corrosion under the Impact of Climate Change. *KSCE J. Civ. Eng.* **2020**, *24*, 131–142. [CrossRef]
32. Gonçalves, A.F.; Ferreira, M.J.E.; Ribeiro, A.B. The New Lnec Specifications on Reinforced Concrete Durability. In Proceedings of the International RILEM Workshop on Integrated Service Life Modelling of Concrete Structures, Guimarães, Portugal, 5–6 November 2007; pp. 131–140.
33. *UNE-EN 1992-1-1*; Eurocode 2: Design of Concrete Structures—Part 1–1: General Rules and Rules for Buildings. European Committee for Standardization: Brussels, Belgium, 2013.
34. *prEN 1992-1-1*; Eurocode 2: Design of Concrete Structures—Part 1–1: General Rules and Rules for Buildings, Bridges and Civil Engineering Structures. European Committee for Standardization: Brussels, Belgium, 2022.
35. MITMA. *Código Estructural. Volumenes 1 y 2*; MITMA: Madrid, Spain, 2021.
36. fib-International Federation for Structural Concrete. *Bulletin 34: Model Code for Service Life Design*; fib: Lausanne, Switzerland, 2006.
37. Fonseca, E.T.; Vellasco, P.C.G.D.S.; Vellasco, M.M.B.R.; de Andrade, S.A.L. A Neuro-Fuzzy Evaluation of Steel Beams Patch Load Behaviour. *Adv. Eng. Softw.* **2008**, *39*, 558–572. [CrossRef]

38. Sanchez, M.; Alonso, M.C. Repassivation of Steel Rebars after an Electrochemical Chlorides Removal Treatment by Simultaneous Application of Calcium Nitrite. In *Concrete under Severe Conditions: Environment and Loading, Proceedings of the 6th International Conference on Concrete under Severe Conditions, CONSEC'10, Yucatán, Mexico, 7–9 June 2010*; CRC Press/Balkema: Leiden, The Netherlands, 2010; Volume 2, pp. 1109–1114.
39. Alonso, M.C.; García Calvo, J.L.; Hidalgo, A.; Fernández Luco, L. 10—Development and Application of Low-PH Concretes for Structural Purposes in Geological Repository Systems. In *Geological Repository Systems for Safe Disposal of Spent Nuclear Fuels and Radioactive Waste*; Ahn, J., Apted, M.J., Eds.; Woodhead Publishing Series in Energy; Woodhead Publishing: Sawston, UK, 2010; pp. 286–322. ISBN 978-1-84569-542-2.
40. Andrade, C. Corrosion Of Steel Reinforcement. *WIT Trans. State-of-the-Art Sci. Eng.* **2007**, *28*, 185–216. [CrossRef]
41. *UNE EN 206-1:2008*; Concrete —Part 1: Specification, Performance, Production and Conformity. Asociación Española de Normalización y Certificación (AENOR): Madrid, Spain, 2008.
42. Hernández, L.; Gil-Martín, L.M. A Comparison of the Most Important Properties of Structural Concrete: European Standards versus American Standards. *Eur. J. Environ. Civ. Eng.* **2022**, 1–14. [CrossRef]

sustainability

Article

Numerical Modeling and Design Method for Reinforced Polyvinyl-Alcohol-Engineered Cementitious Composite Beams in Bending

Qiao-Ling Fu [1], Shao-Bo Kang [2] and Dan-Dan Wang [2,*]

[1] Faculty of Architectural Engineering, Chongqing Water Resources and Electric Engineering College, Chongqing 402171, China; ginkgoling@163.com
[2] School of Civil Engineering, Chongqing University, Chongqing 400044, China
* Correspondence: wangdd0914@cqu.edu.cn

Abstract: The polyvinyl-alcohol-engineered cementitious composite (PVA-ECC) is a superior cementitious material when used for tension and flexural loading. The utilization of PVA-ECC in the tension zone can prevent the development of wide cracks and increase the flexural resistance of reinforced PVA-ECC members. In this paper, a nonlinear finite element model is established to simulate the behavior of PVA-ECC beams in bending. In the model, the constitutive models for PVA-ECC in compression and tension are employed by simplifying them as piece-wise linear models, and the bond between the reinforcing bar and PVA-ECC is also considered. The load–deflection curve and failure mode of beams can be obtained from the finite element model. Comparisons between numerical and experimental results show that the developed numerical model can estimate the ultimate load and failure mode of beams with reasonably good accuracy. After evaluating the accuracy of the finite element model, parameter analysis is conducted to investigate the effects of the reinforcement ratio, steel strength grade, and mechanical properties of PVA-ECC on the flexural behavior of reinforced PVA-ECC beams. The numerical results conclude that the effects of reinforcement ratio on the peak load, stiffness, and deflection are obvious while the influence of steel grade is mainly on the peak load. The tensile localization strain of PVA-ECC mainly affects the ductility of the beam. Furthermore, a design method is proposed based on the plane-section assumption to calculate the ultimate load of reinforced PVA-ECC beams, in which the contribution of PVA-ECC to the moment resistance of beam sections is considered. Comparisons between existing design methods and the proposed method indicate that the ultimate load of beams can be predicted more accurately by considering the tensile strength of PVA-ECC in the tension zone.

Keywords: finite element analysis; bond–slip model; load–deflection relationship; failure mode; design method

check for updates

Citation: Fu, Q.-L.; Kang, S.-B.; Wang, D.-D. Numerical Modeling and Design Method for Reinforced Polyvinyl-Alcohol-Engineered Cementitious Composite Beams in Bending. *Sustainability* **2023**, *15*, 10130. https://doi.org/10.3390/su151310130

Academic Editors: Mahdi Kioumarsi and Vagelis Plevris

Received: 2 June 2023
Revised: 19 June 2023
Accepted: 22 June 2023
Published: 26 June 2023

1. Introduction

The polyvinyl-alcohol-engineered cementitious composite (PVA-ECC) is a superior cementitious material for its excellent mechanical properties under tension. Combined with discontinuous fiber, the ECC can mitigate the development of major cracks and attain a tensile strain capacity higher than 3% [1–3]. As a bendable concrete suitable for sustainable and resilient infrastructure, ECC has been studied by many researchers and has been applied in high buildings, bridge structures, and high railways [4].

Due to its superior tensile properties, the application of ECC in the tension zone of flexural members has been investigated by many researchers [5]. Ge et al. [5] studied the effect of reinforcement type and ECC thickness on the flexural behavior of reinforced ECC-concrete beams. The moment capacity and bending stiffness were enhanced and the crack width was well controlled when ECC was used to replace the concrete with the same thickness. Yuan et al. [6] investigated the flexural behavior of

ECC/concrete composite beams and confirmed the enhancement from ECC to the flexural resistance of reinforced concrete beams. Apart from the application of ECC in the tension zone, flexural properties of the whole ECC beam have also aroused great interest. Even though the influence of concrete type and reinforcement ratio on reinforced concrete beams has been investigated [7–10], its effect on PVA-ECC beams showed different results. Shao et al. [11–13] proposed two failure paths of PVA-ECC beams, namely, failure after a dominant crack or after gradual strain hardening of a reinforcement, depending on the reinforcement ratio. According to the test results of twelve simply supported beams, the failure paths were mainly affected by the reinforcing ratio and steel type. A simplified flexural strength prediction method was also proposed to predict the load-bearing capacity of ECC beams. The shear behavior of reinforced ECC beams has also been investigated by researchers from different countries [14–16].

Besides experimental results, finite element simulations were also utilized to study the behavior of reinforced PVA-ECC beams. Zheng et al. [17] adopted ABAQUS/Standard software to estimate the shear capacity of RC beams strengthened by ECC composite layers. Shanour et al. [18] tested twelve reinforced ECC beams with different ratios of PVA and polypropylene fibers. Nonlinear element analysis was also conducted to predict the load–deflection curves and crack patterns of ECC beams. However, the calculated initial stiffness from the numerical results was greater than that from the experimental results. Selim et al. [19] developed three-dimensional finite element models for beams under four-point loading by using ABAQUS. The numerical results were in good agreement with the test results at the initial loading stage, followed by significant differences at the peak load. Shao et al. [20] adopted a new ECC compression model to simulate the gradual compression softening behavior of flexural members by using nonlinear element analysis software DIANA. The numerical results concluded that the proposed model with the new compression model and a hybrid-rotating/fixed-crack model predicts the beam behavior, especially the failure mode and drift capacity of beams, with good accuracy.

To predict the flexural properties of reinforced PVA-ECC beams, a nonlinear finite element model without initial flaws is proposed in this paper. Six four-point bending specimens from the literature [21] are simulated to evaluate the accuracy of the proposed model. Parameter analyses are conducted to investigate the effect of the reinforcement ratio, steel strength grade, and mechanical properties of PVA-ECC on the load–deflection behavior. In addition, a design method is developed to consider the tensile strength of PVA-ECC in calculating the moment capacity of beams. Comparisons are made between different design methods and test data to show the effect of the tensile strength of PVA-ECC in beams.

2. Numerical Model

2.1. Finite Element Models

In reference [11], singly reinforced PVA-ECC beams were tested in four-point bending. The PVA-ECC used for beams was designed by using cement, fly ash, sand, PVA fibers, and superplasticizer. Cement, fly ash, and sand were put into a mixer and mixed for about five minutes, and then fibers were put in the cement paste and mixed for three minutes to ensure uniform dispersions. Nonlinear finite element analysis is conducted to model the flexural behavior of the PVA-ECC beam. Software DIANA FEA [22] is employed to simulate the load–deflection curves and failure modes of the beam. Figure 1 shows the geometry, test setup, and finite element model for the beam. A two-dimensional model with a quadrilateral mesh is used for the beam. Simple supports are defined at the two ends of the beams. The plane stress element is defined for PVA-ECC and the mesh size is 10 mm × 10 mm with a 150 mm thickness. The height of the cross-section is 250 mm. Stirrup and longitudinal reinforcements are embedded into the PVA-ECC. As for longitudinal reinforcements, the bond–slip behavior between PVA-ECC and the reinforcement is taken into consideration. A displacement-control load is applied at the middle point of steel block, with 0.1 mm at each step in the simulation.

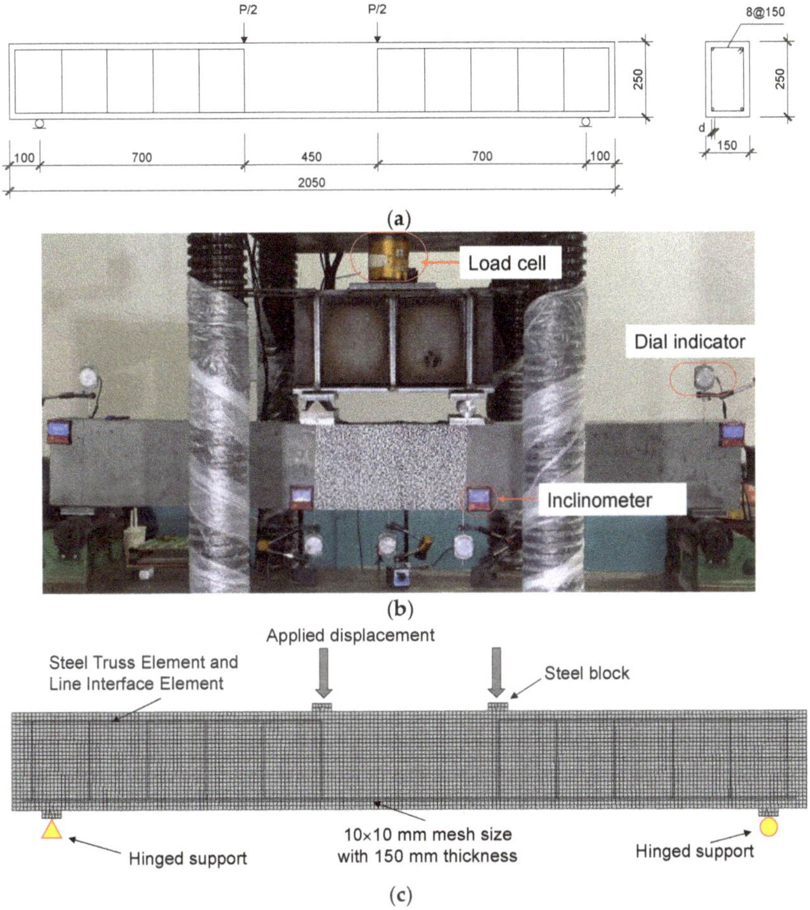

Figure 1. Finite element model for simply supported beams. (**a**) Geometry of beams. (**b**) Experimental setup. (**c**) Finite element model.

The mechanical model of PVA-ECC used in the simulations is based on tension and compression tests conducted by Shao [12]. As shown in Figure 2a, four stages are observed from the compressive stress–strain curve, namely, the elastic stage, nonlinear riding stage, linear descending stage, and plateau stage. Table 1 summarizes the mechanical properties of PVA-ECC in tension and compression. In the table, f_r represents the residual compressive stress and is suggested to be 30% of the peak stress in this paper, $\varepsilon_{c,p}$ represents the peak strain when the compressive strength is reached, $\varepsilon_{c,1/2}$ is obtained using a compressive stress equal to 0.5 times the peak stress divided by the elastic modulus, and $\varepsilon_{c,u}$ is the residual strain when the residual stress reaches 30% of the peak stress.

The constitutive model for ECC in tension is simplified as four stages, namely, the elastic stage (OA), linear rising stage (AB), plateau stage (BC), and linear softening stage (BC), as shown in Figure 2b. Table 1 shows values of tensile stress and strain at different stages in the numerical simulation. In the table, f_t is taken as 1.96 MPa according to four-point flexural tests; $f_{t,cr}$ represents the tensile cracking strength and is recommended to be 1.86 MPa; $f_{t,cr}$ is assumed to be 95% of the tensile strength f_t [20]; $\varepsilon_{t,cr}$ represents the tensile strain when the stress is 0.95 times the peak strain; $\varepsilon_{t,1}$ is the tensile cracking strain when obvious cracking forms; $\varepsilon_{t,p}$ is the tensile strain when a principal crack forms, namely, the localization strain. The steel constitutive model adopts the elastoplastic model and detailed data are shown in the literature [21].

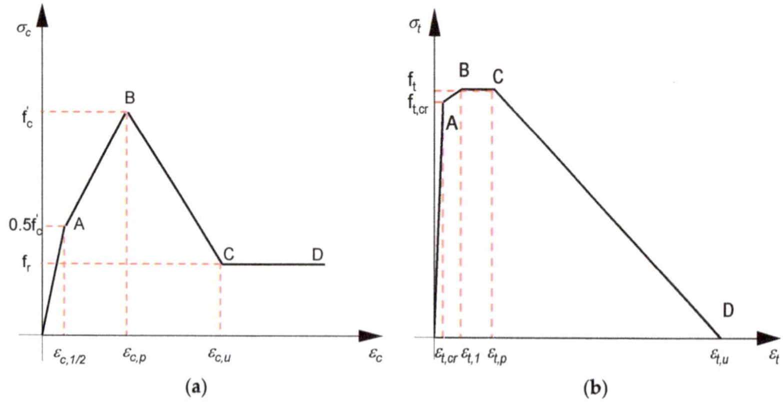

Figure 2. Mechanical properties of PVC-ECC: (**a**) compressive constitutive model; (**b**) tensile constitutive model.

Table 1. ECC properties in numerical model.

Parameter	Symbol	Unit	ECC
Compressive strength	f_c'	MPa	38.5
Compressive strain at peak strength	$\varepsilon_{c,p}$	mm/mm	0.0034
Compressive strain at 0.5 times the peak strength	$\varepsilon_{c,1/2}$	mm/mm	0.0011
Compressive strain at the residual strain	$\varepsilon_{c,u}$	mm/mm	0.005
Elastic modulus	E_c	GPa	17.1
Tensile strength	f_t	MPa	1.96
Tensile cracking strength	$f_{t,cr}$	MPa	1.86
Tensile strain when the stress is 0.95 times of peak stress	$\varepsilon_{t,cr}$	mm/mm	0.00011
Tensile cracking strain	$\varepsilon_{t,1}$	mm/mm	0.0011
Tensile localization strain	$\varepsilon_{t,p}$	mm/mm	0.0073
Maximum tensile strain of ECC	$\varepsilon_{t,u}$	mm/mm	0.035

The longitudinal reinforcement is embedded in PVA-ECC. The embedded reinforcement element coupled with concrete elements deforms together as a whole and contributes to the stiffness. In DIANA software, the bond–slip model of the reinforcement can be defined based on the embedded reinforcement element without the definition of an interface element, which can save time and improve modeling efficiency. The bond–slip relationship between ECC and the reinforcement is shown in Figure 3. Equation (1) shows the equation of the bond–slip curve proposed by Shao et al. [20] and Bandelt et al. [23]. The bond–slip curve consists of three stages, including a nonlinear ascending stage, a linear descending stage, and a plateau stage resulting from the presence of a frictional bond.

$$\frac{\tau}{1.17 \cdot \tau_{\max}} = \begin{cases} s^{0.28} & s \le s_1 \\ 1 - 0.06 \cdot (s-1) & s_1 < s \le s_2 \\ 0.3 & s_2 < s \end{cases} \tag{1}$$

where $\tau_{\max}$ is the peak bond stress and can be taken as $1.2\sqrt{f_c'}$; s_1 and s_2 are quantified to be 1 and 12.7, respectively.

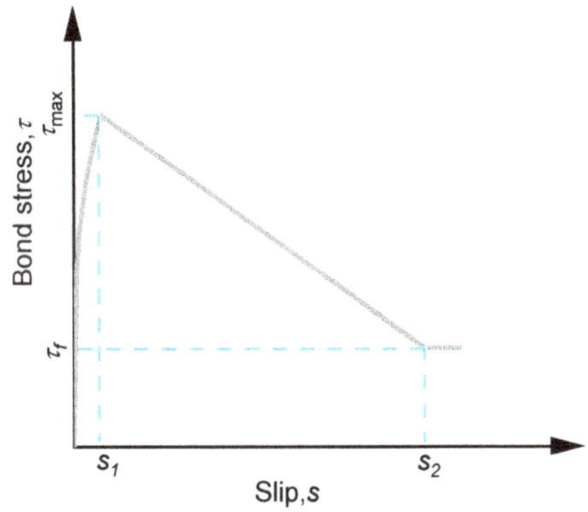

Figure 3. Bond–slip model between ECC and steel bars.

2.2. Validation of Proposed Model

Before simulations, a mesh sensitivity is performed to calibrate the finite element model with different mesh sizes for specimen ECC-16-1.072, namely, 25 mm, 10 mm, and 5 mm. It can be seen from Figure 4 that the load–deflection curves are identical to each other for three mesh sizes before peak load is reached. However, the load–deflection relationship after the peak load is lower than that of the experimental results when the mesh size is increased to 25 mm. Even though the results with a mesh size of 5 mm are closer to the test result, the simulation is time-consuming, taking more than 24 h, which is not appropriate. Therefore, a mesh size of 10 mm is suitable for numerical modeling.

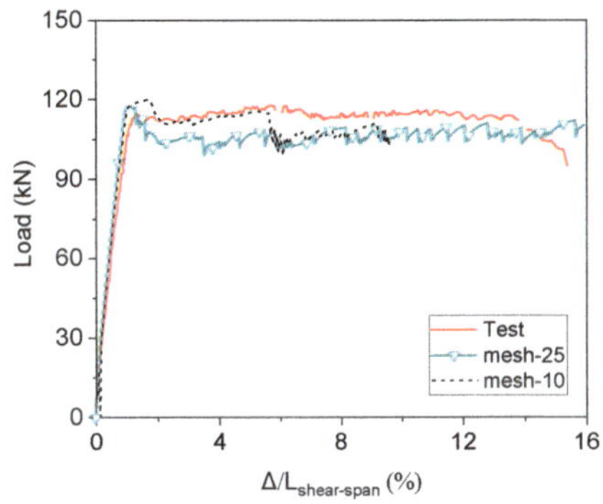

Figure 4. Parameter analysis on mesh size for specimen ECC-16–1.072.

Figure 5 compares the load–deflection curves of three PVA-ECC beams obtained in the experimental tests and numerical models. Note that the blue cross denotes the point where the fracture of rebars occurs. It can be observed from the figure that the load–deflection curves of the numerical results are in good agreement with the experiment results at the initial and ultimate stages. The maximum load capacity of experimental and numerical results is shown in Table 2. The ratio of numerical results to experimental results of the peak load ranges between 1.01 and 1.07, with a mean value of 1.04. The mean ratio of the beam deflection at the fracture of rebars is 0.96, with a coefficient of variation of 6.7%, as shown in Table 2. It indicates that the ultimate load of beams is slightly overestimated by using the developed numerical model, whereas the deflection at the fracture of reinforcement is underestimated. The overestimation of the load capacity might come from the difference between the actual and measured tensile strengths of PVA-ECC. In reinforced PVA-ECC beams, the presence of flaws in PVA-ECC can reduce its tensile strength.

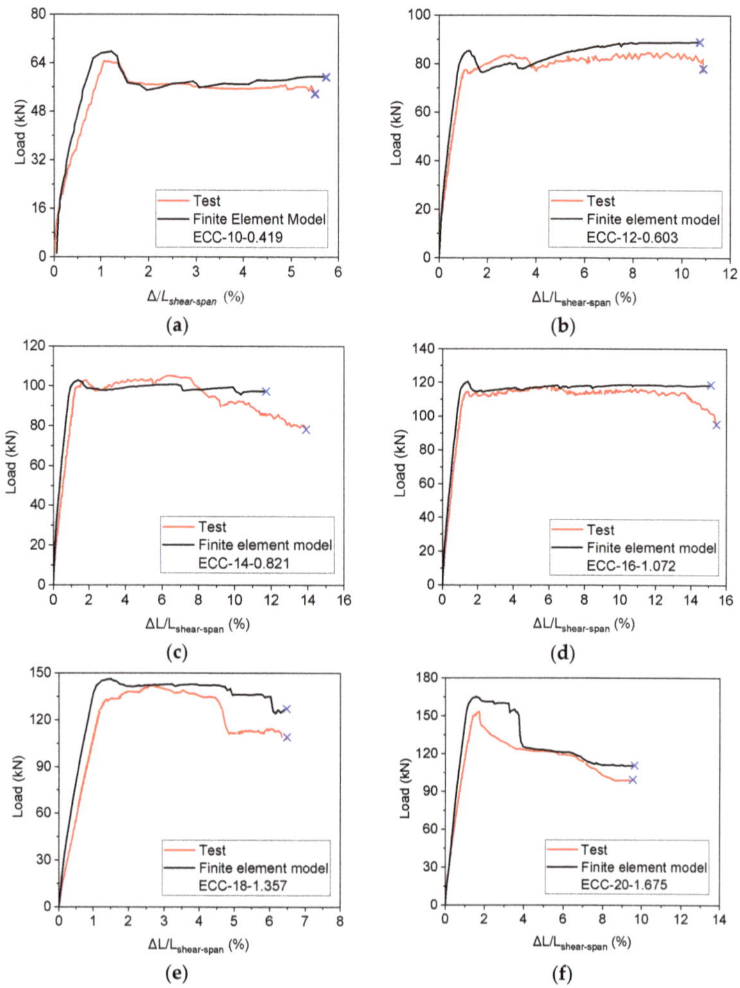

Figure 5. Experimental and numerical results of load deflection curves: (**a**) ECC-10-0.419; (**b**) ECC-12-0.603; (**c**) ECC-14-0.821; (**d**) ECC-14-0.821; (**e**) ECC-14-0.821; (**f**) ECC-14-0.821.

Figure 6 shows the principal tensile strain contours of the three specimens. Similar to the experimental results, the failure model of ECC-10–0.419 is due to the fracture of reinforcement in the tension zone, as shown in Figure 6a, and then the load value decreases sharply. When the reinforcement ratio is increased to 1.075% (see Figure 6b) and 1.675% (see Figure 6c), the purple zone, namely, the compression zone, is extended and the crack length across the section decreases. This indicates that the failure model is changed from the fracture of steel reinforcement to the crushing of the compression zone. Thus, the failure mode of reinforced PVA-ECC beams depends highly on the reinforcement ratio at the bottom side. When the reinforcement ratio is low, fracture of the tensile reinforcement occurs near the loading points, leading to the failure of beams. However, if the reinforcement is high, crushing of PVA-ECC may occur in the compression zone first, resulting in significant reductions in the applied load.

Besides the principle tensile strain, the strain of steel reinforcement can also be obtained from the numerical model. The maximum stress of the beam bottom reinforcement is located at the midspan of beams, with a value of 509.0 MPa for specimen ECC-10-0.419.

With the increase in reinforcement ratio, the maximum steel stress decreases from 509.0 MPa to 463.4 MPa. With regard to the top reinforcement, the compression stress increases from 35.8 MPa to 125.5 MPa when the reinforcement ratio increases from 0.419% to 1.675%.

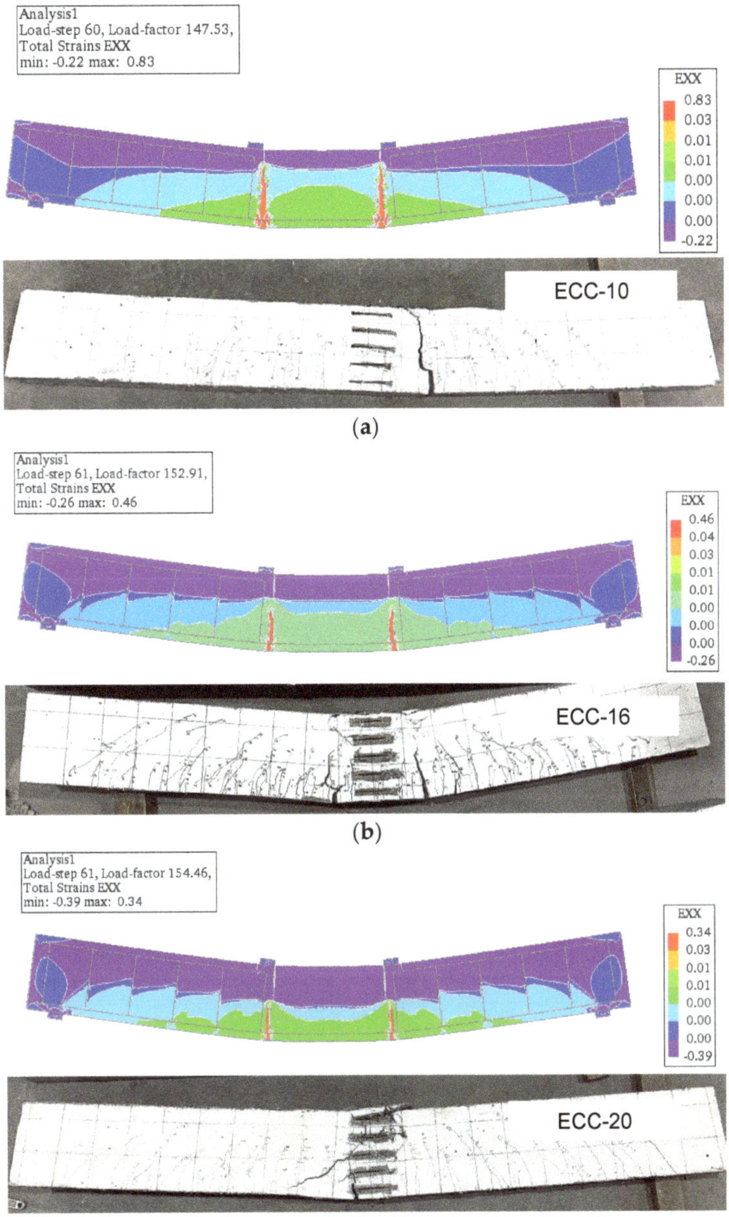

Figure 6. Principal tensile strains of beams in bending: (**a**) ECC-10-0.419; (**b**) ECC-16-1.075; (**c**) ECC-20-1.675.

Table 2. Comparison of test results and finite element results.

Specimen	Ultimate Load			Deflection at Fracture of Rebars		
	Experimental Results P_{Exp} **(kN)**	**Numerical Results** P_{FE} **(kN)**	P_{FE}/P_{Exp}	**Experimental Results** D_{Exp} **(%)**	**Numerical Results** D_{FE} **(%)**	D_{FE}/D_{Exp}
ECC-10-0.419	64.7	68.2	1.05	5.94	5.97	1.01
ECC-12-0.603	84.7	88.7	1.05	10.99	10.34	0.94
ECC-14-0.821	102.8	103.5	1.01	13.91	11.71	0.84
ECC-16-1.072	117.7	120.5	1.02	15.55	15.11	0.97
ECC-18-1.357	142.4	146.3	1.03	6.62	6.46	0.98
ECC-20-1.675	153.4	164.3	1.07	9.53	9.63	1.01
Average			1.04			0.96
Coefficient of variation			2.2%			6.7%

2.3. Parameter Analysis

With the verified numerical model, three beams with different reinforcement ratios of 0.269%, 0.343%, and 0.419% are modeled to study the effect of reinforcement ratio on the flexural behavior of beams. Other parameters of the beam remain the same as those of ECC-10-0.419. Figure 7a shows the load–deflection curves with different reinforcement ratios. It can be observed that specimen ECC-10-0.419 develops the highest load after the development of a principal crack. The effects of reinforcement ratio on the peak load, stiffness, and deflection are obvious when the specimen fails. By reducing the reinforcement ratio from 0.419% to 0.269%, the ultimate load of the beam is considerably decreased. The initial stiffness of beams is nearly the same before the cracking of PVA-ECC, whereas the stiffness after the cracking of PVA-ECC shows a remarkable difference due to the development of microcracks. The stiffness following the cracking of PVA-ECC decreases gradually with decreasing reinforcement ratio.

In addition to the reinforcement ratio, the influence of steel grade, maximum tensile strain, tensile localization strain, and bond strength on the flexural behavior of beams is also investigated using the numerical model. Three different steel grades, namely, yield strengths of 335 MPa, 450 MPa, and 500 MPa, are selected in the numerical simulation. Figure 7b presents the load–deflection relationship of beams with different steel strengths. The peak loads of the three specimens are 54.7 kN, 64.3 kN, and 68.4 kN for beams with steel yield strengths of 335 MPa, 450 MPa, and 500 MPa, respectively. This indicates that the ultimate of beams can be enhanced if the reinforcement strength increases. As for the deformation capacity, namely, the deflection at which the load decreases to 85% of the ultimate load, it is increased slightly with increasing steel yield strength, as shown in Figure 7b. The influences of the maximum tensile strain and tensile localization strain on load–deflection curves are also discussed, as shown in Figure 7c,d. It can be seen from these three figures that the ultimate tensile strain of ECC does not have a significant effect on the load–deflection curve of beams when the value varies between 0.001 and 0.035. Nevertheless, when the tensile localization strain varies from 0.005 and 0.010, the deformation capacity of beams is considerably increased, but the ultimate load of beams is not considerably affected.

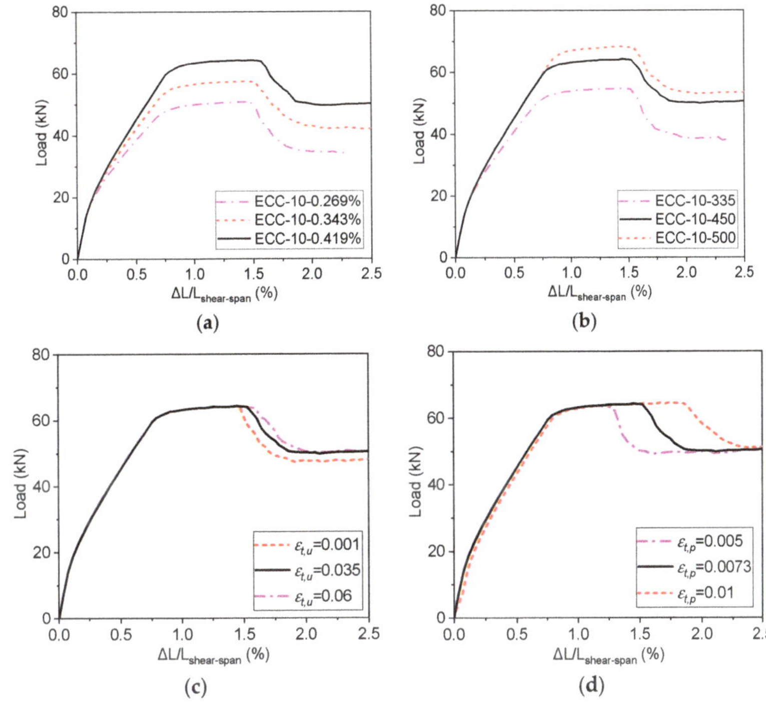

Figure 7. Load–deflection curve of test beam with different reinforcement ratios and steel bar strengths: (**a**) reinforcement ratio; (**b**) steel bar strength; (**c**) maximum tensile strain; (**d**) tensile localization strain.

3. Existing Design Methods

Numerical modeling is used to predict the load–deflection curves, crack width of singly reinforced beams. Even though it is possible to obtain multiple simulation results, the calculation procedure is time-consuming and complicated. To obtain the peak load easily and concisely, existing design methods are used to calculate the ultimate load of reinforced PVA-ECC beams and are then compared with test data to show their accuracy.

3.1. American Concrete Institute and Chinese Code

For conventional concrete beams, American concrete institution and Chinese guidelines propose similar calculation methods for the load-bearing capacity based on the plane-section assumption. They assume that the maximum strain at the extreme concrete compression fiber is equal to 0.003 and the tensile strength of concrete is neglected. The compressive stress in the compression zone is represented by a rectangular compressive stress block. Equations (2) and (3) show the calculation method of moment capacity. Table 3 shows the peak load of reinforced PVA-ECC beams calculated by using GB 50010-2010 [24] and ACI 318-19 [25]. For the design method in GB 50010-2010, it cannot consider the contribution of PVA fibers. However, for reinforced PVA-ECC beams, the tensile strength of PVA-ECC has not been exhausted when the load capacity of the beam is reached. As a result, the peak load is far lower than the experimental values except for specimen ECC-20-1.675. Hence, the tensile strength of PVA-ECC has to be considered. The peak load of all specimens calculated by ACI-318-19 is also lower than the experiment result. Therefore, it is essential to consider the contribution of fibers in the tension zone while calculating the peak load of PVA-ECC beams.

$$M = \alpha_1 \beta_b f_c b x (h_0 - x/2) \tag{2}$$

$$x = \frac{f_y A_s}{\alpha_1 \beta_b f_c b} \tag{3}$$

where β_b represents the simplified coefficient and is 0.85 and 0.8 in ACI-318-19 [25] and GB 50010-2010 [24], respectively.

3.2. *fib Model Code*

fib Model Code [26] proposes a calculation method to consider the influence of fiber in the tension zone. Flexural failure occurs when one of the following conditions is reached (see Figure 8):

1. Attainment of the ultimate compressive strain of ECC, $\varepsilon_{c,u}$;
2. Attainment of the ultimate tensile stain in the steel, $\varepsilon_{s,u}$;
3. Attainment of the ultimate tensile strain of ECC, $\varepsilon_{t,u}$'

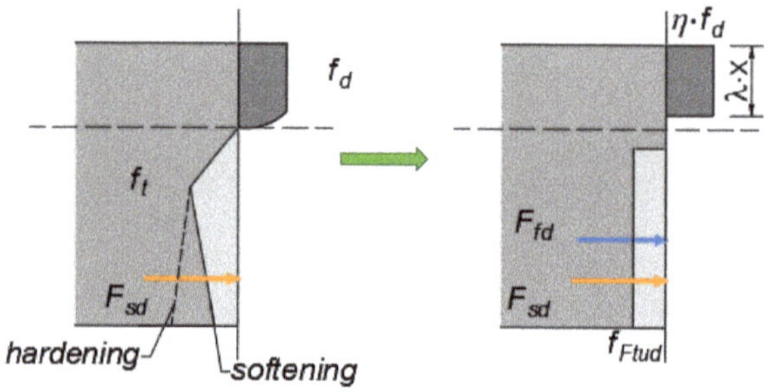

Figure 8. Calculation of bending moment from *fib Model Code*.

Similar to conventional concrete, it is assumed that the ultimate compressive strain of ECC is achieved. In order to calculate the compressive stress and tensive stress conveniently, a rectangular zone is used in the compression and tension zones, as shown in Figure 8. The second condition is neglected in the calculation, as the steel bar is at the post-yield stage when PVA-ECC beams achieve the peak load. Coefficients λ and η adopt 0.8 and 1.0 from *fib Model Code*, respectively [26]. In Figure 8, f_t represents the tensile strength of PVA-ECC, and f_{Ftud} represents the residual tensile strength and is assumed to be $1/3\ f_t$.

The calculation results are presented in Table 3. The mean value of P_{fib}/P_{Exp} is 0.84, with a coefficient of variation of 0.19. *fib Model Code* adopts the Prisco model [27] and the basic principles governing the structural design of fiber-reinforced concrete elements. Even though the contribution of fiber-reinforced concrete in tension is considered, the calculated load is still lower than the experimental value, but it is higher than the value calculated using GB 50010-2010 and ACI 318-19 for conventional concrete.

3.3. *Fehling*

Fehling [28] extended the design principles for ultra-high-performance concrete by taking fibers into consideration. In the design method, the plane-section assumption is used to define the strain profile, and the stress in the compression and tension zones is calculated from the corresponding constitutive models. However, the tensile stress of ultra-high-performance concrete is assumed to be uniformly distributed in the tension zone, which is different from the actual distribution of tensile stresses across a beam section. The stress–strain curve of ultra-high-performance concrete is defined by a stress–crack-width relationship. Figure 9a shows the stress distributions and internal forces acting on the cross-section. The concrete compressive stresses are simplified as a triangle and the resulting stress F_{cd} lies at the centroid of the triangle. The distribution of tensile stresses is parabolic, which is directly in accordance with the minimum crack width. In this section, σ_{cf0d} is assumed to be f_t, as shown in Figure 9a. To simplify the calculation, the tensile stress distribution in Figure 9a can be converted to an equivalent stress block (see Figure 9b).

Table 3. Comparison of experimental ultimate load with values calculated using different codes and models.

Specimen	Experimental Results	GB 50010-2010 [24]		ACI 318-19 [25]		*Fib Model Code* [26]		UHPC [28]		Proposed Method [20]	
	Load (kN)	Load (kN)		Load (kN)		Load (kN)		Load (kN)			
	P_{Exp}	P_{GB}	P_{GB}/P_{Exp}	P_{ACI}	P_{ACI}/P_{Exp}	P_{fib}	P_{fib}/P_{Exp}	P_{UHPC}	P_{UHPC}/P_{Exp}	P_{Po}	P_{Po}/P_{Exp}
ECC-10-0.419	64.7	44.8	0.69	40.1	0.62	46.3	0.72	60.6	0.94	65.7	1.02
ECC-12-0.603	84.7	61.2	0.72	54.7	0.65	62.7	0.74	75.7	0.89	79.2	0.93
ECC-14-0.821	105.1	80.3	0.76	71.6	0.68	81.7	0.78	93.4	0.89	94.5	0.90
ECC-16-1.072	117.7	105.9	0.90	94.1	0.80	107.3	0.91	117.8	1.00	114.7	0.97
ECC-18-1.357	142.4	127.4	0.89	112.8	0.79	128.7	0.90	138.6	0.97	130.9	0.92
ECC-20-1.675	153.4	155.6	1.01	137.1	0.89	156.7	1.02	173.9	1.13	151.4	0.99
Mean value			0.83		0.74		0.85		0.97		0.95
Coefficient of variation			0.20		0.28		0.19		0.09		0.06

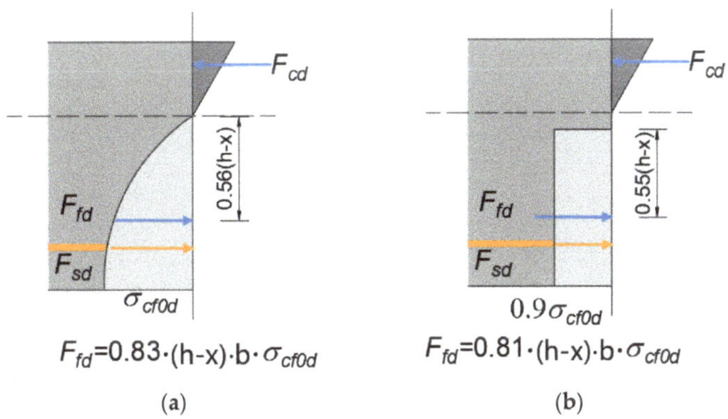

Figure 9. Assumed stress distribution and resultant tensile force at ultimate limit state: (**a**) realistic stress distribution; (**b**) simplified diagram.

Table 3 shows the comparison between the experimental result and that calculated by using the method proposed by Fehling. It can be observed that the mean load ratio is 0.97, with a coefficient of variation of 0.09. Thus, the calculated result is in good agreement with the experimental result. However, the tensile stress distribution in the tension zone is a bit different from that of ECC in tension and should be properly modified in the subsequent study.

4. Proposed Calculation Method

Wang et al. [20] suggested a calculation method to compute the moment capacity and peak load of singly reinforced PVA-ECC beams. The constitutive model of ECC in tension is modified in the design method, especially at the formation stage of microcracks, namely, the plateau stage of tensile stresses. Two limiting reinforcement ratios are developed based on different flexural behavior in Equations (4) and (5), namely, the minimum reinforcement ratio and the balanced reinforcement ratio. When the reinforcement ratio is lower than the minimum reinforcement ratio, the failure mode of flexural beams is characterized by the fracture of reinforcement following the formation of a principle crack at the mid-span of beams. Once the reinforcement ratio is higher than the minimum reinforcement ratio, the failure mode changes from major cracking to crushing of PVA-ECC in the compression zone. The balanced reinforcement ratio represents the case in which the ultimate tensile strain of PVA-ECC in the tension zone and the ultimate compressive strain in the compression zone are achieved simultaneously.

$$\rho_{min} = \frac{0.16 f_t}{f_u - f_y} \tag{4}$$

$$\rho_t = \frac{\left[f_c'\left(0.75\varepsilon_{c,p} - 0.5\varepsilon_{c,1/2}\right) - f_t\left(\varepsilon_{t,p} - 0.5\varepsilon_{t,1}\right)\right]}{f_y\left(\varepsilon_{t,p} + \varepsilon_{c,p}\right)} \tag{5}$$

where ρ_{min} is the minimum reinforcement ratio and ρ_t is the balanced ratio.

With increasing reinforcement ratio, it is possible that the reinforcement does not yield when the ultimate compressive strain of PVA-ECC beams is obtained. Therefore, the yield reinforcement ratio when the yielding of tensile reinforcement and crushing of compressive PVA-ECC occur simultaneously is suggested in Equation (6).

$$\rho_y = \frac{\left[f_c'\left(0.75\varepsilon_{c,p} - 0.5\varepsilon_{c,1/2}\right)(h - a_s) - f_t\left(\varepsilon_y h + \varepsilon_{c,p} a_s - 0.5\varepsilon_{t,1}(h - a_s)\right)\right]}{f_y\left(\varepsilon_y + \varepsilon_{c,p}\right)h} \tag{6}$$

Based on different calculation equations of reinforcement ratio, all the values of ρ_{min}, ρ_t, and ρ_y for different rebar diameters are calculated and presented in Table 4.

Table 4. Minimum, balance, and yield reinforcement ratio for different steel diameters.

Diameter (mm)	Actual Reinforcement Ratio (%)	Minimum Reinforcement Ratio (%)	Balanced Reinforcement Ratio (%)	Yield Reinforcement Ratio (%)
10	0.419	0.297	0.390	1.806
12	0.603	0.201	0.407	1.772
14	0.821	0.199	0.416	1.903
16	1.072	0.202	0.405	1.741
18	1.357	0.183	0.419	1.945
20	1.675	0.200	0.414	1.911
Mean value		0.214	0.409	1.846

It can be seen from Table 4 that the minimum reinforcement ratio, balanced reinforcement ratio, and yield reinforcement ratio are close to each other for different rebar diameters. The average values for ρ_{min}, ρ_t, and ρ_y are 0.214%, 0.409%, and 1.846% for the HRB400-deformed steel bar, respectively. The actual reinforcement ratio in reference [21] was between the average balanced reinforcement ratio of 0.409% and the yield reinforcement ratio of 1.846%. Hence, the tensile reinforcement is in the post-yielded stage, while the compression strain at the edge of the compression zone reaches the peak compression strain when the beam is at the peak load.

A calculation method is also proposed to calculate the ultimate moment capacity of reinforced PVA-ECC beams in bending. Figure 10 shows the proposed method for the calculation of ultimate load, which is controlled by the crushing of PVA-ECC in the compression zone, namely, compression-control failure. Compared with the design method by Fehling [28] for ultra-high-performance concrete, the constitutive model in the tension and compression zone of beams is simplified bilinear, as shown in Figure 10. The plane-section assumption and force equilibrium are adopted in the calculation procedures.

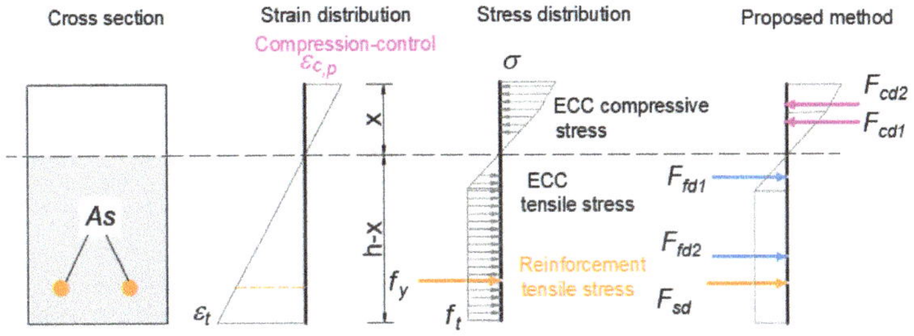

Figure 10. Proposed methods for calculation of ultimate load of beams in bending.

Table 3 lists the comparisons of the calculated load capacity and the experimental loads. The mean ratio and coefficient of variation of the calculated load capacity to the experimental loads are 0.95 and 0.06, respectively. Figure 11 shows the comparison among experimental results and calculated results based on different methods. It can be observed that the design methods in GB 50010-2010 [24], ACI 318-19 [25], and *fib Model Code* [26] significantly underestimate the ultimate load of reinforced PVA-ECC beams, as shown in Figure 11a, particularly when the diameter of longitudinal reinforcement is small, as the tensile strength of PVA-ECC is not considered in the design methods. When the bilinear constitutive model for PVA-ECC is considered in the numerical model and the proposed method, the ultimate load of beams can be predicted with good accuracy, as shown in Figure 11b.

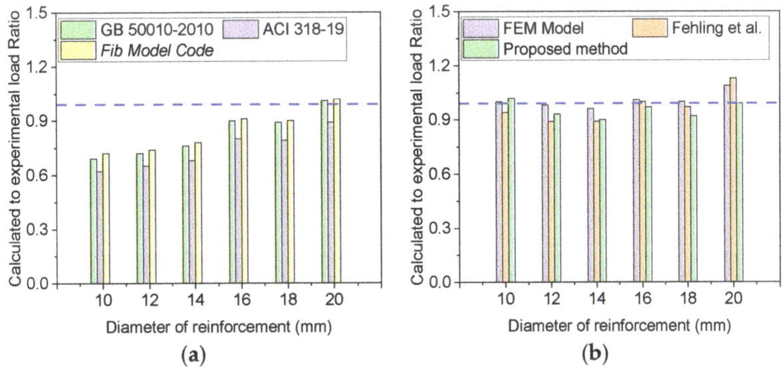

Figure 11. Ratio between analytical and experimental results with different calculation methods. (**a**) Results using design standards. (**b**) Results using FEM, Fehling et al., and proposed method.

5. Conclusions

In this study, a nonlinear finite element model is proposed for singly supported PVA-ECC beams in bending to simulate the load–deflection curve and failure mode. In the numerical model, the constitutive models for PVA-ECC are represented by simplified piece-wise linear curves, and the bond–slip behavior between the embedded reinforcing bar and surrounding PVA-ECC is considered in the numerical model. Calculation methods in different design codes are used to calculate the ultimate load of PVA-ECC beams. Moreover, a design method is also proposed to consider the tensile strength of PVA-ECC in calculation. The following conclusions can be obtained from the numerical study and design method.

(1) By considering the piece-wise constitutive models for PVA-ECC in tension and compression and the bond–slip behavior between PVA-ECC and steel reinforcement in the numerical model, the numerical results show a good agreement with the experiment results in terms of the load–deflection curves and failure modes. The ultimate load and deformation capacity of reinforced PVA-ECC beams can be predicted by using the numerical

model with good accuracy. The average numerical-to-experimental ratio of load capacities is 1.04, and the mean ratio of the numerical to experimental deflection at the fracture of reinforcement is 0.96.

(2) Parameter analyses show that increasing the reinforcement ratio of reinforced PVA-ECC beams significantly increases the stiffness after cracking and the ultimate load of beams. However, the increase in reinforcement strength grade has a limited effect on the initial stiffness of beams, whereas the ultimate load-bearing capacity of beams is effectively increased. The tensile localization strain of PVA-ECC affects the deformation capacity of beams. The deformation capacity of reinforced PVA-ECC beams is increased by increasing the tensile localization strain.

(2) The ultimate load of PVA-ECC beams is calculated using three different design codes, ACI 318-10, GB 50010-2010, and *fib Model Code*. All design methods underestimate the ultimate load of PVA-ECC beams due to the neglection of the tensile strength of PVA-ECC in the tension zone. By considering the tensile strength of PVA-ECC in the tension zone, the method proposed by Fehling can yield more accurate predictions of experimental results.

(3) When the tensile stress–strain relationship is considered, the proposed design method for reinforced PVA-ECC beams can predict the ultimate load of PVA-ECC beams accurately by dividing the failure modes into tension-control and compression control.

Author Contributions: Conceptualization, S.-B.K. and Q.-L.F.; methodology, D.-D.W.; software Q.-L.F.; validation, D.-D.W. and S.-B.K.; formal analysis, D.-D.W.; investigation, Q.-L.F.; writing—original draft preparation, Q.-L.F. and D.-D.W.; writing—review and editing, S.-B.K.; supervision, S.-B.K.; project administration, D.-D.W.; funding acquisition, Q.-L.F. All authors have read and agreed to the published version of the manuscript.

Funding: This research was funded by the Natural Science Foundation of Chongqing, China (Grant No. cstc2021jcyj-msxmX0357) and the Graduate Research and Innovation Foundation of Chongqing, China (Grant No. CYB21030), the Chongqing Commission of Education (Grant No.: KJQN202103802 and KJQN202103801), and the Chongqing Water Resources and Electric Engineering College (Grant No.: K202214).

Institutional Review Board Statement: Not applicable.

Informed Consent Statement: Not applicable.

Data Availability Statement: The data presented in this study are available on request from the corresponding author.

Conflicts of Interest: The authors declare no conflict of interest.

Notation

a_s	thickness of the concrete cover	P_{ACI}	peak load calculated by ACI 318-19
A_s	cross-sectional area of reinforcement	P_{Exp}	peak load of experimental results
b	width of cross-section	P_{FE}	peak load of numerical results
d	diameter of reinforcement	P_{fib}	peak load calculated by *fib Model Code*
D_{Exp}	experimental deflection at ultimate load	P_{GB}	peak load calculated by GB 50010-2010
D_{FE}	numerical deflection at ultimate load	P_{Po}	peak load calculated by proposed method
E_c	elastic modulus of ECC	P_{UHPC}	peak load calculated by Fehling's method
E_h	hardening modulus of steel bars	s_1	slip at peak bond stress
E_s	elastic modulus of steel bars	s_2	slip at the onset of frictional bond stress
f'_c	compressive strength	$\varepsilon_{c,1/2}$	compressive strain at 50% of peak strength
f_d	compressive stress in the compression zone that adopts the compressive strength	$\varepsilon_{c,p}$	compressive strain at compressive strength
f_{Ftud}	ultimate strength, assumed as $1/3f_t$	$\varepsilon_{c,u}$	compressive strain at 50% of compressive strength
f_t	tensile strength	$\varepsilon_{s,u}$	ultimate tensile stain of steel
$f_{t,cr}$	cracking tensile strength	$\varepsilon_{t,1}$	cracking strain of ECC
f_u	ultimate strength of reinforcement	$\varepsilon_{t,cr}$	tensile strain when the stress is 95% of peak stress
f_y	yield strength of reinforcement	$\varepsilon_{t,p}$	tensile localization strain

f_r	residual stress of PVA-ECC	$\varepsilon_{t,u}$	ultimate tensile strain of ECC
F_{cd}	force in the compression zone	ρ_{min}	minimum reinforcement ratio
F_{cd1}	force in the initial compression zone	ρ_t	balanced reinforcement ratio
F_{cd2}	force in the second compression zone	ρ_y	yield reinforcement ratio
F_{fd}	force of ECC in tension zone	β_b	simplified coefficient
F_{fd1}	force of increasing stage in tension zone	σ_{cf0d}	tensile stress at the bottom of the tension zone, assumed to be f_t
F_{fd2}	force of plateau stage in tension zone	τ_{max}	maximum bond stress
F_{sd}	reinforcement force	τ_f	frictional bond stress
M	moment of beams in bending	λ	a coefficient that adopts 0.8
h	height of cross-sections	η	a coefficient that adopts 1

References

1. Jin, H.; Li, F.; Hu, D. Research on the flexural performance of reinforced engineered cementitious composite beams. *Struct. Concr.* **2021**, *23*, 2198–2220.
2. Maalej, M.; Li, V.C. Introduction of Strain-Hardening Engineered Cemenetitious Composites in Design of Reinforced Concrete Flexural Members for Improved Durability. *ACI Struct. J.* **1995**, *92*, 167–176.
3. Yang, E.; Li, V.C. Strain-Hardening Fibre Cement Optimization and Component Tailoring by Means of a Micromechanical Model. *Constr. Build. Mater.* **2010**, *24*, 130–139.
4. Victor, C.L. *Engineered Cementitious Composites (ECC): Bendable Concrete for Sustainable and Resilient Infrastructure*; Springer: Berlin/Heidelberg, Germany, 2022.
5. Ge, W.; Ashour, A.F.; Cao, D.; Lu, W.; Gao, P.; Yu, J.; Ji, X.; Cai, C. Experimental study on flexural behavior of ECC-concrete composite beams reinforced with FRP bars. *Compos. Struct.* **2019**, *208*, 454–465.
6. Yuan, F.; Pan, J.; Leung, C.K.Y. Flexural Behaviors of ECC and Concrete/ECC Composite Beams Reinforced with Basalt Fiber-Reinforced Polymer. *J. Compos. Constr.* **2013**, *17*, 591–602.
7. Dabiri, H.; Kaviani, A.; Kheyroddin, A. Influence of reinforcement on the performance of non-seismically detailed RC beam-column joints. *J. Build. Eng.* **2020**, *31*, 101333.
8. Behnam, H.; Kuang, J.S.; Samali, B. Parametric finite element analysis of RC wide beam-column connections. *Comput. Struct.* **2018**, *205*, 28–44.
9. Onuralp, O.Y.; Karalar, M.; Aksoylu, C.; Beskopylny, A.N.; Stel, S.A.; Shcherban, E.M.; Qaidi, S.; Pereira, I.D.S.A.; Monteiro, S.N.; Azevedo, A.R.G. Shear performance of reinforced expansive concrete beams utilizing aluminium waste. *J. Mater. Res. Technol.* **2023**, *24*, 5433–5448.
10. Ozkılıç, Y.O.; Basaran, B.; Aksoylu, C.; Karalar, M.; Martins, C.H. Mechanical behavior in terms of shear and bending performance of reinforced concrete beam using waste fire clay as replacement of aggregate. *Case Stud. Constr. Mater.* **2023**, *18*, e02104.
11. Shao, Y.; Billington, S.L. Flexural performance of steel-reinforced engineered cementitious composites with different reinforcing ratios and steel types. *Constr. Build. Mater.* **2020**, *231*, 117159.
12. Shao, Y. *Improving Ductility and Design Methods of Reinforced High-Performance Fiber-Reinforced Cementitious Composite (HPFRCC) Flexural Members*; Stanford University: Stanford, CA, USA, 2020.
13. Shao, Y.; Billington, S.L. Predicting the two predominant flexural failure paths of longitudinally reinforced high-performance fiber-reinforced cementitious composite structural members. *Eng. Struct.* **2019**, *199*, 109581.
14. Suryanto, B.; Nagai, K.; Maekawa, K. Modeling and Analysis of Shear-Critical ECC Members with Anisotropic Stress and Strain Fields. *J. Adv. Concr. Technol.* **2010**, *8*, 239–258.
15. Suryanto, B.; Nagai, K.; Maekawa, K. Smeared-Crack Modeling of R/ECC Membranes Incorporating an Explicit Shear Transfer Model. *J. Adv. Concr. Technol.* **2010**, *8*, 315–326.
16. Varela, S.; Saiidi, M.S. *Shear Behavior of Engineered Cementitious Composite Structural Members, Second Conference on Smart Monitor, Assessment and Rehabilitation of Civil Structures*; ITU: Istanbul, Turkey, 2013.
17. Zheng, Y.-Z.; Wang, W.-W.; Mosalam, K.M.; Fang, Q.; Chen, L.; Zhu, Z.-F. Experimental investigation and numerical analysis of RC beams shear strengthened with FRP/ECC composite layer. *Compos. Struct.* **2020**, *246*, 112436.
18. Shanour, A.S.; Said, M.; Arafa, A.I.; Maher, A. Flexural performance of concrete beams containing engineered cementitious composites. *Constr. Build. Mater.* **2018**, *180*, 23–34.
19. Selim, A.H.; Mahmoudi, F.; Abdalla, J.A.; Hawileh, R.A.; Abed, F.; Kyaure, M. Finite Element Modeling of Engineered Cementitious Composite (ECC) Prisms and Beams. In Proceedings of the 2022 Advances in Science and Engineering Technology International Conferences (ASET), Dubai, United Arab Emirates, 21–24 February 2022; IEEE: Piscataway, NJ, USA; pp. 1–6.
20. Shao, Y.; Hung, C.-C.; Billington, S.L. Gradual Crushing of Steel Reinforced HPFRCC Beams: Experiments and Simulations. *J. Struct. Eng.* **2021**, *147*, 04021114.
21. Wang, D.-D.; Li, J.-Z.; Qin, S.-D.; Liu, C.-X.; Long, B.; Kang, S.-B. Experimental tests and analytical model for reinforced polyvinyl alcohol-engineered cementitious composite beams in bending. *Struct. Concr.* **2023**, *24*, 3520–3544.
22. DIANA. DIANA Users' Manual Release 10.2, DIANA FEA bv. 2017. Available online: https://manuals.dianafea.com/d102/Diana.html (accessed on 21 June 2023).
23. Bandelt, M.J.; Frank, T.E.; Lepech, M.D.; Billington, S.L. Bond behavior and interface modeling of reinforced high-performance fiber-reinforced cementitious composites. *Cem. Concr. Compos.* **2017**, *83*, 188–201.

24. 50010–2010; Code for Design of Concrete Structures. National Standard of The People's Republic Of China: Beijing, China, 2015.
25. ACI 318–19; Building Code Requirements for Structural Concrete. Americon Concrete Institute: Farming, MI, USA, 2019.
26. *fib Model Code for Concrete Structures 2010*; Fédération Internationale du Béton: Laussane, Switzerland, 2013.
27. Prisco, M.D.; Colombo, M.; Dozio, D. Fibre-reinforced concrete in fib Model Code 2010: Principles, models and test validation. *Struct. Concr.* **2013**, *14*, 342–356.
28. Fehling, E.; Schmidt, M.; Walraven, J.; Leutbecher, T.; Frohlich, S. *Ultra-High Performance Concrete UHPC: Fundamentals, Design, Examples*; Ernst & Sohn GmbH & Co.: Berlin/Heidelberg, Germany, 2014.

MDPI
St. Alban-Anlage 66
4052 Basel
Switzerland
www.mdpi.com

Sustainability Editorial Office
E-mail: sustainability@mdpi.com
www.mdpi.com/journal/sustainability

www.ingramcontent.com/pod-product-compliance
Lightning Source LLC
Chambersburg PA
CBHW061550210926
51850CB00009B/2055
* 9 7 8 3 7 2 5 8 0 4 4 6 7 *